An Introduction to SCIENTIFIC RESEARCH METHODS in GEOGRAPHY

To Violet and Élan, for their love,
support, and relentless pursuit of our improvement.

An Introduction to SCIENTIFIC RESEARCH METHODS in GEOGRAPHY

Daniel R. Montello
University of California, Santa Barbara

Paul C. Sutton
University of Denver, Colorado

SAGE Publications
Thousand Oaks ▪ London ▪ New Delhi

For information:

Sage Publications, Inc.
2455 Teller Road
Thousand Oaks, California 91320
E-mail: order@sagepub.com

Sage Publications Ltd.
1 Oliver's Yard
55 City Road
London EC1Y 1SP
United Kingdom

Sage Publications India Pvt. Ltd.
B-42, Panchsheel Enclave
Post Box 4109
New Delhi 110 017 India

Printed in the United States of America.

Library of Congress Cataloging-in-Publication Data

Montello, Daniel R., 1959-
An introduction to scientific research methods in geography /
Daniel R. Montello, Paul C. Sutton.
p. cm.
Includes bibliographical references and index.
ISBN 1-4129-0286-X (cloth: alk. paper) — ISBN 1-4129-0287-8
(pbk. : alk. paper) 1. Geography—Research. I. Sutton, Paul C. II. Title.
G73.M58 2006
910.72—dc22

2005028149

This book is printed on acid-free paper.

06 07 08 09 10 9 8 7 6 5 4 3 2 1

Acquiring Editor: Robert Rojek
Production Editor: Jenn Reese
Typesetter: C&M Digitals (P) Ltd.
Indexer: John Roy
Cover Designer: Michelle Kenny

Contents

Preface

In this text, we provide a broad and integrative introduction to the conduct and interpretation of scientific research in geography. We cover both conceptual and technical aspects of research as they apply to all topical areas within geography, including human geography, physical geography, and geographic information science. We have attempted to produce a comprehensive text that discusses all parts of the research process, including scientific philosophy; basic research concepts; generating research ideas; quantitative and qualitative data collection; sampling and research design; data analysis, display, and interpretation; reliability and validity; using geographic information techniques in research; communicating research and using library resources; and ethical conduct in research. The text is intended for English-language undergraduate and graduate courses on research methods in geography and related disciplines, such as environmental studies. In addition, we hope it will have some value as a primer or reference work for students, faculty, and other research professionals who want an integrated yet concise introduction to scientific research methods in geography.

In this text, we apply the research philosophy and methods of the social and natural sciences to research in geography. At the same time, we recognize and respect the diversity and heterogeneity of geography, and avoid simplistic conceptions of scientific geography as narrowly "positivistic," "objective," or "quantitative." In this way, the text attempts to promote rigor and progressiveness in geography, helping to build bridges among the various subfields of geography and the other social and natural sciences, while avoiding some of the limiting meta-theoretical conflicts that have characterized geography in recent decades.

Geography is a very heterogeneous discipline. Not every subdiscipline of geography accepts the intellectual preferences characteristic of scientists (discussed in Chapter 1) or practices systematic empirical research to the same degree. In particular, geography also includes humanities, engineering, and craft approaches that we make no attempt to cover thoroughly in this text. But make no mistake—our focus on a scientific approach still leaves quite a bit of room for conceptual and methodological pluralism in geography. For example, we make it clear that "subjective" mental constructs, such as beliefs and attitudes, can be validly measured and studied scientifically; so can many of the theoretical ideas and concepts that stem from various critiques of strict positivism in geography, such as unequal power relationships and constructed worldviews. At the same time, we do not naively treat phenomena such as symbolic meaning, so important to human experience and activity, as if they were no different whatsoever from temperature or chemical composition. However, we believe the differences can at least partially be conceptualized within a common scientific framework, thereby allowing the study of diverse geographic subject matters to be approached with a largely common set of intellectual and analytic tools.

Overview of Chapters

We have organized the chapters of this text to correspond roughly to the order of tasks one would encounter while actually carrying out research. However, we have written the chapters to be largely self-contained, so they can to a degree be read in different orders. However, we do recommend that, if you are a novice to research methodology, you read the entire book before embarking on research. Each chapter starts with a short list of **learning objectives**, a set of questions that serves to briefly summarize and preview the chapter's major topics and to stimulate your thought processes by getting you to begin actively pondering what you are about to read. Similarly, each chapter ends with a set of **review questions** designed to remind you of important general ideas within the chapter and help you consolidate what you have read. Important terms—we call them **key terms**—are in **boldface** throughout the text; each chapter's terms are listed and defined in glossaries at the end of the chapters. The end of each chapter also contains a **bibliography** that lists books, articles, and other resources that have shaped our thinking about research methods and deserve to be recommended as further sources for the interested reader.

Chapter 1 is entitled **Introduction: A Scientific Approach to Geography**. It introduces scientific research and methodology with a discussion of an example research project. We discuss problems with the conduct and interpretation of this project; we intend this to provide a concrete example of the scope of issues that are part of scientific research methodology. We introduce the logic and philosophy of science, both by defining its approach and goals, and by considering "characteristic beliefs" held by most scientists *as* scientists. We also consider the very definite limits of science. Science is not just a search for truth or the application of rational thought, but an appreciation of the value of empirical observation *coupled with* rational thought as a way to increase our understanding of the world. As such, many of the most important questions that the humanities address cannot or should not be answered scientifically, although they may be informed by scientific results. Furthermore, scientists are humans, of course, and scientific institutions are human institutions. Human shortcomings are certainly a part of science. We finish this chapter by reviewing the history and conceptual systems of the discipline of geography. We note the distinction between regional and systematic approaches (our concern in this book is mostly with systematic geography), and go over the basic areas of human and physical geography, as well as research on geographic information techniques.

Chapter 2 is **Fundamental Research Concepts**. In it, we introduce a series of basic concepts that are fundamental to the conduct and interpretation of scientific research. We classify these as "idea concepts" (theory, hypothesis, causality, model, construct) or "empirical concepts" (case, variable, measurement, data, measurement levels, continuity versus discreteness, accuracy and precision). We go over the key concept of scale in geographic research and thought, including phenomenon scale, analysis scale, and cartographic scale. We finish the chapter by going over some ways to generate research ideas, including a concise "plan of action" for generating good research questions and designing studies to address them.

In Chapter 3, **Data Collection in Geography: Overview**, we discuss the difference between primary and secondary sources of data. We introduce the various types of data collection in geography that we cover in subsequent chapters, grouped into just five types. Finally, we provide an introduction to the distinction between "quantitative" and "qualitative" methods in research, rooting it in its historical development in different social-science and humanities disciplines. We interpret the distinction in terms of data collection and analysis methods, although we do not favor drawing the distinction in an excessive way. Some of the distinctions people occasionally make between the two are misleading or not useful, however, and we think they contribute to the unnecessary philosophical and methodological chasm one sometimes sees between scientific research in human and physical geography.

Chapter 4 is entitled **Physical Measurement.** In it, we discuss a broad type of data collection that is very popular in geography—in fact, the most common way data are collected in most areas of physical geography. Physical measurements are collected by recording physical properties of the earth surface, the ocean, the atmosphere, and the biota. Geographers measure a great variety of physical properties, using a variety of tools and techniques. Physical properties include size and number, temperature, chemical makeup, moisture content, texture and hardness, the reflectance and transmissivity of electromagnetic energy (including optical light), air speed and pressure, and more. Physical measurements are often made via aerial and satellite remote sensing, which we discuss in Chapter 12. Human geographers also make physical measurements, often by observing the "physical traces" left behind by human behavior or activity. For example, geographers observe house design and agricultural patterns.

In Chapter 5, entitled **Behavioral Observations and Archives**, we discuss two types of data collection that human geographers often use. We place these two together mostly for convenience; they are not that closely related, but neither requires an entire separate chapter. Behavioral observation is the systematic observation of people's actions. Archives are records of events or characteristics that were generally collected for purposes other than scientific research (financial records or crime statistics, for instance). Behavioral observation and archives do share the fact that as measures of human activities, both are often relatively "nonreactive" as compared to the explicit reports of the next chapter. That is, they often do not require the people being studied to realize that they are being studied, a realization that can change people's actions. Finally, both types of data collection often produce "records" that do not serve directly as data but must be coded (classified) to serve as scientific data. We discuss how to do coding.

Chapter 6 is entitled **Explicit Reports: Surveys, Interviews, and Tests.** In this chapter, we cover one of the most flexible ways of collecting data in human geography.

Explicit reports are written surveys (questionnaires) and oral interviews that assess beliefs, attitudes, activities, and demographic characteristics. They also include tests, which assess factual knowledge. Explicit reports are usually verbal but can be designed in a number of other formats as well. They may be administered in a variety of ways, via a variety of administration media, including the Internet. We discuss ways to structure report items, especially the distinction between closed-ended and open-ended items. We give specific instructions for designing different types of closed-ended items. We also discuss ways to generate items—how do you create questions for a survey, for instance? We also cover secondary report data, particularly the very important secondary source of U.S. census data. We conclude with some interesting, more conceptual, issues about the limits of explicit-report measures, such as the fallibility of memory, the impenetrability of certain aspects of mental activity, and the imperfect relationship of mind to behavior.

In Chapter 7, **Experimental and Nonexperimental Research Designs**, we discuss the topic of research designs—the structure of manipulations/comparisons in a study that determines which questions can be asked of data, and with what degree of validity they can be answered. We extensively consider the distinction between true experiments and nonexperimental studies, as this is such a fundamental distinction with respect to the validity of causal conclusions in research. We discuss three ways of increasing the validity of causal conclusions in research: physical, assignment, and statistical control. Assignment control is the manipulation of variables, which is at the heart of the experimental/nonexperimental distinction. In this light, we introduce the concepts of control groups, random assignment, independent and dependent variables, and quasi-experiments. We explain the distinction between lab and field research. We present some specific research designs, both between-case and within-case. We present developmental, or temporal-change, designs. We also discuss single-case and multiple-case designs. We finish with a section on computational modeling, an approach to research design that is also a type of data collection and method of analysis.

Chapter 8 is entitled **Sampling**. We define a population as the entire set of entities you wish to generalize to; a sample is any incomplete subset of that. What constitutes a population and what constitutes a sample therefore depends on your research goal—what you intend to generalize to in a particular situation. We make it clear that obtaining entire populations would always lead to more certain, and therefore better, research conclusions. In most basic-science research, however, the population is a hypothetical entity that cannot be obtained in entirety, if for no other reason than that it includes entities that no longer exist and entities that do not exist yet. We also explain that researchers do not just sample cases (people, rivers, cities, soil regions, and so on), but virtually all other components of research (such as times and places of measurement, survey questions, sensors) as well. We introduce the important concept of the sampling frame, which is the set of entities from which samples are actually drawn. The sampling frame is usually a subset of the population to which one wishes to generalize. We describe basic sampling procedures, the way sampling frames are chosen from populations, and then the way individual entities are chosen from sampling frames. We organize our description of procedures in terms of nonprobability and probability sampling. Although nonprobability sampling technically does not allow the derivation of probabilities to be used in inferential statistics, it is nonetheless quite commonly done in scientific research. We go over various specific types of probability sampling, including simple random, stratified, and cluster sampling. We discuss the implications of particular sampling frames and procedures. We also consider the possible effects of nonparticipation and volunteer bias. Then we go over some of the unique implications of sampling entities distributed over space and time. Finally, we consider the issue of sample size, including recommendations for determining adequate sample size and an introduction to formal power analysis.

Chapter 9 covers the topic of **Statistical Data Analysis**. We explain why data analysis in geography is typically statistical (probabilistic) in nature—primarily because it is conceptualized as resulting from a sampling procedure and measured with error, but also because of the complex multivariate nature of most geographic phenomena. We describe the two components of most data analysis: (a) describing properties of data sets and (b) inferring properties of population data from properties of sample data. We introduce basic statistical concepts and focus on giving a clear explication of basic statistical logic. In particular, we carefully present the conceptual basis for hypothesis (significance) testing. However, we do not present a detailed tutorial on statistical analysis in this chapter; other courses exist to cover this important and rich topic, and readers should take at least a couple of them. We finish by reviewing some of the special properties of geographic data that arise from their spatiality. These "spatial" properties often lead to some interesting consequences for data analysis in geography, either because the properties are the subject of geographic theories or because the properties strongly influence the validity with which otherwise nonspatial data are interpreted.

Chapter 10 covers **Data Display: Tables, Graphs, Maps, Visualizations**. We discuss the purposes of data displays, including the initial examination of data, the interpretation of their meaning, and the communication of data and their meaning to other people. All of these purposes boil down to effective communication as the guiding principle in designing and using data displays. We go over several types of displays, including tables, graphs, maps, and computer visualizations. We discuss maps relatively briefly; mapping is so important that geographers should take at least one separate course on it. We provide advice on choosing the right display type and designing it so it is effective at its purpose. We finish by briefly discussing emerging trends and technologies in visualization, including animations, sonifications, spatializations, and virtual reality.

Chapter 11 is entitled **Reliability and Validity**. In it, we discuss the two fundamental concepts of reliability and validity that are germane to all scientific research. Reliability is the "repeatability" of scores or measured values of variables. High reliability occurs when you measure something twice and get the same value each time, assuming the thing you are measuring is actually the same at the time of the two measurements. Low reliability is caused by random errors of measurement and has a variety of detrimental effects on research outcomes. We go over techniques for assessing and increasing reliability. Validity is the "truth-value" of research results and interpretation. Given that research is largely an attempt to increase the truthfulness of our understanding of the world, validity is a core concern for researchers. Following an influential typology, we discuss four classes of validity: internal, external, construct, and statistical conclusion. We finish by discussing certain special classes of validity threats that arise because researchers are human, and in human geography, their research subjects are human as well. These "participant and researcher artifacts" can produce biased and distorted data that reflect various expectancies or beliefs people have about research situations, or about classes of people being tested or doing the testing.

Chapter 12 concerns **Geographic Information Techniques in Research**. We describe geographic information and define geographic information techniques to include geospatial analysis, cartography, remote sensing, and computer geographic information systems (GISs). We cover the first two in Chapters 9 and 10, so this chapter focuses more on the fundamental nature of geographic information and research applications of the last two techniques, remote sensing and GIS. The data collected via remote sensing qualify as physical measurement, covered in Chapter 4; in this chapter, we consider aspects of physical measurement especially relevant to aerial and satellite sensing. We also summarize some characteristics of our planet earth and the representation of its surface in globes and digital databases. We explain some basics of coordinate systems and Global Positioning Systems (GPSs). We finish by overviewing basic GIS operations.

Chapter 13 is about **Scientific Communication in Geography**. In it, we discuss both formal and informal scientific communication in geographic research, both to other geographers and to lay audiences. We provide a list of forms of scientific communication, including journal articles, books, grant applications, and oral presentations. We describe the "peer review system" for academic publishing and grant applications. We give specific instructions for how to put together an empirical research paper, including its parts, general structuring, and so on. We cover some aspects of writing style, although we recognize that geography has neither a single accepted style nor a universally accepted style manual. We also discuss aspects of giving effective oral presentations. We then turn to a discussion of the most important storehouse of scientific communications—the library, whether "brick-and-mortar" or electronic. We include detailed advice on how to perform scientific literature searches.

Our final chapter, Chapter 14, is a discussion of **Ethics in Scientific Research**. Ethics is the study of moral or proper action. Research ethics involves a variety of rights and responsibilities, including those of the researcher, the people who work with the researcher (including human research subjects), and society at large. In fact, many people now accept that ethical concerns extend to the animals, plants, places, and cultures that play a role in the geographer's research. We go over guidelines for the ethical treatment of places encountered on research field trips. We also cover specific rules and guidelines for the ethical treatment of human and animal research subjects, including rules that geographers who work with human or animal subjects must follow, typically as part of the requirements of "Institutional Review Boards."

Acknowledgments

We would like to thank everyone who gave us input and advice on this book. Melanie V. Gray went above and beyond the call of collegiality and friendship in this respect. She read all the chapters, providing many suggestions as to content, grammar, and writing style. Most of the faculty members in the Geography Department at UCSB provided some direct input, and the rest provided indirect input. Oliver Chadwick, Sara Fabrikant, Catherine Gautier, Michael Goodchild, Phaedon Kyriakidis, Joel Michaelsen, Dar Roberts, Dave Siegel, Stuart Sweeney, Waldo Tobler, and Libe Washburn deserve specific mention. We thank several of our students over the years for the sounding board they provided and feedback that helped shape our ideas; this feedback often came in the form of questions like "What in the world are you talking about?" We also thank the many teachers over the years who inspired and informed us, especially about logical, statistical, and methodological issues. We appreciate the people who reviewed earlier versions of portions of this text and provided helpful comments, including Donald Friend, Kathy Graham, Juliana Maantay, Rod McCrae, and several anonymous reviewers. We are very grateful to Sage Publications for the opportunity they have provided to write this book and have it published. Special thanks go to Robert Rojek at Sage, who solicited the proposal for the book and championed it through to fruition. Finally, we wish to express our admiration and appreciation for Reg Golledge. Not only does he continue to teach us much about geographic research, but it was he who "admitted" both of us to UCSB back in 1992, eventually making this book and our friendship possible.

CHAPTER 1

Introduction

A Scientific Approach to Geography

Learning Objectives:

- What is a scientific approach to geography?
- How is science both an individual and a social activity?
- What are several metaphysical beliefs characteristically held by scientists?
- What are four goals of scientific activity?
- What are the relationships of natural science, social science, and the humanities to the study of geography, currently and throughout its history?

John[1] was pursuing his Masters degree in geography. He was interested in geographical factors that contribute to causing social ills, such as violent crime, in inner cities. Having read some of the literature on this subject, John had discovered the concepts of "associative" and "dissociative" institutions. The first are thought to create community identity and social cohesion; churches might be an example. The second are thought to destroy community identity and social cohesion; crack houses might be an example. John theorized that, "social decay in the inner city is caused by a prevalence of dissociative, rather than associative, institutions." To test his theory, John looked at the city of Milwaukee (it was convenient for him). He got data from the police department on the number of suicides and homicides that had occurred in the previous 10 years in Milwaukee. He also looked in the phone book yellow pages for the Milwaukee Metropolitan Area, which

[1]Although our real experiences over the years inspired John's story, he is fictitious and does not refer to any single real person.

includes suburbs and peripheral areas as well as the urban core of Milwaukee. From the phone book, John counted the number of liquor stores, noting their addresses. He then organized his data into census tract units. Census tracts, created by the U.S. Census Bureau, include regions in cities at about the size of neighborhoods where between 3,000 and 8,000 people reside. John thus assigned each census tract two numbers, the number of "wrongful deaths" and the number of liquor stores. He calculated a Pearson correlation coefficient on these two variables, a statistical index that identifies linear patterns of relationships between two metric-level variables. He found a positive correlation of .31, which suggests that census tracts with more liquor stores in his data set were somewhat more likely to have more wrongful deaths, at least within the previous 10 years. John concluded that he had proven that dissociative institutions cause social decay in inner cities, and he recommended getting rid of liquor stores in inner-city areas.

Should we accept John's conclusions and agree with his recommendation? Probably not. There are numerous problems with the way he conceived, conducted, and interpreted his study. For instance, the yellow pages lists most businesses, but not all. Why look only at liquor stores and not bars? In Wisconsin, people often purchase alcohol in grocery stores and small markets. Shouldn't John have looked at other potentially dissociative institutions, such as adult clubs, gambling parlors, or criminal organizations? His theory is about the presence of dissociative institutions relative to associative institutions, but he didn't even look at associative institutions. What about other indicators of social ills besides murder and suicide, such as assault or rape? Are there other factors that we might expect to be related to the incidence of murder and suicide that vary considerably across census tracts in Milwaukee? Good candidates are socioeconomic status (SES), age, residential density and housing style, housing tenure (ownership status), ethnic makeup, citizenship, and immigration history. John used census tracts as the unit of analysis because of convenience, but are census tracts the proper unit of analysis for the concepts that interested him? And why Milwaukee in the first place? Are there any special characteristics of Milwaukee that make it less representative of cities, including inner cities?

Our story about John's research and its faults and limitations provides a concrete introduction to the topic of this book: **scientific research methods**. Scientific research methods (or methodologies) are the suite of techniques and procedures for empirical scientific investigation, along with the logic and conceptual foundations that tie scientific investigations together and connect them with substantive theory. The topic of research methods clearly touches on many issues important to researchers in all natural- and social-science disciplines, including geography. Research methods concern which problem domains are studied; which specific ideas within the domain are investigated; what entities are studied; what is observed or measured about the entities; how they are observed; where, when, and how many observations are collected; how the observations are analyzed (including graphing, mapping, statistical analysis, or simple tallying); what patterns are in the observations and whether the patterns can be generalized to some larger population of entities, times, or places; what explains the patterns in the observations; and even what the observations say, if anything, about solving practical problems. This is an impressive list. What's more, all of these issues are potentially relevant not only to how we carry out our own research but to how we interpret others' research. The study of research methods is thus central to deciding what conclusions we can draw about the meaning of research, the contexts in which these conclusions hold, and the degree of confidence we have in these conclusions. In other words, you cannot competently carry out or critique scientific research without considerable knowledge of methods.

Overview of the Logic and Philosophy of Science

Let's consider what makes an activity scientific research. What is a **scientific approach**? There is no precise answer to this question. Like art or cheeseburgers (does it count when the "meat" is soy protein?), science is a somewhat vague concept that includes clear central examples but also many examples that most people would agree are more or less scientific, rather than clearly and definitely examples of science or not. That said, we can start with this simple and fairly inclusive definition: *Science is a personal and social human endeavor in which ideas and empirical evidence are logically applied to create and evaluate knowledge about reality.* Let's consider a few components of this definition. Science is a personal and social human endeavor because it is something humans do, as individuals and as social groups. Individual scientists learn from other scientists, work with colleagues and assistants, and act within various cultural and institutional contexts. **Empirical**[2] evidence is derived from systematic observation of the world via the senses, often aided by technology. The systematic nature of scientific empiricism crucially distinguishes it from the observations we all make informally every day. Because science aims for stable and publicly consensual truth, scientific empiricism strives to be repeatable, accumulable, and publicly observable. A necessary reliance on empirical evaluation is, to a large extent, the hallmark of scientific activity, as opposed to other human enterprises that strive to understand the world (more on this below). It helps differentiate science from intuition, authority, anecdote, profitability, physical or political power, the need for happy endings, and other approaches and motivations. Ideally, scientists apply ideas and evidence according to certain formal and informal logical principles in science. It is not possible to give a finite and complete list of these principles, but they certainly include such things as the following: (1) One must avoid contradictions, (2) our confidence in a phenomenon increases as our observations of it increase, and (3) past regularities will probably recur in the future.

The relationship in science between ideas and evidence deserves further comment. Scientists use ideas to design **studies**—units of focused observation or data collection—and to interpret their results. Scientists explain patterns in their empirical observations by reference to ideas about reality. But they also understand that

---

[2]By the term "empiricism," we are not referring to the philosophical position of Empiricism that holds that all knowledge is ultimately derived from experience after birth.

any empirical observations can potentially be explained not just by ideas about reality but also by ideas about the way they obtained or interpreted the observations. That is, scientists consider that a pattern of observations may reflect such empirical factors as biased instruments, idiosyncratic testing environments, unusual samples, and so on—not just the phenomenon under study.

Notice that our definition does not restrict science to just the physical or biological world. Science is also concerned with the world of human activity, artifact, and institution. There are **natural** (biophysical) **sciences** and **social** (human) **sciences**. This is especially important to recognize in a discipline as broad as geography, which involves both biophysical and human sciences; as we discuss below, it involves humanities, engineering, and craft as well. This text deals with scientific methodology for all of geography, including the biophysical and human domains. Therefore, we always use the generic term "science" inclusively in this text to mean both natural and social sciences.

Our definition also avoids claiming that science restricts itself to one specific approach to logic. The history of debates about the proper way to do science includes numerous claims that, for instance, "real" science applies hypothetico-deductive reasoning, in which scientists use prior hypotheses to deduce observational consequences that they can then compare to empirical evidence. Others have claimed that science is inductive in nature, relying on initial observations to generate hypotheses about reality.[3] But scientists use both **deductive** and **inductive** approaches; we find it misleading to claim that one is generally more common or appropriate than the other.[4] In fact, although our definition highlights logical thinking, it makes no claim that scientists think exclusively in a logical manner. Like artists and other nonscientists, scientists gain insights and create new ideas in any number of different ways that would not be considered strictly logical, including intuition, fantasy, and the like (we look at these further in Chapter 2 when we discuss how to generate research ideas). Clearly, scientists often come to understand a phenomenon through a process of insight, an inferential process that seems to leap from premises to conclusions with no conscious systematic reasoning plan. This form of reasoning is sometimes called **abductive**.

Finally, although our definition points out that science includes both an idea part and an empirical part, it does not claim that every individual scientist or laboratory must engage in both parts equally. Although a science such as physics has become so specialized that some physicists describe themselves as "theoretical" and others as "empirical," all physicists recognize that the full activity of physics includes both theory and empirical observation. Scientists believe that they produce empirical observations in order to evaluate and generate ideas about reality, and they

---

[3]In 17th–19th century philosophical debates, Rationalists and Empiricists championed idea-first and observation-first approaches to science, respectively.

[4]Although deduction is sometimes defined as deriving specific truth from general truth, and induction as deriving general truth from specific truth, the distinction actually refers to the certitude of inferences one makes with each type of logic—deduction definitely leads to true conclusions, whereas induction only probably leads to true conclusions.

believe that someone, eventually, needs to empirically evaluate the ultimate truth of ideas about reality. To look at it another way, ideas about reality suggest studies to conduct and ways to explain the observations that result from those studies. "Theoretical" scientists may not collect and analyze empirical observations, but they believe it is important that someone does. In other words, as we stated above, the dual components of science describe a *social* activity, not just an individual enterprise.

Characteristic Metaphysical Beliefs of Scientists

In addition to our short definition of science, we believe it is useful to identify a set of **characteristic metaphysical beliefs** or intellectual preferences held by most scientists. As we said above, it is probably impossible for anyone to give a strict definition of the scientific enterprise that actually succeeds at including all instances of scientific activity (past, present, future) while excluding all activities that are not scientific. Delimiting the meaning of a concept like science depends on the nature of human conceptualization, not just the actual reality to which the concept refers. Furthermore, the human activity called science has evolved over the centuries (if not longer) in a somewhat haphazard way—no overarching "creator of science" defined and implemented it. Thus, over time, fairly different activities have been considered better or worse examples of scientific activity. However, we believe that we can identify characteristic beliefs that help us understand what is more scientific than not, and when someone probably is or is not carrying out science. These are metaphysical because they concern beliefs about the ultimate nature of reality (**ontology**) and how a scientist can know about it (**epistemology**). We think a majority of practicing scientists hold these beliefs but don't consider them essential to the definition of science. For example, we would not claim that a person who does not believe in the existence of a world independent of sentient minds could not be doing scientific research. We also call these beliefs "preferences" because they are just that—personal preferences or intuitions. They are unproven, and they are unlikely to be provable, to be the best possible avenues to truth; that is, they are elements of faith!

*1. **Realist philosophy.*** Nearly all scientists at least implicitly accept a philosophy of *realism.* They believe the universe actually exists, independently of *sentient* (thinking, feeling) beings, as matter and energy patterned in space and time. The matter and energy coheres into meaningful pieces (entities and events) but is also organized into meaningful pieces by the sentient beings.

*2. **Only continuously connected and forward causality.*** This might be thought of as an extension of the belief in realism, but we find it valuable to note it separately. Scientists tend to insist that causes and effects are continuously connected in space and time, and only in a temporally forward direction. That is, cause A can only bring about effect B if A's influence can move forward "densely" in space and time; the space and time between A and B is continuously filled with causal connections that transmit the cause to the effect. Put another way, the patterned matter and energy that is the physical instantiation of causal influence cannot get from A to B without traveling continuously between the space and time separating the two. In Chapter 2, we discuss philosophical and scientific ideas about causality in greater detail, including issues surrounding its forward and continuously connected nature.

3. Simplicity. Scientists prefer the simplest explanation that is adequate. This is often called the principle of **parsimony**[5]; it is also largely captured by the notion that scientists like ideas that are "elegant." Because of their preference for simplicity, scientists prefer general-purpose truths to idiosyncratic truths. We return to this issue in Chapter 7 in our discussion of "nomothetic" (general, lawlike) versus "idiographic" (specific, idiosyncratic) approaches, but we note here that extremely idiographic approaches are essentially nonscientific, however true they may be. We must stress that parsimony still requires explanations to be adequate, in terms of fitting with observations and other ideas that are already accepted. Some people mistakenly believe that a preference for parsimony is a blind requirement to pick the simplest idea in all cases. We also note that although relative simplicity is often fairly obvious (a model with three parameters is simpler than one with five parameters), in other cases it may be a deep intellectual question as to what constitutes greater simplicity.

4. Skepticism. Although scientists are searching for truth, they doubt they will find it in absolute form. Theories, for instance, are considered provisional even when widely and repeatedly supported by empirical evidence. Partially because of their skepticism, scientists dislike ideas that cannot potentially be falsified by evidence or inconsistency with other ideas.[6] Also in line with their skepticism, scientists typically entertain *chance* as a first explanation for patterns in their observations; they must first discard chance before more substantive explanations warrant attention. This logic of falsification and the entertainment of chance are fundamental to the statistical analysis of data, to which we return in Chapter 9.

5. Quantitative thinking. Scientists apply observations and logical thought in order to achieve understanding. To this end, they like precision, of both ideas and observations (we define and compare precision and accuracy in Chapter 2). To increase

---

[5]The principle of parsimony is often referred to by the charming name "Occam's Razor." William of Ockham was a medieval English philosopher and Franciscan monk who favored minimalism in life, a view expressed in his famous dictum that "plurality should not be posited without necessity."

[6]The 20th-century philosopher Karl Popper offered extensive arguments for why we need to rely on falsifying wrong ideas and not on confirming true ones. His work is part of a larger tradition of philosophical debate about scientific epistemology. As a point of history, however, we do not believe scientists have ever or will ever stick only to falsification or disconfirmation as an epistemological strategy, nor do we believe they need to. Nonetheless, whatever logical value falsification has over confirmation, we believe their preference for skepticism leads scientists to focus more on disproving than proving.

the precision of ideas and observations, scientists often turn to mathematics and computation. When feasible, they often express theories as mathematical equations. They also attempt to carry out observations of the world very carefully, avoiding distorting effects as much as possible. One way they satisfy these preferences is to develop new technologies of observation (procedures, tools) that can extend the ability of the senses to observe the world. Such new technologies have historically extended the reach of scientific observation and hence advanced scientific ideas; they include the telescope, the computer, the chromatograph, and the methods of psychophysics (by which one can quantify people's perceptual responses). We note, however, that although the use of mathematics and technology is desirable to scientists, it is not required in order to classify something as scientific. Less-developed disciplines or those whose problem domains are more complex may still be scientific even if they rely on relatively little sophisticated mathematics and technology.

Nonscientific Ways of Knowing

We can contrast our definition of science and list of characteristic beliefs of scientists with various nonscientific ways of knowing. At most liberal arts colleges or universities, the major nonscientific approaches to knowing are applied in the **humanities**, traditionally including history, philosophy, languages and literature, art history, and so on; below, we discuss the fact that much geographic research is carried out in the tradition of the humanities. The humanities are like science in their logical application of ideas in order to understand reality, specifically the reality of human existence, but for the most part they do not employ systematic empirical observation of reality as does science; instead, they often informally analyze texts and other symbolic artifacts of human thought, activity, and culture (unlike the systematic coding of archives discussed in Chapter 5). The humanities are thus rarely mathematical in their work.[7] Perhaps scholarship in the humanities is even more distinctive because of a difference in the type of understanding for which it strives. Scientists want general truth, as reflected in their faith in simplicity as a guiding principle, whereas scholars of the humanities want specific truth about people or societies in particular places and times. In addition, humanities scholars are often concerned with promoting ideas about human values and morality that they cannot easily justify objectively or measure quantitatively. We return to these stylistic differences in Chapter 7.

A variety of other approaches to knowing are nonscientific because they do not pursue general knowledge of reality, do not apply systematic empiricism, or strongly oppose one or more of the characteristic beliefs of scientists. Artists (in the visual arts, music, dance) arguably aim for general knowledge but are not systematically empirical in their methods, nor are they likely to endorse many of the characteristic

---

[7]The huge exception is mathematics itself, which in traditional form is quantitative logic rather than science. Mathematics is a common language of science but is itself primarily a branch of the philosophy of logic (some recent approaches in mathematics are empirical, however).

metaphysical beliefs of scientists. Various crafts and vocations do not have general knowledge as their aim; they are typically about doing something or producing tangible products, rather than knowing something. As with the arts, a variety of approaches to knowing that might be called *spiritual* (religion, mysticism) do not typically employ systematic empiricism, nor do they embrace the characteristic beliefs of scientists. In particular, they tend to take explicit issue with the realist philosophy and skepticism of most scientists. Finally, practitioners of the "paranormal" (astrology, tarot, extrasensory perception) often eschew systematic empirical evidence, but even when they welcome it, they tend to lack skeptical attitudes about their beliefs. However, psychic phenomena have been the subject of genuine scientific research. Evaluations of the meaning of this research are mixed, and skepticism about it is quite strong, perhaps mostly because many paranormal ideas so clearly violate principles of forward connected causality.

However, we must appreciate these claims about the nature of nonscientific ways of knowing in the context of certain limits we see to scientific understanding. First is that our description of science is an ideal that we rarely attain in practice. Scientists are human beings acting within social, institutional, and cultural contexts. They have imperfect personalities and are sometimes motivated by greed, egotism, or prejudice. Although scientists as a group endorse skepticism, individual scientists sometimes fail to apply adequate skepticism to their own ideas. But here again is where the social nature of science is critical. Social mechanisms, such as peer review of scientific reports (see Chapter 13), serve to blunt the distorting effects of individual human fallibilities on scientific research.

However, we see another limit to scientific understanding as more fundamental. That is simply that one can accept the value of a scientific approach without believing that science is the *only* valid and useful way of knowing. Being nonscientific does not mean that an approach to knowing is necessarily wrong or useless or irrelevant. Many of the most important questions of interest to humans cannot or should not be answered scientifically, although in some cases scientific results may inform them. What is the meaning of human existence? Why is it wrong to hurt people? What is beauty? Is there a God? What is the best form of government? Why should I get out of bed in the morning? Do I want chocolate or vanilla? We believe that some overzealous promoters of a scientific approach might occasionally fail to stress this adequately. We also believe that some critics of scientific approaches, especially when applied to the study of humans, fail to appreciate that reasonable promoters of a scientific approach recognize that it has limits. We think of science as an interesting and useful way to grasp some truth about reality, including human reality, and we recognize that we pursue it in part because we personally enjoy scientific thinking. We do not, however, consider it the royal road to all truth and enlightenment.

Goals of Science

According to our definition, the purpose of science is "to create and evaluate knowledge about reality." We can elaborate on this purpose in terms of four goals toward which different sciences and scientists strive, to various degrees. The goals are intellectually progressive, in that goals lower on the list presuppose some mastery of those above them. The goals have also largely been historically progressive, in that scientific disciplines have tended to focus more on goals farther down the list as their ideas and empirical techniques developed over time; more mature sciences thus tend to focus more on "lower" goals. The four goals, ordered progressively, are

1. Description. Whatever their domain of interest, scientists must distinguish and describe the basic phenomena (entities and events) within that domain. This is essentially the intellectual act of classification (categorization) common to all sentient creatures, but scientists often carry it out especially systematically.

2. Prediction. Given that they know something of the content of their domain, scientists want to be able to predict phenomena about which they cannot learn simply by direct observation. These predictions are often about the future, but can also concern facts about phenomena from the present or the past that are as yet unknown. The most powerful tools for prediction available to scientists are inferences (both extrapolations and interpolations) from patterns of observations; these inferences take advantage of mathematical precision while exploiting the logical principle that observed regularities will probably hold in other situations not yet observed. We discuss the logic of prediction further in Chapter 9.

3. Explanation. Once scientists can describe and then predict, they want to explain *why* some described and predicted pattern exists. This requires the explication of causal relations among entities and events. As we mentioned above, we discuss the logic and philosophy of causality more in Chapter 2; in Chapter 9, we consider the relation of causality to prediction in the context of data analysis; in Chapters 7 and 11, we discuss how research designs and techniques strengthen or weaken conclusions about causality in empirical studies.

4. Control. Finally, being able to describe, predict, and explain phenomena within their domain of interest, scientists (and those who fund scientists) typically want to apply this knowledge in order to control the phenomena—to bring about desired changes in the phenomena. Now that I understand erosion, can I prevent it? Now that I understand the development of a globalized economy, can I make sure it happens in a way that preserves economic fairness and environmental health?

A distinction is often made between **basic** and **applied** scientific research. Basic research focuses on understanding reality for its own sake; it is primarily an expression of human curiosity and the desire for intellectual mastery. In terms of the goals of science, basic research is very concerned with description, prediction, and explanation, but less with control. In contrast, applied research focuses on control, in addition to the first three goals, for the purpose of making some object or procedure that will help meet specific practical needs or solve specific problems. Although engineering is often contrasted with science, because it is concerned more with making something work (or work better) than with understanding how something works, we can see that engineering might accurately be considered applied science. Similarly, much medical and educational research is applied science. Both basic and applied foci are prevalent in geographic research. Like the definition of science itself, however, the distinction between basic and applied science is somewhat vague and should not be overstressed. Many scientists work in both arenas to various degrees. Optimally, there is interplay between basic and applied science in which the results and needs of each inform and motivate the other.

Before leaving our discussion of the goals of science, we need to mention a few caveats. The first is that scientists sometimes violate the progressive quality of the goals. Although later goals presuppose earlier goals, this need only be partially true. For instance, explanation requires prediction, but not anything like perfect prediction—it only has to be at least better than chance prediction. Or to take another instance, a certain amount of practical control can be exerted over phenomena without having a complete explanation for them; applied sciences like engineering often focus on successful control to the point of happily applying trial-and-error approaches that can lead to control without understanding. Our second caveat about the goals is that scientists may be able to predict phenomena at some scale of analysis but not at others that are smaller or larger, despite feeling like they have a fairly complete explanation of the phenomena. This is especially relevant to geography, concerned as it is with phenomena that exist and interact at a wide range of scales (we discuss the concept of scale in Chapter 2 and the problems of analysis related to scale in Chapter 9). Finally, an understanding of the ultimate limits of prediction was one of the great intellectual achievements of the 20th century, when we recognized that very small events have the power to radically alter the future (the "butterfly wing effect"). Thus, prediction in complex systems has ultimate limits because of the possibility the system will enter into "chaotic" states that we cannot predict, even with complete prior knowledge. Our ability to predict weather will apparently always be limited in this way, for example. A related intellectual achievement of the 20th century concerns ideas developed by quantum physicists about limits to the traditional notion of causality (more in Chapter 2).

History and Philosophical Systems of the Discipline of Geography: Natural Science, Social Science, and Humanities

We finish this chapter with a short overview of the history and philosophical systems of the discipline of geography, and their relation to scientific and other approaches to knowing. Traditionally one might define **geography** as *the study of the earth as the home of humanity* (the word's literal meaning is "earth writing"). A more modern and impressive-sounding definition is that geography is *the study of the distribution of human and natural structures and processes over the earth's surface, and the role of space and place in understanding these human and natural structures and processes.* Like other disciplines, the domain, methods, and philosophical foundations of geography have changed over the centuries. In fact, geography has arguably gone through even more intellectual changes than other traditional disciplines, especially during the 20th century. The result of all this is that geography is an extremely broad and heterogeneous discipline. Many books discuss these changes (see the bibliography at the end of the chapter), and we touch upon them only briefly here. We heartily recommend reading these books and taking a course on the history and philosophical systems of geography.

Geographical thought perhaps began when humans first recognized that different places have different characteristics ("areal differentiation"): The land surface varies, plants vary, people look and sound different, and so on. Surely this occurred long before writing first appeared. A more formal study of geography probably began in the ancient worlds of Africa, the Middle East, and the Far East, as part of astronomy and land surveying. From these early days, military activity was also a major impetus for the development of geographic knowledge of all kinds, including the measurement of the earth (**geodesy**), and the description of its human and natural variation; the military motivation for geography continues to this day. Trade was another early motivation for accumulating geographic knowledge. And the logs and diaries kept by travelers and explorers over the centuries provided a rich source of descriptions (occasionally accurate) of faraway places. These early intellectual endeavors provided the seeds for the diverse approaches of modern geography. Thus, geographers from the beginning applied a mixture of linguistic, graphic (including cartographic), and mathematical approaches as part of their intellectual activity. Although the relative mixture of the three has shifted over the history of the discipline, geographers still apply all three today.

By the time geography emerged as a separate academic discipline in the 19th century, it had developed a venerable tradition of characterizing places and regions in terms of the totality of their natural (geomorphological, climatological, botanical, and so on) and human (cultural, economic, political, and so on) characteristics. This approach is called **regional geography**. Regional geography is still a part of the discipline of geography, of course, and it is perhaps what most lay people think primarily constitutes the subject matter of geography; it is sometimes called the "*National Geographic* approach." However, during the 19th century, as academic specialization flowered, a different approach began within geography. This approach focused on particular topical areas or "systems" within the domain of geography, trying to describe and, even more, explain the workings of these systems wherever they found expression on the earth. A practitioner of **systematic geography** might therefore study river systems or urban structure anywhere they occur.

Many scholars who championed the systematic approach during the early 20th century thought it made geography look more like other sciences, which were continuing to develop depth, perhaps at the expense of breadth. A penchant for applying mathematics and the application of a strict interpretation of positivist philosophy also characterized this quest for scientific respectability. It also contributed to the division in geography between those who specialized in the natural aspects of the earth and those who specialized in the human aspects. This all culminated in the so-called quantitative revolution of the mid-20th century. Particular scholars and departments championed statistics, geometry, calculus, computers, airplane and satellite remote sensing, and then (a little later) geographic information systems (GIS) as the "right" way to do geography.

Almost as soon as the quantitative revolution occurred, however, a counter-revolutionary response criticized the supposed limits of quantitative approaches. To shorten a complicated story, these criticisms charged especially that **positivist** geographers oversimplified human experience and activity to the point of caricature. Instead of the clean and precise abstractions of scientific modeling and analysis, these critics called for approaches that recognized a messier, more subjective, and self-willed geographic reality. Furthermore, according to these critics, geographic reality was often the expression of unequal power relations among various stakeholders. These **post-positivist** critiques came in a large variety of flavors during the later 20th century, including phenomenology, Marxism, feminism, social theory, deconstructionism, and postmodernism. There are important differences among these positions, of course; the bibliography at the end of the chapter provides some relevant readings.

The situation today is that of a pluralistic geography. Both regional and systematic approaches are evident. Geographers apply linguistic, cartographic, and mathematical methods in a bewildering array of combinations. This is especially true within human geography, where the perspectives of a plethora of disciplines, both social science and humanities, inform the study of human experience, activity, society, and culture. Across the breadth of geography, scholars study an enormous assortment of specific topics. Human geographers investigate transportation, migration, population, cultural distribution and diffusion, communication, economic activity (production, consumption, buying and selling), regional development, recreation and tourism, place perception and identification, spatial and environmental thought, urban structure and change, and resources and hazards. Physical geographers investigate landform formation and change, soils and minerals, lakes and rivers, groundwater, climate and atmosphere, plant and animal distribution, glaciers and ice fields, and ocean and coastal processes. Yet other geographers specialize in the refinement and development of new geographic information methods and techniques that cut across the human/physical distinction, including GIS, database design, cartography and visualization, remote sensing, spatial theory and analysis, and geostatistics.

Given its very broad subject matter and pluralistic nature, geography in the early 21st century is remarkably **multidisciplinary** and **interdisciplinary**.[8] Physical geography overlaps with most of the physical and life sciences, especially the earth and environmental sciences of geology, biology, ecology, oceanography, hydrology, climatology, and atmospheric science. Human geography overlaps with most of the social and behavioral sciences, especially sociology, economics, anthropology, psychology, and political science. Alternatively, a great deal of human geography overlaps with the humanities, especially history, literature, philosophy, art, and cultural studies. Various technical specialties within geography overlap with engineering of several kinds, as

[8]The distinction being the degree to which a collection of multiple disciplines, each with its own concepts, vocabulary, and methods, is integrated into a single hybrid "interdiscipline."

well as mathematics and computer science. Last but not least, some areas of geography focus on the expression of "geographic craft." It is still, however, the case today that for many of us, the real promise of the field is the integration of the natural, human, and technical aspects of "the study of the earth as the home of humanity."

Review Questions

- To what does the phrase "scientific research methods" refer, and why is attention to methods important for conducting and interpreting research?

Overview of the Logic and Philosophy of Science

- What are some characteristics of a scientific approach to geography? What is scientific empirical observation, and how does it differ from everyday, informal empirical observation?
- What are the following "characteristic metaphysical beliefs" held by scientists: realism, continuously connected and forward causality, simplicity, skepticism, quantitative thinking?
- What are some common types of nonscientific ways of knowing, and how are they nonscientific?
- What are some important limitations of a scientific approach to knowing?

Goals of Science

- What are the four scientific goals of description, prediction, explanation, and control, and how do they relate to each other?
- What are basic and applied science, and how is this distinction relevant to geographic research?

History and Philosophical Systems of the Discipline of Geography

- What is the focus of geography as a scholarly discipline, and how has this changed historically? What are the regional and systematic approaches to the discipline of geography?
- What are major developments in ideas and approaches of the discipline of geography during the 20th century?

Key Terms

abduction: a type of implicitly logical reasoning that can lead to true conclusions without systematic reasoning from explicit premises

applied science: a style of scientific research that focuses on understanding reality in order to control it

basic science: a style of scientific research that focuses on understanding reality for the sake of understanding

characteristic metaphysical beliefs of scientists: beliefs or intellectual preferences commonly held by scientists about the ultimate nature of reality and how it can be understood; they help to understand the concept of scientific activity

control: the most mature of the four goals of science; being able to bring about desired changes in the phenomena within a scientific domain for practical purposes

deduction: a type of explicitly logical reasoning in which premises definitely lead to true conclusions

description: the least mature of the four goals of science; distinguishing and characterizing the phenomena within a scientific domain, typically by classifying

empirical: evidence derived from systematic observation of the world via the senses, often aided by technology

epistemology: the philosophical study of how people, including scientists, can acquire knowledge about reality; together with ontology, it makes up the study of metaphysics

explanation: the third most mature of the four goals of science, before control; explicating causal relations in order to answer the question of why some phenomenon is the way it is

geodesy: the theory and technology of measuring the size and shape of the earth and the distribution of features on its surface

geography: the study of the earth as the home of humanity; it literally means "earth writing"

goals of science: four specific ways that scientists strive to attain their ultimate goal of understanding reality, including description, prediction, explanation, and control; the goals are intellectually progressive from least to most mature

humanities: nonscientific disciplines that study the human world of individual and social activity, artifact, and institution; they include such disciplines as history, philosophy, languages and literature, art history, and much of human geography

induction: a type of explicitly logical reasoning in which premises probably lead to true conclusions

interdisciplinary: an approach to scholarship that combines two or more traditional disciplines by integrating their concepts, vocabularies, or methods into a new hybrid discipline

multidisciplinary: an approach to scholarship that combines two or more traditional disciplines without integrating their concepts, vocabularies, or methods

natural sciences: scientific disciplines that study the natural, or biophysical, world; they include such disciplines as atmospheric science, biology, chemistry, geology, oceanography, physics, and physical geography

ontology: the philosophical study of the ultimate nature of reality; together with epistemology, it makes up the study of metaphysics

parsimony: a belief widely held by scientists that the simplest adequate explanations are the best

positivism: a philosophical crystallization in the late 19th and 20th centuries of much of traditional scientific belief, explicitly advocating the rationality of such things as mind-independent reality, publicly observable truths that are objectively measurable, and so on

post-positivism: various diverse philosophies developed in the mid and late 20th century that criticize aspects of positivist philosophy as a model of how science is done and should be done

prediction: the second least mature of the four goals of science, after description; guessing unknown phenomena within a scientific domain at better than chance level

realism: a belief widely held by scientists that the universe actually exists, independently of sentient beings

regional geography: a traditional approach to geographic inquiry in which places and regions are studied in terms of the totality of their natural and human characteristics; in contrast to systematic geography

scientific approach: a personal and social human endeavor in which ideas and empirical evidence are logically applied to create and evaluate knowledge about reality

scientific research methods: the suite of techniques and procedures for empirical scientific investigation, along with the logic and conceptual foundations that tie scientific investigations together and connect them with substantive theory

sentient: entities that think and feel, including at least humans and many other animals; sentience has implications in scientific research for how we collect, interpret, and communicate data, and for various ethical considerations

social sciences: scientific disciplines that study the human world of individual and social activity, artifact, and institution; they include such disciplines as anthropology, communications, economics, political science, psychology, sociology, and much of human geography

studies: units of focused observation or data collection

systematic geography: an approach to geographic inquiry that emerged in the 19th century in which geographers study particular topical areas or "systems" within geography wherever they operate on the earth; in contrast to regional geography

Bibliography

Abler, R. F., Marcus, M. G., & Olson, J. M. (1992). *Geography's inner worlds: Pervasive themes in contemporary American geography.* New Brunswick, NJ: Rutgers University Press.

Craig, E. (Ed.) (1998). *Routledge encyclopedia of philosophy.* London: Routledge.

Geography Education Standards Project (1994). *Geography for life: National geography standards 1994.* Washington, DC: National Geographic Research & Exploration.

Laudan, L. (1977). *Progress and its problems: Toward a theory of scientific growth.* Berkeley: University of California Press.

Martin, G. J., & James, P. E. (1993). *All possible worlds: A history of geographical ideas* (3rd ed.). New York: Wiley.

National Research Council (1997). *Rediscovering geography: New relevance for science and society.* Washington, DC: National Academy Press.

CHAPTER 2

Fundamental Research Concepts

Learning Objectives:

- What are the major idea concepts and the major empirical concepts in science?
- What is causality, and what role does it play in scientific inquiry?
- What are the four levels of measurement, and why are they important?
- What does the concept of scale mean in geography?
- What are some systematic ways to generate research ideas?

In Chapter 1, we discussed the fact that scientists create and evaluate ideas, via logic and empirical observation. In this chapter, we flesh out a variety of basic scientific concepts that are fundamental to the conduct and interpretation of scientific research. We classify these as **idea concepts** and **empirical concepts**.

Idea Concepts

Given that scientific research ultimately concerns developing valid ideas about reality, we first consider the idea concepts: theory, law, hypothesis, causality, model, and construct. Probably the central idea concept is **theory**. Defined narrowly, as we prefer, a theory is an idea or conjecture about a causal relationship in reality. It answers the question of "why" something is the way it is by identifying its antecedent causes. This narrow use that we prefer is in contrast to the term's broad use in everyday speech to refer to a conjecture about any aspect of reality. For example, I could "theorize" that door number 2 has a goat behind it, which of course is not a guess about *why* there is a goat there. Scientists also use the term broadly in some cases; they sometimes speak of "theorizing" about a description or prediction of reality, rather than an explanation. For instance, chemists might speak of the "theory of atomic structure," which posits electrons, protons, and so on. Although the theory does not necessarily posit that the components of the atomic structure have a causal relationship to each other, it does posit that their structure is the cause of patterns of data that result from observing atoms. Similarly, we may understand the "theory of gravity" as a description or prediction of the behavior of bodies with mass; in and of itself, the theory does not explain *why* bodies attract. Mathematical expressions of relationships that we expect to hold precisely in an "ideal" world, like the formula for the force of gravitational attraction between two bodies, are often called **laws** rather than theories. For the most part, laws have been identified only in the physical sciences.

We prefer to use "theory" in the narrow sense of a conjecture about causality in order to recognize explanation as the ultimate goal of basic scientific research, as we discussed in Chapter 1. We also prefer the narrow use of "theory" in order to distinguish it from another common idea concept, that of **hypothesis**. A hypothesis is a conjecture about a pattern of observations of the world. The difference between theory and hypothesis is in fact fairly subtle, and as we have suggested, some people use the terms virtually synonymously. But the term hypothesis tends to refer more to a specific and directly testable idea, often times a specific idea about a particular pattern of data in some population that may or may not imply anything directly about causality. We may think of the relationship between theory and hypothesis in terms of conditional ("if-then") logic: If theory A is true, then one hypothesizes that data pattern B will hold.

Let's consider the idea concept of **causality** a bit more. There is a long history of philosophical analysis of causality, notably by the 18th-century British Empiricist philosophers, particularly David Hume. Causality is the apparent fact that the occurrence of an event A (state A, entity A) brings about or determines the occurrence of event B; A is a reason B occurred, or the occurrence of B depends in some sense on the occurrence of A. A is the **cause** and B is the **effect**. For instance, lightning is a cause of wildfires. Hume offered three principles in his analysis of causality:

a. Covariation between cause and effect (they co-occur)

b. Temporal precedence of cause (the cause comes first)

c. Controlling the cause will control the effect.

The 20th century witnessed a great deal of critical discussion of the concept of causality, especially in the context of quantum and relativistic physics. Is causality in the mind or in the world? Can causality occur simultaneously between two spatially separated entities? Can causality move backward in time? In spite of these discussions, however, it is still true that most everyday scientists who focus on reality above the atomic size (including geographers) accept the importance of causality as a concept and the meaning of theories as attempts to explicate causality. That is, notions like quantum causality apparently have little or no relevance to the concerns of most geographers.

But it is clearly naive to think of causality simply as one perfect pool ball hitting another perfect pool ball on the felt-covered slate of reality. In most areas of geography, both physical and human, causality is complex and multivariate. Furthermore, it is typically **probabilistic** (**stochastic**) in nature rather than **deterministic**. That is, it is only required that causes *probably* bring about effects, not definitely each and every time the cause occurs. In fact, a few problem areas in geography, particularly physical geography, do employ deterministic rather than probabilistic analyses. We discuss this further in Chapters 7 and 9 but note here that probabilistic processes occur in systems in which causality is complex, and partially outside of our ability or interest to observe or conceptualize.

The notion of probabilistic causality is related to an old distinction between **necessary** and **sufficient causes**. An effect cannot happen if a necessary cause does not occur, but it may not happen just because the necessary cause occurs. Droughts are pretty much necessary causes of wildfires; generally speaking, a wildfire will not take place without dry conditions, but wildfires do not have to take place just because it is dry. An effect happens if a sufficient cause occurs, but it may happen even if the sufficient cause does not occur. Given dry conditions, a lightening strike constitutes a sufficient cause of wildfires; an arsonist can take the place of the lightning, however.

Another distinction is between **mechanistic** and **functional** causality. Mechanistic causality is the classic idea that causes move forward "densely" in space and time—the continuously connected causality we discussed in Chapter 1 as being widely preferred by scientists. Let's consider this preference in greater detail with an example. Consider the scenario of hitting a light switch to cause a bulb to illuminate a room. Starting somewhat arbitrarily at the moment when the metal piece connected to the switch provides contact between the power source and the wire leading to the light bulb (we could start, for example, at the moment we desire to turn on the light), causality moves continuously along the wire, hits the base of the bulb, and is transmitted to the filament, which causes electrical stimulation, making the wire emit electromagnetic photons in the visible range of the spectrum. Then the photons quickly move out continuously over the space of the room, directly impinging on the retina or bouncing off surfaces before hitting the retina, then travel along the optic nerve and tract to the visual cortex.[1] That's mechanistic causality.

Alternatively, many theories in and out of science posit a different type of causal explanation, one that places the cause after the effect by focusing on the cause as functional or purposeful—one that considers the cause as a goal. For example, Darwinian evolution is often thought of as "survival of the fittest," as if the traits of organisms evolve because of their function in improving the organism's future reproductive fitness. In economic geography, individuals and firms are often said to locate themselves in order to "maximize profit," a future state of affairs that apparently motivates the present or past. Functional causality may seem to violate scientists'

---

[1]The question of how causality is then transmitted continuously to brain areas responsible for the conscious experience of light is somewhat understood, although the conscious part is still largely a scientific mystery. As a *scientific* mystery, however, we assume the ultimate answer would not violate continuously connected causality.

antipathy to disconnected and backward-acting causality. In fact, we can make functional causality compatible with mechanistic causality by rephrasing functional explanations ultimately in mechanistic terms. Evolution does not occur because of an attraction to a future perfect form; genetic changes are random or haphazard, but some traits are simply more likely to lead to survival and reproduction in particular environments, which leads to the creation of more organisms with the new genetic material. Likewise, a firm does not locate because of future profit; it locates because of the current *anticipation* of future profit. We can find functional causes in scientific theories, but we should understand them as useful heuristic devices rather than literal truths.

Closely related to the concept of theory is that of **model**. A model is a simplified representation of a portion of reality, expressed in conceptual, physical, graphical, or computational form. A model is essentially a complex information-bearing assemblage that conveys a set of interrelated theories about structures and processes of a system of interest in the world—it expresses the parts and their causal interactions. There are many examples of models from all areas of systematic geography. For instance, the Huff model, an example of so-called gravity models in economic geography, predicts and explains store choice by consumers according to the distance the consumer must travel to each store and the relative attractiveness of each store compared to the others in the comparison set. Another instance is the Davisian model, a geomorphologic model of land forms, which says that the physical shape of the earth's surface terrain is a function of the geological composition and elevation of the land (due to mountain-building forces), processes of soil and rock denudation (wearing away and transport), and the passage of time. These two examples are relatively simple models; sometimes models reach great levels of complexity in geography. We discuss models and their logic in more detail in Chapter 7, particularly those expressed in mathematical or computational form. But it is worth emphasizing here that when we say models are "simplifications of reality," we want to stress that they are *necessarily* simplified and that their usefulness to researchers stems a great deal from their simplified nature.

Our final idea concept is that of **construct**. Actually, construct is essentially another word for a scientific concept. Just as a theoretical statement is an elementary component within a model, a construct is an elementary component within a theory. Constructs are pieces of the idealized world that compose the subject matter of theories, and they are the hypothetical entities that we attempt to measure when we perform our systematic empirical observations. For example, in John's research discussed in Chapter 1, he attempted to measure the construct of "dissociative institutions" by counting liquor stores in the phone book. Biogeographers use the construct of "plant communities" in some of their theories and attempt to measure them by observing and recording the locations of various plant species, their soil and climate conditions, and such properties as elevation and angle to the sun. Discussing constructs in this way may seem to imply that they are rare, exotic, or the sign of a scientist who doesn't really understand what he or she is studying. On the contrary, constructs are utterly commonplace, even ubiquitous, among all scientists, including physical scientists. For example, the idea that a table has length treats length as a construct; we realize that generating a number by laying a ruler end to end is not the table's *actual* length but the result of one attempt to *measure* its length. Such a measurement is at least slightly inaccurate and could be very inaccurate. The construct of a table's length cannot be in error this way.

Thus, it is critical to note that constructs are abstract idea entities that scientists care about, but they are not observed directly—their effect upon measurements is observed. Early in the 20th century, psychologists attempted to finesse the difficult problem of defining intelligence by saying it was "what intelligence tests measure." But intelligence is *not* what intelligence tests measure—it is what intelligence tests *try* to measure. Thought of this way, a construct may be called a **latent variable**, whereas its expression as a set of observations may be called a **manifest variable**. We try to observe as well as we can; that is, we try to measure accurately and precisely and completely (more below).[2] But in fact, measurements (manifest variables) are always imperfect reflections of the constructs (latent variables) they attempt to capture. In Chapter 11, we discuss issues of validity that arise from the relationship of constructs to their measurement.

Empirical Concepts

Empirical concepts include cases, variables, measurement, measurement levels, discrete versus continuous variables, and accuracy versus precision of measurement. A **case** is the thing or entity studied. Synonyms include unit of analysis, entity, element, individual, research subject, and respondent. Cases in physical geography are sometimes called "samples," as in a "soil sample" or a "water sample." This usage is a bit confusing, however, insofar as the entire set of entities we measure in a study is called the sample (Chapter 8)—not just one of the entities. In human geography, the case is often an individual person or group of people, such as a family; a city block; a city, county, state, or country; a census tract or other census region; an industry or corporation; or a society or cultural group. Examples of cases in physical geography include water bodies, such as lakes, rivers, marshes, estuaries, or oceans; mountains or mountain chains; air masses; forests or other vegetation communities; soil profiles; and ecosystems. These examples show that cases in geography, especially physical geography, are often units of time or space carved out of a continuous reality. We will see in Chapters 8 and 9 that this creates some very intriguing issues in geographic data sampling and analysis but also some rather special difficulties.

But we do not study cases directly—we study attributes or properties of cases. We don't study mountains; we study their formation or mineral content. We don't study cities; we study their economic base or percentage of senior citizens. Because they vary from case to case, or within cases over time, these properties are called

---

[2]Defining a variable in our research by describing the techniques (operations) that are used to measure it is called an **operational definition**. The Methods section of a research report would always describe these operations, so that a reader can understand and evaluate what was done in the research (see Chapter 13).

variables, as opposed to **constants**. To say that variables vary is to say that they take on multiple values when observed across cases; thus, the simplest variable possible is a **dichotomous variable** with two values.

The process by which we observe cases and determine their values on our variables of interest is called **measurement**. Or put more formally, measurement is assigning numbers to cases to reflect their values on a variable; in the case of nominal classification, nonnumerical symbols may be used. To return to our table example, we may measure the length (the variable) of a table (the case) by laying a ruler end-to-end (measurement procedure) to determine that it is 2.27 meters long. The measured numbers are the **data** or data set.[3] Measurement of many different types occurs in geography, as we discuss in greater detail in Chapter 3 and several of the subsequent chapters. A mountain's elevation in meters comes from a global positioning system (GPS). The average number of vehicles that travel through a busy intersection each hour comes from time-stamped digital signals recorded whenever anything of sufficient weight goes over a cable laid across the road. The age of trees of a particular species that grow in an area comes from a count of their rings exposed in a tree core. A resident's attitude about a toxic waste dump, expressed as a numerical value on a rating scale, comes from a survey he or she fills out.

As these examples make evident, measurement varies in terms of what type of case is measured, what variable is measured, and how it is measured. Of central importance, however, is characterizing the quantitative content that results from measuring particular things in particular ways. This is called the **measurement level** of a variable—the degree and nature of quantification implied by a measurement. It's important because of its implications for the way we choose and interpret techniques of data analysis and display, as we discuss in Chapters 9 and 10. There are four levels of measurement, starting with the least quantitative content and ending with the most. Each level expresses the quantitative content of all levels above it:

1. ***Nominal.*** Nominal measurement is not quantitative at all. It is simply assigning numbers (or letters or any other symbols) to distinguish one case's value on a variable from that of another case. Nominal measurement most often expresses *classification*—the placement of a case into a class or category that has qualitatively different properties than other classes: for example, the species of each plant in a particular ecosystem. Sometimes, however, nominal measurements simply name, distinguishing one case from another: for example, the case number assigned to each tree in a database that serves as a distinguishing label. Whether classifying or only naming, numbers used to record nominal measurements have no quantitative meaning, although of course numbers used to indicate classes do express qualitative meaning

[3]"Data" is a plural word in its original Latin, so traditional use dictates we speak of data in the plural, such as "The data show that . . ." rather than "The data shows that . . ." "Data set" is singular for the entire collection of data, and a single measured score is a "data point" or even "datum" (Latin). Recently there have been indications that editors and other gateway tenders will allow "data" to be used singularly, but we maintain the traditional plural use in this book.

(a white oak is different than a black oak). Other examples of nominal measurement in geography include soil type, sex (gender), and type of primary industrial activity.

*2. **Ordinal.*** Ordinal measurement is minimally quantitative. It is assigning numbers to distinguish the relative order, or rank, of the value of one case on a variable from that of another case; for example, the oldest tree gets a "1," the next oldest gets a "2," and so on. Notice that ordinal variables do express "more" and "less," which are quantitative properties, but they do not specify *how much* more or less one case is than another. The second-oldest tree might be 10 years or 100 years younger. In geography, examples of ordinal measurement include ranking cities in terms of importance in the urban hierarchy or ranking streams in terms of their position within a watershed. Many textbooks in geography and other disciplines recommend that scores from rating scales of attitudes or preferences be treated as ordinal data; that is, that rating preference for the state of Ohio as a "5" and the state of Maine as a "9" represents four "ranks of liking" rather than an interval of four "units of liking." In Chapter 6, we discuss the use of rating scales in explicit reports and argue that rating-scale data should not be treated as merely ordinal.

*3. **Interval.*** Interval measurement expresses not only the ranks of cases on some variable but the quantitative lengths of intervals between the cases: for example, the relative locations of trees in a stand. Although interval measurements contain information about lengths between data scores, they do not contain information about a *true zero*. That is, an interval variable does not express a value of "nothing"—no amount of the variable. The classic example is temperature expressed in Celsius or Fahrenheit; 0° does not mean no temperature or no heat but simply represents another value like the rest do.[4] In our example, none of the trees can be said to have "no" location. In fact, spatial location is an important example of an interval variable in geography; consider location expressed in a spatial coordinate system like latitude-longitude (0° latitude is not "no" latitude).

*4. **Ratio.*** Ratio measurement expresses not only the lengths of intervals between cases on some variable but also the lengths of intervals relative to a true zero. For example, the widths or heights of trees are ratio variables. Because a ratio variable does express a value of "nothing," comparisons between its score values can be validly conveyed as a ratio. That is, a tree that is 0.8 meters wide is *twice* as wide as a tree that is 0.4 meters wide (a ratio of 2:1). Notice how interval variables cannot validly be placed in a ratio like this; 70° F is *not* twice as hot as 35°, notwithstanding that we have heard weather reporters say this. Examples of ratio variables in geography are very common, and include amounts of rainfall, distances between places, and family incomes. Ratio and interval measurement, taken together, are known as

[4]0° kelvin (–273° C) is a theoretical abstraction that represents absolutely no heat. Of course, 0° C (slightly above 0°, to be precise) does have the special relevance of corresponding to the phase change between liquid water and ice. To many people, 0° F has the special psychological relevance of indicating a temperature below which it is preposterously cold to go outside.

metric, to reflect their important property of expressing quantitative distances between values.[5]

Related to the concept of measurement level is the concept of whether variables are **discrete** or **continuous.** This concept too has implications for appropriate data analysis and display. Discrete variables have a limited set of distinct possible values. For instance, the number of states bordering a given state in the conterminous U.S. is 1, 2, 3, 4, 5, 6, 7, or 8 (can you guess which has 8?[6]). It is not possible for a state to border a fractional number of other states, because any contact, even at a point, is considered one whole contact. Similarly, a city may contain 123,488 or 123,489 people, but not 123,488.3 people (a discrete variable like this is called **countably infinite** because it could take on any arbitrarily large value). Between any two values of a continuous variable, in contrast, there are potentially an infinite number of additional values. Between a snow pack depth of 1.52 and 1.53 meters could be a depth of 1.524 meters. Thus, continuous variables essentially map onto the real number line, or a piece thereof. Only measurement precision (discussed below) limits the number of possible values of a continuous variable. Because measurement must necessarily have finite precision, any actual data always consist of discrete values, although the number of different values may be very large and include values with several digits past the decimal. How does the discrete-continuous distinction relate to levels of measurement? There is a partial overlap. Nominal and ordinal variables are necessarily discrete; interval and ratio variables may be either discrete or continuous. Put conversely, discrete variables may be any of the four levels, but continuous variables must be interval or ratio.

The final empirical concept we introduce here is the distinction between **accuracy** and **precision of measurement.** Accuracy refers to the correctness of a measurement—how close the measured value is to the true value of the thing being measured. Precision refers to the sharpness or resolution of a measurement—how small the units are with which a value is measured. To understand this distinction more clearly, it may help to consider an analogy to a cluster of darts thrown at once toward a target (Figure 2.1). Think of the resulting cluster of five darts as a single measurement, with the bull's-eye as the true value of the thing being measured. The distance from the spatial center (centroid) of the darts to the bull's-eye is accuracy; the spread of the five darts around their centroid is precision. As Figure 2.1 shows, a tight cluster of darts may be centered near the bull's-eye or far from it. A wide cluster may be centered right over the bull's-eye or some distance away; of course, a wide cluster would likely have some darts off the board if it were inaccurate in this manner. Similarly, a digital bathroom scale may measure in a precise manner

---

[5]Many writers reserve the term "measurement" exclusively for ordinal and metric measurement, or even just metric. We find it useful to use the term to refer to all situations in which numbers or other symbols are assigned to cases to represent their value on variables, even if that variable is a set of classes. In other words, all empirical studies in geography involve some form of measurement, even if it is "qualitative" measurement.

[6]Both Tennessee and Missouri border eight states. Only Maine borders one state.

Box 2.1 Zeno's Paradoxes: Space, Time, and Theme as Discrete or Continuous

The distinction between discrete and continuous is not as straightforward or as mundane as it might seem. Yes, it has implications for data collection, analysis, and display. More than this, however, the distinction is in fact a major intellectual enigma that can be fascinating to ponder. In Chapter 8 and again in Chapter 12, we learn that geographers conceptualize phenomena as being continuous fields (for example, atmospheric temperature) or discrete objects (for example, lakes). But this is not always easy to decide. We noted in this chapter that measurement precision is necessarily finite, which ultimately forces all data to be discrete. Arguably, however, all real phenomena are actually continuous. Even nominal variables like soil type and sex are categorical simplifications of multivariate and continuously valued possibilities of reality (for example, mixed alphisols-mollisols, hermaphrodites). Seemingly discrete entities like clouds, lakes, and mountains actually have vague boundaries that are difficult to identify precisely and change over time; their very existence as objects is debated by trained scientists, never mind lay people. (We once attended an entertaining talk at a national geography conference that concerned itself with the question of whether the "mountains" of West Virginia are really tall enough to deserve that name; the speaker concluded they are not.) But then again, one can also make a good argument that all real phenomena are actually discrete. As the physical sciences have apparently shown us over the centuries, all reality is really composed of multitudes of tiny discrete entities (atoms, electrons, photons, quarks . . . ?). So which is reality—continuous or discrete?

The Greek philosopher Zeno, in his famous paradoxes about space and time, touched upon this mystery more than two millennia ago. Zeno of Elea was a contemporary of Socrates and Plato who believed, with Parmenides, *"that there is only one thing, and it does not change or move, never came into existence and will not cease to exist."* (We don't even want to think about the implications of this for our lives.) Zeno presented four logical arguments in support of this that can largely be understood to rest on the nature of space and time as discrete or continuous substrates for reality. These four "paradoxes" attempt to prove that change and motion are impossible. A paradox is a seemingly contradictory statement that may nonetheless be true; ultimately, the contradiction of a paradox is only apparent. Zeno's four paradoxes are as follows:

1. **The motionless runner.** The first paradox concerns a runner trying to get from a start location at A to a finish at B. The paradox argues that motion is impossible because in order to get from A to B, the runner must first get halfway between A and B. Before the halfway point, the runner must get one fourth of the way, and so on. This paradox rests on discretizing space into an infinite number of points, all of which must be reached in an infinite number of moments of time. Zeno argues that this is forever.

2. **Achilles and the Tortoise.** The second paradox posits that Achilles can never catch, let alone pass, a tortoise that is given a head start in a race. Here Zeno discretizes time and space in a similar manner to the runner paradox by arguing that *when* Achilles reaches the place where the tortoise started, the tortoise will have moved on some distance; Achilles will then have to catch up to the new location of the tortoise. Thus, Achilles will have to exist in an infinite number of points in time, all of which take place while Achilles is behind the tortoise.

3. **The arrow.** Imagine an arrow in flight. At any given *instant* of time the arrow rests at a specific location in space. An arrow cannot move in an instant, so how does it change its spatial location during an infinite sum of instants?

4. The stadium. Zeno's fourth paradox describes a person standing still in a stadium at point A. Two other people are running at the same speed toward A from opposite sides, west or east. To each other, the runners appear to be traveling at twice the speed they appear to the stationary person (which is "impossible").

These paradoxes, especially the first three, rest in part on the apparent incommensurability of discrete and continuous reality (the fourth actually anticipates Einstein's 20th-century arguments about the intrinsic dependence of space and time). We find them intellectually enriching and entertaining to ponder. They point to some of the deep conceptual and philosophical questions inherent in the subject matter of geography, even subject matter that most nongeographers probably think is obvious and not at all controversial. Just how many lakes are there in Minnesota?

because it reads off weight to the nearest tenth of a pound but measure inaccurately because, unknown to you, it reads off weight 30 pounds too heavy (for some of us, such inaccurate scales are apparently to be found everywhere).

Accuracy and precision are thus, in an important sense, separate issues, but they are intimately related. Accuracy is the correctness of measurement *at a given level of precision.* It is perfectly accurate to say that the average adult man weighs 200 pounds, as long as you recognize that the statement is precise to the nearest 100 pounds; that is, the measurement can be expressed only in units of 100 pounds. On the other hand, precision is the smallest resolution of measurement *that produces accurate digits in the measured score.* It is not more precise to say that the Nile River is 6,652.327 km long rather than 6,652 km, unless the .327 km can be measured accurately (it cannot be). Such false or **spurious** precision is unfortunately common, probably because of the tendency of computers to output numbers with many more decimal places than are actually warranted by the quality of the data. Perhaps some people also hold a mistaken belief that greater numerical precision is necessarily a sign of more "scientific" work. In any case, one should not report a data value with greater precision than the measurement procedure warrants—the precision it produces that is accurate. When working with summary indices (see Chapter 9), such as the mean or variance, acceptable precision is typically considered to be one digit more precise than the precision of the original data values.[7]

[7]This **rounding** advice is based on the typical situation in which the most precise digit of a measured score is halfway between the lower and upper ranges of possible values. For example, a mountain that is accurately described as 3,528 meters high is actually somewhere between 3,527.5 and 3,528.5 meters high. You should not apply our rounding advice if this situation does not hold. For example, we note that when people report their age in years, they do not follow the usual measurement convention. A person who accurately says he or she is "46" really means "between 46.0 and 47.0" not "between 45.5 and 46.5." You could treat a 46 as a 46.5 for averaging, or better yet, request date of birth.

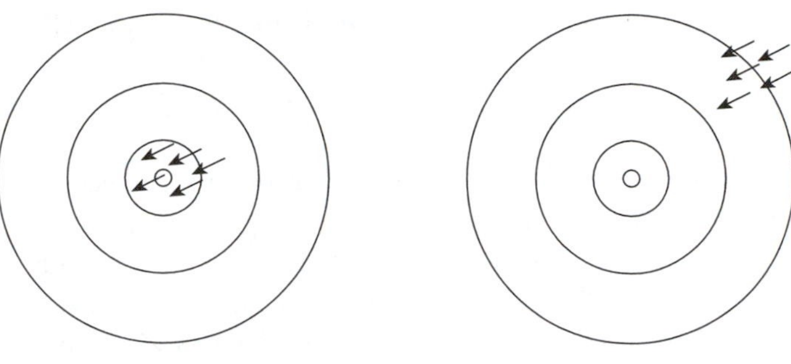

a. High accuracy and high precision.

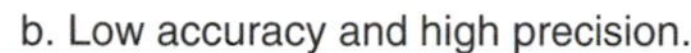

b. Low accuracy and high precision.

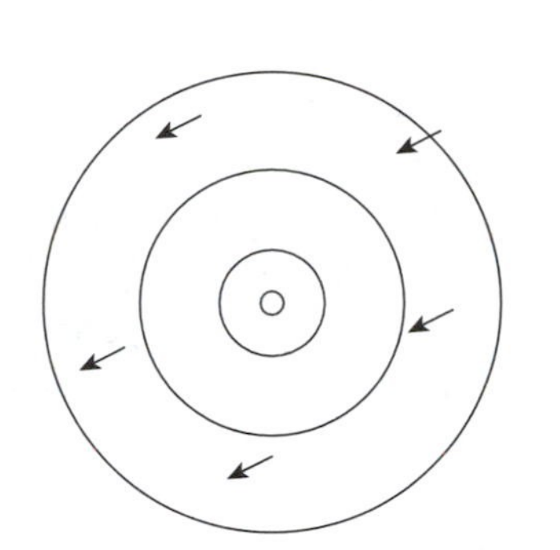

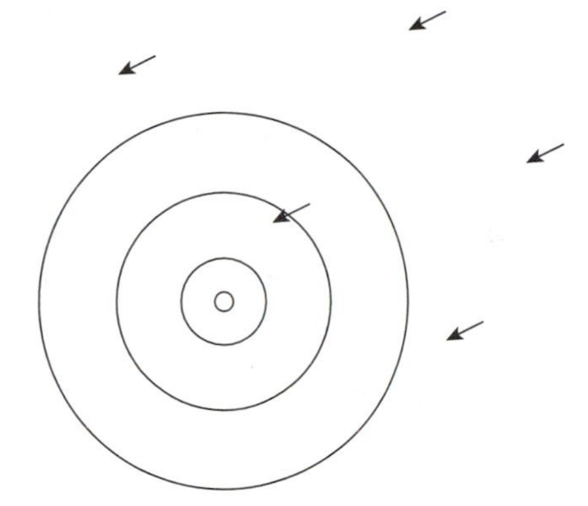

c. High accuracy and low precision.

d. Low accuracy and low precision.

Figure 2.1 Measurement accuracy and precision depicted by analogy to darts thrown at a target. The cluster of five darts thrown at each target should be thought of as one measurement, and the bull's-eye is the actual value of the construct being measured. The distance from the spatial center (centroid) of the darts to the bull's-eye is accuracy; the spread of the five darts around their centroid is precision. The game of darts is but one of many situations where high accuracy and high precision together make a winning combination.

The Concept of Scale in Geography

There is one concept that is especially central to geographic inquiry—the concept of **scale.** Although scale is relevant to all natural and social science disciplines, perhaps no discipline is more sensitive to its implications than geography is. Scale is both an idea concept and an empirical concept. Scale has many implications for research in geography, and we discuss specific aspects of it in several chapters of this book. But it is so fundamental to geography that it deserves to be introduced here.

Scale is about size, either relative or absolute. Scale is relevant not just to space in geography, but also to time and theme (**themes** are the nonspatial and nontemporal characteristics of human and natural phenomena that geographers measure and map as variables). Scale has several meanings in geography, which can be confusing. We group these meanings into three categories: **phenomenon scale, analysis**

scale, and **cartographic scale**. Phenomenon scale refers to the size at which human or physical earth structures or processes actually exist, regardless of how they are studied or represented. A lake is larger than a pond (in the English language anyway), and a city is larger than a city block. Analysis scale refers to the size of the unit at which some problem is analyzed. Data at the state level are at a larger scale than data at the county level. Finally, cartographic scale refers to the depicted size of a feature on a map relative to its actual size in the world. (Because cartographic scale is expressed in terms of map size relative to earth size, a "small-scale" map shows a large earth area; most people who are not cartographers find this a little confusing.) Although the three meanings of scale are frequently treated independently, they are in fact interrelated in important ways that are relevant to all geographers and the focus of research for some geographers. We examine several of these interrelations throughout the rest of this book.

It is widely recognized that various scales of geographic phenomena interact, or that phenomena at one scale emerge from smaller- or larger-scale phenomena. This is captured by the notion of a **hierarchy of scales**, in which smaller phenomena are nested within larger phenomena. For example, local economies are nested within regional economies, which are in turn nested within the global economy. Conceptualizing and modeling such scale hierarchies and "couplings" across scales can be quite difficult; for this reason, much geographic work continues the practice of focusing on a single scale.

Many geographers have claimed that we can partially define the discipline of geography—the study of the earth as the home of humanity—by its focus on phenomena at certain scales, such as cities or continents, and not other scales. The range of scales of interest to geographers is often summarized by the use of terminological continua such as "local/global" or "micro-, meso-, and macroscale." Not everyone shares the view that geographers must restrict their focus to particular ranges of scales, however, and advances have and will continue to occur when geographers stretch the boundaries of their subject matter. Nonetheless, few would argue that subatomic or interplanetary scales are properly of concern for geography.

Generating Research Ideas

We finish this chapter by discussing some approaches to generating research ideas. A beginning researcher may wonder where scientists get their ideas. The short answer to this question is . . . *anywhere!* Although we offer some tips here that can help, generating research ideas is a creative component of scientific activity that cannot and should not be entirely formularized. You can get ideas from your intuition, your dreams, from movies or books, or from your personal experiences. You can get them from your neighbor, or your aunt or uncle. You can get them from eating too many chili peppers. However, we can identify a few more systematic approaches[8] that may help you generate research ideas, especially if you are new to designing research:

[8]Adapted from McGuire, W. J. (1973). The yin and yang of progress in social psychology: Seven koan. *Journal of Personality and Social Psychology, 26*, 446–456.

1. Intensive case study. Look closely at a particular marsh, including its shape, size, and depth; its water temperature, clarity, and chemical makeup; and its flora and fauna. This could lead to research on the functioning of aquatic ecosystems.

2. Paradoxical incident. Notice that families often return to hazardous areas after a disaster such as a flood or fire. This could lead to research on the variables that influence residential choice or responses to risky events.

3. Analogical extension. Identify an analogy between people's tendency to shop at closer stores over more distant stores and the greater pull that close planets have over more distant planets (of the same mass). This could lead to research on the "social gravity model" of spatial interaction.

4. Practitioner's rule of thumb. Examine the choices professional mapmakers make when designing topographic maps. This could lead to research on which cartographic variables are more or less effective at communicating relief.

5. Account for conflicting results. Observe that plants of a certain species grow on the sunny face of mountains in one part of the world but on the shady side in another. This could lead to research on the factors that affect plant growth other than insolation, like soil or wind.

6. Reduce complexity to simpler components. Break down a person's daily activity patterns into components like work, shopping, recreation, and so on. This could lead to a model of how commuters organize their travel at different times of the day.

7. Account for exceptions to general findings. This is useful to pursue whenever you find it.

Of course, we don't just want to get a research idea, we want to get a good one—interesting, novel, relevant, and feasible. Identifying ideas like this requires the kind of expertise that hopefully develops over time with experience as a research scientist. But we can help get you started by offering a plan of action for generating and pursuing *good* research ideas, and implementing them as research:

1. Find a research area; focus on what interests you.
2. Generate research ideas, first on your own; avoid groupthink and staleness by not referring to literature or experts right away.
3. Link with other knowledge you already have; is your idea plausible?
4. Check existing literature; ask experts.
5. Formulate your idea as one or more specific hypotheses.
6. Design research to address your hypotheses.

Review Questions

Idea Concepts

- What are the idea concepts of theory, hypothesis, causality, model, and construct?
- What are some historical and contemporary ideas within the philosophy of causality?
- What are the distinctions between probabilistic and deterministic causality; necessary and sufficient causality; mechanistic and functional causality?

Empirical Concepts

- What are the empirical concepts of case, variable, measurement, data, accuracy, and precision?
- What are the four measurement levels for variables, and why is the measurement level of a variable important?
- What is the distinction between discrete and continuous variables, and what are some ways the distinction is important?
- What is spurious precision, and what is a general rule for reporting data with appropriate precision?

The Concept of Scale in Geography

- To what does the concept of scale in geography refer?
- What are phenomenon, analysis, and cartographic scale?

Generating Research Ideas

- What are some general strategies for generating research ideas?
- What are steps of a plan of action for developing good research ideas?

Key Terms

accuracy: the correctness of values resulting from a particular measurement process, at a particular level of precision

analysis scale: the size of the unit at which some problem is analyzed

cartographic scale: the depicted size of a feature on a map relative to its actual size in the world; unlike other meanings of scale, a "small-scale" map shows a large earth area, and a "large-scale" map shows a small earth area

case: the thing or entity a scientist studies; synonyms include unit of analysis, entity, element, individual, research subject, respondent, and (in physical geography) sample

causality: the concept that the occurrence of one state or event can bring about another state or event

cause: antecedent state or event that brings about an effect

classification: grouping entities into classes or categories based on some type of similarity of class members to each other or to a standard, and some type of dissimilarity between class members and nonmembers

constant: attributes or properties of cases that do not vary from case to case but take on a single value; in contrast to variables, which vary across cases

construct: concept that is a piece of the idealized world comprising the subject matter of theories; the hypothetical entities that we attempt to measure when we perform our systematic empirical observations

continuous variable: variable that can take on an infinite number of possible values between any two values (assuming unlimited measurement precision)

countably infinite variable: discrete variable that can take on an infinite number of possible values but not between any two values, only toward positive or negative infinity

data: the values obtained by measurement that constitute empirical evidence in a study; data are analyzed, interpreted, and displayed by scientists

deterministic: causal processes that necessarily bring about effects, or relationships that always hold, at every occurrence

dichotomous variable: simplest possible variable, it takes on only two values across cases

discrete variable: variable that can take on only a limited set of distinct possible values (even assuming unlimited measurement precision)

effect: a subsequent state or event brought about by a cause

empirical concepts in science: scientific concepts that directly refer to empirical observations of reality; they include cases, variables, measurement, measurement levels, discrete versus continuous variables, and accuracy versus precision of measurement

functional causality: the idea that causes can follow effects, providing goal states for the effects; often used heuristically by scientists but not literally

hierarchy of scales: the fact that geographic phenomena at different scales often interact, existing in nested and nesting relationships to one another

hypothesis: idea or conjecture about a pattern of observations of the world; similar to theory but more specific and directly testable

idea concepts in science: scientific concepts that directly refer to ideas about reality; they include theory, law, hypothesis, causality, model, and construct

interval measurement: the third of the four levels of measurement; it expresses quantitative distance between scale values, but not an absolute zero point

latent variable: hypothetical entity that we attempt to measure when we perform our systematic empirical observations; synonym of construct

law: mathematical expression describing a quantitative relationship that is expected to hold precisely in an "ideal" world, but is not explanatory in and of itself; the law of gravity is an example

manifest variable: actual entity expressed by our measurements when we perform systematic empirical observations; synonym of measured variable

measurement: assigning numbers or other symbols to cases to reflect their values on a variable

measurement level: typology of types of variables based on the quantitative content they express—the degree and nature of quantification implied by a measurement; from the least to the most quantitative, the four levels are nominal, ordinal, interval, ratio

mechanistic causality: the idea that causes move forward "densely" in space and time, with continuously connected causes and effects

metric measurement: either interval or ratio measurement, which express quantitative distance between scale values

model: simplified representation of a portion of reality, expressed in conceptual, physical, graphical, or computational form

necessary cause: cause that must be in place for the effect to occur, but by itself it may not be enough to make the effect occur

nominal measurement: the first of the four levels of measurement; it expresses no quantitative information at all, only classification or naming

operational definition: defining a variable by describing the techniques (operations) used to measure it

ordinal measurement: the second of the four levels of measurement; it expresses only rank order

phenomenon scale: the size at which some human or physical earth structure or process actually exists, regardless of how it is studied or represented

precision of measurement: the sharpest or highest resolution of accurate values resulting from a particular measurement process

probabilistic: causal processes that sometimes bring about effects, or relationships that sometimes hold, but not at every occurrence; same as "stochastic"

ratio measurement: the fourth of the four levels of measurement; it expresses quantitative distance between scale values and an absolute zero point, which allows for the valid creation of ratios among scale values

rounding: reduction in the precision of measured or calculated values that is done in order to avoid excessive or unnecessary levels of precision

scale: both an idea concept and an empirical concept that concerns size, either relative or absolute; besides spatial scale, temporal and thematic scale are also relevant to geography

spurious precision: precision in measured or calculated values that exceeds the accurate precision actually present

sufficient cause: cause that by itself will make the effect occur, but it may not need to be in place for the effect to occur

themes: the nonspatial and nontemporal characteristics of human and natural phenomena that geographers measure and map as variables

theory: idea or conjecture about a causal relationship that provides an answer to the question of "why" something is the way it is; sometimes used broadly to refer to any conjecture

variable: attributes or properties of cases that researchers measure and study; their value varies from case to case, in contrast to constants, which take a single value across cases

Bibliography

Clugston, M. J. (Ed.) (2004). *The new Penguin dictionary of science* (2nd ed.). London: Penguin.

Cook, T. D., & Campbell, D. T. (1979). *Quasi-experimentation: Design and analysis issues for field settings.* Boston: Houghton Mifflin.

Haines-Young, R. H., & Petch, J. R. (1986). *Physical geography: Its nature and methods.* London: Paul Chapman.

Hudson, J. C. (1992). Scale in space and time. In R. F. Abler, M. G. Marcus, & J. M. Olson (Eds.), *Geography's inner worlds: Pervasive themes in contemporary American geography* (pp. 280–298). New Brunswick, NJ: Rutgers University Press.

McGraw-Hill (1997). *McGraw-Hill encyclopedia of science & technology* (8th ed.). New York: McGraw-Hill.

Montello, D. R. (2001). Scale, in geography. In N. J. Smelser & P. B. Baltes (Eds.), *International encyclopedia of the social & behavioral sciences* (pp. 13501–13504). Oxford: Pergamon.

Rosenthal, R., & Rosnow, R. L. (1991). *Essentials of behavioral research: Methods and data analysis* (2nd ed.). New York: McGraw-Hill.

CHAPTER 3

Data Collection in Geography

Overview

Learning Objectives:

- What is the distinction between primary and secondary data sources?
- What are the five major types of data collection in geography?
- What are some of the ways geographers and others have made a distinction between quantitative and qualitative methods, and how do they relate to scientific and humanistic approaches in geography?

In the previous chapter, we explained that the empirical part of scientific research involves systematically observing cases in order to record measurements of variables that reflect properties of those cases. Researchers analyze the resulting set of data (usually numbers) graphically, verbally, and mathematically in order to learn something about the properties of the cases. Data collection efforts do not generally go on continuously but are grouped into periods of activity focused on particular research issues or questions. Such a focused period of data collection and analysis is a **study** (in Chapter 7, we learn that there are two major categories of scientific studies, experimental and nonexperimental). In this chapter, we introduce some basic characteristics of data collection in geography, including the distinction between primary and secondary data sources, the five major types of data collection, and the distinction between quantitative and qualitative methods.

Primary and Secondary Data Sources

One way to characterize data in geography concerns whether they were collected specifically for the purpose of a researcher's particular study. If so, we call the data **primary**. An example would be a geographer who interviews people about their attitudes toward bioengineered agriculture. If, instead, the data have been collected for another purpose, usually by someone other than the researcher, we call it **secondary**. An example of that would be a geographer who uses *Landsat* imagery to study landslides on the California coast. The imagery was not collected by that researcher, and it was not collected primarily so he or she could study landslides.

The major asset of primary data is that they are collected in a way specifically tailored to a particular research question, which means they are probably the data best suited to answering that question. In our attitude example above, the geographer would design the survey specifically to address the issue of attitudes toward bioengineering and agriculture, including customizing it to fit the people answering the survey and the place where they live. But all of this takes considerable time and effort to do well. In contrast, the major asset of secondary data is that they are sometimes the only data available to address a particular research question that are even moderately suited to that question. Also, secondary data are almost always less expensive than primary data (in terms of money, time, and effort). In our landslide example, the geographer gets a very large amount of free data obtainable in something like an hour or less, depending on the geographer's units of analysis, but that geographer has to accept the way the Landsat satellite collects imagery. This includes the extent of earth surface coverage, the time the satellite passes over, the spatial resolution of the imagery, and the spectral bands recorded.

Some geographers use mostly primary data, whereas others use mostly secondary data. This depends mostly on the geographer's topical area of research. However, compared to many other scientific disciplines, both human and physical geographers use a great deal of secondary data. This is probably because they so often study phenomena at large spatial and temporal scales, where it is typically so difficult and costly to collect data that a single study does not warrant it. The fact that secondary data are not tailored to the geographer's specific research question influences the nature of many geographers' research. Problems addressed by census data, for example, are the subject of more geographic research than is necessarily warranted from an intellectual or applied perspective. Especially characteristic of much geographic research in this respect is that researchers study problems at the analysis scale of the available data set, which is often not exactly the scale at which the phenomena operate (see Chapter 2). We consider the characteristically geographic problem that results from this "data-driven" approach to science several times in the rest of the text but especially in Chapter 9.

Types of Data Collection in Geography

We can characterize data in geography more precisely than just distinguishing primary from secondary. Geographers collect and analyze many different kinds of data

Table 3.1 Types of Data Collection in Geographic Research

1. Physical measurement (Chapter 4)
2. Observation of behavior (Chapter 5)
3. Archives (Chapter 5)
4. Explicit reports (Chapter 6)
5. Computational modeling (Chapter 7)

in their studies, based on many different variables and collected with many different techniques. However, we can group all of these data collection methods into just five types (Table 3.1). The first is very popular in geography, especially physical geography. **Physical measurements** consist of data collected by recording physical properties of the earth or its inhabitants. Physical properties include size and number, temperature, chemical makeup, moisture content, texture and hardness, the reflectance and transmissivity of electromagnetic energy (including optical light), air speed and pressure, and more. We discuss these physical measurements at some length in Chapter 4. One of the key innovations of 20th-century geography is the use of aerial and satellite remote sensing as ways to efficiently record large amounts of physical measurement data. We discuss physical measurement via remote sensing in Chapter 12. Human geographers often observe the "physical traces" left behind by human behavior or activity (biogeographers might study the physical traces of nonhuman animal activity). They include the house designs in different neighborhoods or cultural regions, crops that have been planted in different fields, or patterns of clear-cut forests left by different harvesting techniques.

The second type of data collection is based on the fact that human geographers also observe and record human behavior directly (again, biogeographers can observe animal behavior). **Behavior** is the overt and potentially observable actions or activities of individuals or groups of people. It is not their thoughts, feelings, or motivations, although very often behavioral observations provide the data that allow geographers to study thoughts, feelings, and motivations scientifically. Geographers make **behavioral observations** in person or with the aid of a variety of recording media. Importantly, records of behavior do not in themselves constitute data; they must be "coded" into categories to become data. As we discuss in Chapter 5, behavioral observations vary greatly in the degree to which they involve people's explicit awareness that they are being studied. We also consider in that chapter the important fact that people's behaviors are not always based on their explicit choices and decision-making.

A third type of data collection practiced by geographers is the use of existing records that others have collected primarily for nonresearch purposes, at least not the geographer's research. These secondary records are known as **archives**. Examples of archives used by geographers include financial records, birth and death records, newspaper stories, industry and business records, museum records, historical documents, diaries, letters, and more. Often, archives must also be coded in order to produce usable data; for this reason, we discuss them in Chapter 5 along with behavioral observations.

The next type of data collection is quite popular in human geography. **Explicit reports** are beliefs people express about things—about themselves or other people, about places or events, about activities or objects. Actually, explicit reports are also observations of behavior; answering a question on a survey is behaving, for instance. But we distinguish reports as distinct types of data collection because they always involve explicit recognition by people that researchers are studying them, and because research participants' explicit beliefs and choices determine the data collected with explicit reports. Explicit reports such as surveys and interviews often consist of questions that have no right or wrong answers, or at least the correctness of the answer is not of chief interest to the researcher. When the explicit report consists of questions that do have right or wrong answers, and the correctness of answers *is* of interest to the researcher, we call the explicit report a test. That is, whereas many types of explicit reports are used to study opinions, attitudes, and preferences, tests are used to study knowledge. These measures are called "explicit" reports because people responding to them *know* they are responding to a request for information by a researcher. This turns out to be both an important strength and an important limitation of explicit reports, as we discuss in detail in Chapter 6.

The fifth and final type of data collection is **computational modeling**, applied in both physical and human geography. In Chapter 2, we defined models as simplified representations of portions of reality. We noted that models can be realized in conceptual, physical, graphical, or computational form. Understood in this broad way, models and modeling are pervasive in geography and other sciences. We refer to them frequently in this book, in several different chapters. For instance, in Chapter 9 we discuss statistical models, and in Chapter 10 we discuss graphical models (maps are models). We consider conceptual models in several different chapters, at least implicitly. As a unique approach to data collection, computational modeling is modeling that evaluates theoretical structures and processes expressed mathematically, typically in a computer. We discuss computational modeling in detail in Chapter 7, which covers research designs, because we believe it makes sense to think about modeling as an alternative to standard experimental and nonexperimental approaches. We see in Chapter 7 that we evaluate how well models fit portions of reality by comparing outputs of the model to measurements made on the reality to which the model refers. Alternatively, models are sometimes created and thought about as if they were creations of new realities rather than simulations of existing realities. We consider how this creation of "artificial realities" may or may not be thought of as *scientific* research in Chapter 7.

An Introduction to Quantitative and Qualitative Methods

Geographers, and other natural and social scientists, have been collecting and analyzing all of the types of data we have just discussed for well over a century (of course, many specific techniques and procedures are regularly introduced). Besides geographers, these scientists have included geologists, biologists, oceanographers, hydrologists, atmospheric scientists, anthropologists, psychologists, sociologists, economists, political scientists, and others. Many of these early scientists incorporated a variety of data collection techniques and a variety of data types in order to understand their phenomena of interest. In other words, early scientists of the earth and its people were unabashedly heterogeneous in their empirical methods, using whatever they thought provided insight into their problem domain. We enthusiastically believe that this heterogeneous approach is still the best approach.

During the middle and latter part of the 20th century, characteristics of the varied methodological approaches applied in the sciences, particularly the social and behavioral sciences, were summarized in terms of a distinction between **quantitative** and **qualitative methods**. Like our definition of science in Chapter 1, the quantitative/qualitative distinction is difficult to define in a precise way. The distinction reflects a continuum as much as two sharp categories. There are clear examples of each but also examples that are more-or-less quantitative or qualitative.

A few different factors have been identified that distinguish quantitative and qualitative methods. One concerns the nature of the data recorded and analyzed in a research study. Quantitative data consist of numerical values, measured on at least an ordinal level but more likely a metric level. Qualitative data are nonnumerical, or, as in nominal data, numerical values that have no quantitative meaning. They consist of words (in natural language), drawings, photographs, and so on.

However, the distinction between quantitative and qualitative methods is not just whether a researcher uses numbers or not. Another factor distinguishing the two emphasizes the data collection technique used to create the data, rather than the data itself. According to this, quantitative methods are those that impose a relatively great amount of prior structure on collected data. That is, such methods involve a prior choice of constructs to study, a prior choice of variables with which to measure those constructs, and prior numerical categories with which to express the measured values of those variables. Qualitative methods, in contrast, involve less prior structure on data collection. Data collection that is very clearly qualitative might start with little more than a topic area or a broad research question. The constructs, variables, and especially the measurement values for the variables are determined as observations are made or even afterward. For example, a survey that asks respondents to pick one of a finite number of predetermined categories as a way to measure their attitudes about highway construction would be relatively quantitative in this sense; an interview that asks respondents "how they feel" about highway construction, without any constraints on what they can give for an answer, would be relatively qualitative. Importantly, these examples also show that a single type of data collection, in this case explicit reports (Chapter 6), may be used in a relatively quantitative *or* qualitative way.

Still another factor in differentiating quantitative and qualitative methods focuses on the analysis of data. Either methodological approach may start with relatively unstructured and open-ended responses, such as oral responses in an interview. These can be treated quantitatively, however, by rigorously **coding** the elements in the responses (such as words or phrases) into well-defined categories, which are then tabulated and analyzed statistically; we detail this process in Chapter 5. Alternatively, some researchers choose to treat such records more qualitatively by avoiding rigorous coding and statistical analysis, instead interpreting the records more informally and in a less repeatable manner. Such a qualitative approach usually involves less aggregation of data as well. We return to these issues in Chapter 7, where we contrast single-case and multi-case research designs and further consider the nomothetic/idiographic distinction we touched on in Chapter 1.

We agree with researchers who endorse a "full spectrum" approach that incorporates multiple methods, including both relatively qualitative and quantitative methods. In many research contexts, the two complement each other rather well. Qualitative methods are more applicable when a researcher does not know much to begin with about a particular research topic or domain (that is, no research has been done on it). Such methods are flexible for the researcher, requiring less prior understanding of a phenomenon. Qualitative methods are especially useful in some areas of human geography because they allow a research case that is a sentient human being to "speak in his or her own voice," focusing on what is meaningful or important to himself or herself rather than conforming to the researcher's conceptualization of a situation. However, if qualitative methods are to provide scientifically acceptable evidence about the status of hypotheses and theories, they require difficult and laborious coding. Among other things, this means that researchers working qualitatively typically examine fewer of the cases of interest. Both the relatively low reliability of coding and the typically smaller samples of cases mean that qualitative methods generally do not produce evidence that can be as convincingly generalized as that produced by quantitative methods. For these reasons, we generally favor a research strategy that applies qualitative approaches earlier in an exploratory way and quantitative approaches later, informed by the qualitative results, in a confirmatory way. In many areas of human geography, furthermore, qualitative results can help flesh out and exemplify quantitative results and conclusions.

We make a final observation on the way the quantitative/qualitative distinction has played out in the recent history of the discipline of geography. Like other social scientists, human geographers traditionally took a heterogeneous approach to methods before the late 20th century. However, as part of historical developments in geographic thought during the 20th century, polarized attitudes about proper methodology emerged within the discipline (see discussion and references in Chapter 1). Champions of the quantitative revolution and their predecessors pushed for a "scientific" approach, sometimes interpreted narrowly as relying solely on measuring observables quantitatively and precisely, and avoiding hypothetical constructs, subjective meanings, and other intellectual entities that they thought were too vague for a true positivist science. Various post-positivist critics of this approach during the latter half of the 20th century countered with their own somewhat extreme arguments that a strong positivist approach was not appropriate for understanding human experience, activity, and society. Among other things, this has produced rifts among geographers who focus on the study of human activity and society, and especially between geographers who focus on humans versus those who study the natural earth (unfortunately, the word "versus" sometimes fits all too well).

We believe both of these positions are too extreme. Apparently unlike everything else in the world, humans and some other animals have **agency** (will)—to a degree, they determine when and how they act. Furthermore, because of brain evolution and cultural developments (including language and mathematics), human beings are, in part, **semantic** and **semiotic** entities—meaning and experience, often expressed in symbolic representations, partially guide their activities and explain their geography. Not all human geographers are required to incorporate such constructs in their work, by any means, but anyone who denies their relevance to geography is mistaken. We see no reason that scientific geographers have to ignore meaning and experience, although these constructs certainly create special intellectual and methodological challenges that biophysical scientists do not face. At the same time, there are unequal and unjust power relationships among subsets of people, cultural variations in conceptual structures, and idiosyncratic motivations among individual scientists for doing science. These do not, in our opinion, invalidate scientific approaches to understanding humans, although they certainly have implications for understanding how science should work and does work.

Review Questions

Primary and Secondary Data Sources

- What is the difference between primary and secondary data sources?
- How are primary and secondary data sources used in geographic research?

Types of Data Collection in Geography

- What are the following types of data collection in geography: physical measurement, observation of behavior, archives, explicit reports, computational modeling?

Quantitative and Qualitative Methods

- What are different factors that distinguish between quantitative and qualitative methods?
- What are some strengths and weaknesses of quantitative methods and of qualitative methods, and how do we recommend incorporating the two approaches into research?
- What role have ideas about quantitative and qualitative methods played in the 20th-century history of geography as an academic discipline?

Key Terms

agency: property of humans and some other animals of having at least partial self-determination of when and how to act

archives: type of data collection in which existing records that have been collected by others primarily for nonresearch purposes are analyzed, often after coding

behavior: the overt and potentially observable actions or activities of individuals or groups of people, or other animals

behavioral observation: type of data collection in which ongoing behaviors are recorded and analyzed, often after coding

coding: the process of categorizing qualitative records (such as behavioral recordings, archival records, and open-ended explicit reports) in order to turn them into analyzable data

computational modeling: type of data collection involving the output of a computational model, a model of theoretical structures and processes expressed in mathematical form, typically in a computer

explicit reports: type of data collection in which people's expression of their beliefs about themselves, other people, places, events, activities, or objects are recorded

physical measurement: type of data collection in which physical properties of the earth or its inhabitants are measured and analyzed

primary data: data collected specifically for the purpose of a researcher's particular study

qualitative methods: broad term referring to scientific methods that incorporate some combination of collecting nonnumerical data such as verbal or pictorial records, collecting data using relatively unstructured and open-ended approaches and formats, and analyzing data with nonnumerical and nonstatistical approaches; commonly used only in human geography

quantitative methods: broad term referring to scientific methods that incorporate some combination of collecting numerical data such as metric-level measurements, collecting data using relatively structured and closed-ended approaches and formats, and analyzing data with numerical and statistical approaches; commonly used in both physical and human geography

secondary data: data not collected specifically for the purpose of a researcher's particular study but for another research or nonresearch purpose

semantic: concerning meaning

semiotic: concerning entities or properties that represent or stand for other entities or properties, including signs, codes, symbols, models, and so on

study: unit of data collection and analysis activity focused on addressing a specific question or hypothesis

Bibliography

Abler, R. F., Marcus, M. G., & Olson, J. M. (1992). *Geography's inner worlds: Pervasive themes in contemporary American geography.* New Brunswick, NJ: Rutgers University Press.

Clifford, N., & Valentine, G. (Eds.) (2003). *Key methods in geography.* Thousand Oaks, CA: Sage.

Gaile, G. L., & Willmott, C. J. (Eds.) (2004). *Geography in America at the dawn of the 21st century.* New York and Oxford: Oxford University Press.

Haggett, P. (2001). *Geography: A global synthesis.* Harlow, UK: Prentice-Hall.

CHAPTER 4

Physical Measurements

Learning Objectives:

- What are physical models and how are they used in geography?
- What are representative types of physical measurements made by geographers and other scientists who study the four earth systems of the lithosphere, atmosphere, hydrosphere, and biosphere?
- What is geodetic measurement and what are some types?
- What are nonreactive measurements in human geographic research?
- What are the three actions of accretion, deletion, and modification that create physical traces of human activity, and what are the four types of functions of physical traces?

This chapter, along with Chapters 5, 6, and part of 7, provides an overview of the specific types of data collection in geography we introduced in Chapter 3. We stress that these chapters provide only *overviews* of the various data collection types. In order to conduct research in a specific area, you will likely want to take additional courses and otherwise get additional training more specifically focused on the types of data to use in your particular topical domain, including how best to collect and interpret them. This suggests an important piece of advice for choosing your measurement tools and techniques and using them properly: *Learn what you can about your domain of interest.* That way, not only will you develop an appropriate knowledge base for choosing and applying types of data, types accepted by a particular research community, but you will be able to evaluate the meaning of data other people use to make their scientific case, and you will be able to develop novel types of data that can help answer new questions and old questions in new ways.

One of the most important ways of collecting data in geography is physical measurement: collecting data by recording physical properties of phenomena at or near the earth surface. It is the major type of data collection in physical geography. Physical properties include size and number, temperature, chemical makeup, moisture content, texture and hardness, the reflectance and transmissivity of electromagnetic energy (including optical light), air speed and pressure, and more. Physical measurement is not exclusive to physical geography, however. Human geographers often observe the **physical traces** or "residues" left behind by human behavior or activity. In this chapter, we discuss some basic tools and techniques for collecting physical measurements in both physical and human geography.

Physical Measurements in Physical Geography

As we learned in Chapter 1, physical geographers investigate a diverse array of topics concerning the physical and biological earth as the home of humanity, including landform formation and change, soils and mineral resources, lakes and rivers, groundwater, climate and atmosphere, glaciers and ice fields, ocean and coastal processes, and plant and animal distribution. These are most of the topics of interest to the earth sciences more broadly, although physical geographers have historically been particularly interested in the earth as the home of humanity, like other geographers. They thus focus their interest on the portion of the earth at or near the surface and the implications of these domains for human activity, experience, society, and culture. Traditionally, they do not focus so much on the interior of the earth or on the upper atmosphere, for example.

There are several aspects of physical measurement that apply generally to subfields of physical geography. One is that geographers who collect physical measurements often do so in field settings where their phenomena of interest occur; we contrast lab and field settings in Chapter 7. In many cases, geographers actually make and record (on paper, in computers, and so on) these measurements in the field and analyze them back in an office. In other cases, they actually take samples of the material to be measured from the field and measure them back in a lab; examples include soil or water samples. However, we prefer to reserve the term "sample" for its technical meaning, discussed in Chapter 8, of a set of cases smaller than the entire population of interest. Materials collected in the field are usually cases in a sample, not the entire sample, and in some situations they could be cases in an entire population. To avoid confusion, therefore, we refer to these field samples as **physical materials**. What is important to note here is that collecting physical materials constitutes only part of the process of measurement. They must still be measured back in the lab by procedures that generate symbols, usually numbers, to represent their values on variables of interest. This is analogous to the "records" created by people, discussed in Chapter 5, that are used as sources of archival data, most often by human geographers. They too must be measured—coded—in order to generate interpretable data.

There are several practical issues common to conducting field research in physical geography that deserve comment; to varying degrees, these can apply to field research in human geography too. Field research often goes on in remote places that are difficult to reach. You must often establish the physical accessibility of your site, of course, but you must also establish its social and legal accessibility. Private property requires permission from landowners for its access, and there are restrictions on what activities are appropriate nearly everywhere (we discuss some ethical implications of field research in Chapter 14). When planning field research trips, make sure to be realistic about all the costs involved, including travel to and from sites and loss of or damage to equipment. Will it be feasible to transport any physical material you collect from the site, as well as the equipment required to measure or collect materials both to *and* from the site? Fieldwork can be very hard on equipment, not to mention people. Make sure your equipment is robust to transport, collisions, weather, and so on; make sure you and your assistants are in suitable condition to survive the trip and that you have contingency plans for medical emergencies and the like.

Geographers do not always collect physical measurements in field settings, however. One alternative is to create simulations of field settings in the lab. For instance, the physical or chemical weathering of rocks and other materials has been studied fairly extensively with the use of **physical models** in the lab that expose materials to such weathering agents as solutions of varying chemical composition. Some labs exert "physical control" (Chapter 7) over variables like temperature, pressure, and humidity that influence the performance of their models; the **environmental cabinet** is a device for exerting this type of control. In Chapter 7, we discuss another way to simulate physical processes in the lab: computational models that apply computer programs designed to simulate structural and/or process aspects of your phenomenon of interest. Finally, we note that geographers very often collect physical measurements from the (near) earth surface with the aid of airplane and satellite remote sensing. This is true of physical measurements in human as well as physical geography. We discuss collecting physical measurements via remote sensing in Chapter 12.

Geodetic Measurement

The most fundamental physical measurements in geography are measurements of the spatial structure of the earth itself. In Chapter 1, we defined the theory and technology of measuring the size and shape of the earth and the distribution of features on its surface as the study of geodesy. Geodesy is an ancient field of study; its earliest practice several thousand years ago (by Egyptians, for example) is typically identified with the earliest origins of geography as an intellectual endeavor. Geographic researchers need to have at least a basic understanding of geodesy, insofar as it is usually important that their physical measurements of earth features, such as mountains or rivers, can be accurately located (called "geo-referenced" in Chapter 12). Geographers and others obtain geodetic measurements in one of three ways. First, measurements are made on aerial or satellite imagery; remote sensing is discussed in Chapter 12. Second, measurements are taken from existing geodetic databases. In the past, such data were typically represented in map form; this was an archival function of maps (Chapter 10). In map form, researchers often measured spatial properties off the map as if it were an unmeasured image; for instance, an "opisometer" was a tool like a small wheel on a handle for measuring distances off of straight or curved line segments on maps. Nowadays, these data are frequently stored in numerical database form and accessed via software. Either way, such secondary geodetic data are available from public agencies like the USGS, the British Ordnance Survey, and other national mapping agencies; private companies also have such data for sale. Finally, geographers can measure location and other spatial properties of features directly in the field. This is **ground surveying.**

Geodetic information can tell you about the location of features on the two-dimensional (X-Y) earth surface (**planimetric information**), the third dimension (Z) of **elevation** or **altitude**, or both.[1] Let's consider planimetric information first. Features are typically decomposed into constituent points, as when a rectangle is measured at its four corners, or they are treated directly as points themselves (we further discuss geometric models for features in Chapters 8 and 12). The absolute and relative locations of points on the two-dimensional earth surface can be **fixed** by measuring some combination of distances and/or directions, and subjecting these measurements to geometric and trigonometric calculations. Fixing point locations via directions is called **triangulation**; fixing them via distances is **trilateration.** For example, if I know the location of one point relative to a coordinate system, I can determine the location of another if I know its distance and direction from the first; a line or path segment whose distance *and* direction are known is called a **traverse.** Traditionally, tools such as surveyors' chains or tape measures were used to measure distances over the ground, and various forms of magnetic compasses or more sophisticated **theodolites** were used to measure directions; the theodolite measures angles in horizontal and vertical planes by allowing directions to precisely sighted targets to be measured against internal protractors. Many of these tools are more or less obsolete now, as GPSs (Chapter 12), laser range finders, and other electronic tools have supplanted most of the older mechanical and optical tools. Almost no one uses the opisometer we mentioned above, for example; spatial properties of features can be determined by geographic information systems on digital representations. The supplanting of older technologies by newer ones has always occurred, of course, but it has occurred exceptionally dramatically in the last few decades.

Besides the planimetric properties of distance and direction (and hence area, circumference, and so on), properties of the third spatial dimension are also of interest in many situations. Feature heights can be measured from other features with known heights by using tools like the **clinometer**, a sighting device connected to a gravitational horizontal or vertical index, calibrated by a spirit bubble or plumb bob (the Abney level is another such tool). Theodolites also provide this information, and more precisely. These tools are also useful for measuring slope angles. Conceptually, these traditional approaches are straightforward enough. If I can

[1]Some people use the terms elevation and altitude as perfectly synonymous. Others make the small distinction between elevation as the height of the ground surface above or below sea level (the "geoid"—see Chapter 12) versus altitude as the height of any object or position in the sky. If you make this distinction, a mountaintop has an elevation, for example, whereas a satellite has an altitude.

establish that my sighting tube is precisely at a right angle to gravity, then anything I see in it will be the same height as the tube. If I attach a protractor to the sighting tube, and I am able to level that protractor, I can read off the direction to any target in my sight that is "uphill" or "downhill." For greater elevations, I can use **barometric heighting**, wherein air pressure differences are measured as a function of altitude above or below sea level. When you have a great number of elevations to determine, such as when collecting data for **digital elevation models** (**DEMs**) or sea-surface heights, elevation is determined from **stereoscopic** remotely sensed imagery or satellite range finders.

Physical Measurements of Earth Systems

Since the mid-20th century, physical geographers and other earth scientists have organized their subject matter into major **earth systems,** commonly four of them. Systems are sets of interrelated components; both the components and the system as a whole take on various states as a result of processes operating on them. The four earth systems can be described as follows:

1. *Lithosphere (geosphere).* This is the terrestrial earth surface crust, including bedrock, surface rock, and soil. To nongeographers, the lithosphere is sometimes considered to include the entire solid planet, core and all.

2. *Atmosphere.* This includes the envelope of gases and other materials surrounding the terrestrial surface. Although the atmosphere rises several hundred kilometers above the earth surface, geographers are mostly interested in the lowest band called the "troposphere," which rises to about 18 kilometers above the surface. Most of the matter of the atmosphere is found in this thin layer, including nitrogen, oxygen, and other normal atmospheric gases; gaseous and particulate pollutants; and water vapor and clouds. The bulk of weather processes and the life forms of the atmosphere occur in the troposphere. Of some special interest is the **boundary layer**, the lower portion of the troposphere whose motion is influenced by the frictional influence of the topography of the land surface (boundary layers also occur in the lower layers of water bodies).

3. *Hydrosphere.* This refers to the water bodies, both fresh and saline, on the earth surface. It includes oceans and seas, lakes, rivers, wetlands, groundwater, snow, and ice. It includes water in liquid, gaseous, and solid form (the latter is often distinguished as the **cryosphere**). Movement and transformation of water throughout the earth systems is called the **hydrologic cycle.**

4. *Biosphere.* This is the living earth of plants, animals, fungi, microorganisms, and so on. It extends from a few meters underground (which may be thousands of feet below sea level at the ocean floor) to near the top of the troposphere. Many earth scientists have recently included humans in this earth system; others set them apart in a separate **anthrosphere**. Although humans are certainly part of the natural earth, the methods of the biophysical sciences alone are inadequate for the study of

human geography and other social sciences, mostly because of the semantic and semiotic nature of humans mentioned in Chapter 3 and discussed further in several other chapters.

Some version of this taxonomy has been widely used as an organizing framework in geography and other earth sciences for several decades. The framework is a bit ambiguous, however. It is based largely on location relative to the earth surface but also involves aspects of the physical states of matter (solids, liquids, gases). In fact, we do not believe there is a single consensus version of this framework; if you ask several geographers about its specifics, we think you will probably get some different answers. Is soil really part of the lithosphere, given that it contains water, gases, and living components? Is water vapor part of the hydrosphere or the atmosphere, or both? Are coastal processes such as beach formation and destruction part of the hydrosphere or the lithosphere? In the end, these ambiguities cannot be definitively clarified; they are largely a matter of semantic preference. The solid, liquid, and gaseous components of our world; phenomena above, below, and at the earth surface; and the living and the nonliving are truly interrelated in the complex system that is our planet.

Let's consider some specific types of physical measurement that geographers use to study the four earth systems. A central research focus for geographers has traditionally been the description and explanation of the shape or form of earth surface landforms—**geomorphology**. This involves knowledge of a variety of earth processes, including orogenesis and denudation. **Orogenesis** refers to mountain formation events, including tectonic plate movement and volcanism. **Denudation** refers to any process that degrades landforms, including physical and chemical weathering, mass movement, and erosion. These processes occur via a great variety of mechanisms, including rainfall, river movement, coastal tides and waves, glaciation, and wind ("eolian" processes). In addition, degraded materials are transported by wind and water to be deposited elsewhere, creating a variety of landforms such as river deltas, sand dunes, cave formations, and loess sediments (windblown silty material probably created by glacial grinding). Coral reefs are formed by biological deposition.

Morphology is the study of forms. In geography it refers specifically to the measurement of landforms, at all measurement levels from nominal to ratio. Morphology is carried out in a variety of ways, with a variety of tools. Maps and drawings are sketched, and many of the geodetic tools discussed above, such as theodolites, are used to measure elevations, slopes, and so on. Landform features both above water and underwater are measured to determine their shape or form. For example, the **cross-sectional** and **longitudinal profile** of a riverbed or streambed can be determined via **hydrographic surveying**. A **ranging pole** can be used to measure depths that are not too great; a **sounding line** (or lead line) or **echo sounder** is used for deeper measurements. Chains, wires, boats, and bridges are used to string measuring instruments across river channels.

Another topic that has traditionally been important to geographers is that of **soils**, the dynamic layer of natural material on the crustal surface composed of fine particles of minerals (not pebbles) and organic matter (humus). The soil layer sits atop the rocky bedrock; it is thin or nonexistent in some places, and as deep as hundreds of meters in other places. Soils are especially interesting to geographers because their character influences wild and domesticated (agricultural) plant patterns so strongly, which in turn influences many human patterns (residential density, economic activities). Soils are created by the chemical and physical weathering of rocks we mentioned above, as well as organic and physical decomposition by macro- and microorganisms.

Geographers measure many different aspects of soils. **Soil texture** is the relative mixture of three particle sizes in a given soil: sand (>0.05 mm), silt (0.05–0.002 mm), and clay (<0.002 mm). Soil texture greatly influences the water- and ion-holding qualities of soil, and thus its potential for natural and domesticated vegetation. **Sieve analysis** is one of several methods used to determine soil texture. Several other properties besides texture are important for characterizing soil, including structure, porosity, and moisture content. The moisture content of soil is typically determined by comparing the weight of some physical material before and after it is oven-dried at a little above the boiling point of water until its weight reduces to a stable value. Geographers also assess the chemical makeup of soils. For example, they measure the concentration of mineral ions and soil **pH** (the acidity or alkalinity of a substance). The chemical composition of soils and other materials that geographers study is determined in a large variety of ways. Some examples include electrical conductivity, colorimetric analysis, spectrophotometric analysis, X-ray diffraction, and scanning electron microscopy.

A **soil horizon** is a recognizable layer within the soil, more or less parallel to the surface, that is differentiated from the materials above and below it because of different soil formation processes, such as alluvial (river) deposition, that operate on the different horizons. A cross-section of soil layers from the surface down to the bedrock is known as a **soil profile** (Figure 4.1). The properties of naturally occurring soils are mostly a product of the climatic and vegetative history of a particular place. Humans influence soils in a variety of ways too, mostly by covering it or disturbing it so that it becomes much more susceptible to weathering and erosion. Geographers and other soil scientists summarize the properties of soils with a classification system; for example, the U.S. system describes 11 (recently, 12) naturally occurring **soil orders.**

Turning to the atmosphere, we find that geographers take measurements at and above the earth's surface, usually up to the "tropopause" between the troposphere and the stratosphere. Of major interest is the study of relatively long-term **climate** and, at smaller temporal scales of variability, **weather.** A central climate construct that must be measured is **insolation**, a word derived from "*in*tercepted *sol*ar radi*a*-*tion*." Insolation is the amount of solar energy in a given time period, or its intensity at a given moment, incident on the terrestrial surface of the earth or, sometimes, on an elevated surface such as the top of the atmosphere or the tops of tree canopies.[2] Insolation varies across time and place on the earth surface as a

[2]In many cases, researchers conceive of insolation only as energy directly from the sun; in other cases, researchers include indirect solar energy that has reflected from other surfaces.

Figure 4.1 Measurement of depths of soil horizons in a rainforest soil profile on the Island of Hawaii. (Photograph by Oliver Chadwick. Reprinted with permission.)

function of latitude, surface slope and **aspect**, date of the year (seasonality), time of day, and atmospheric clarity. Various surface materials and structures differ greatly in how much insolation they reflect or absorb, an important property of surfaces called **albedo**.

Temperature, pressure, and precipitation are the major components of weather and climate, and of course, their measurement provides central data to geographers and others who study weather and climate. The three components depend on a variety of factors, including insolation, humidity, the movement of air masses from either land or water (continental versus maritime air), the movement of air masses from either north or south (polar versus tropical air), and so on. We measure these various variables with the aid of ground surface and airborne **rain gauges**, **thermometers**, **barometers**, and **hygrometers**, which measure relative humidity. **Doppler radar** detects the direction of water droplets, thereby measuring wind speed and direction. Also important to climate and weather is the material composition of the atmosphere, including especially the presence of clouds, which are three-dimensional regions of liquid or frozen water droplets, often having condensed around tiny particles of matter. These tiny solid or liquid particles dispersed uniformly within the gaseous atmosphere are called **aerosols** (cloud droplets themselves are too large to be considered aerosols). Airflow at a variety of scales is also quite important to patterns of climate and weather, including local winds and breezes and larger-scale Westerlies, trade winds, and high-altitude jet streams. Air and other atmospheric constituents move around the earth in a variety of stable and variable patterns, in response to temperature and pressure gradients, and in response to the rotation of the planet (the "Coriolis force").

Geographers and other earth scientists are interested not only in the climate of the present and the future but of the past, a topical field of study known as **paleoclimatology**. These scientists use a variety of approaches in attempting to "measure" the climate of the past. We put quotes around "measure" because such measurements of course cannot be done directly on past temperature, precipitation, and so on. Instead, various **proxy measures** are taken; these are physical measurements (or even archival documents—see Chapter 5) that are based on a present trace of past climate conditions. For example, trees add a seasonal layer of growth each year, and the layer grows thicker during wet years. **Dendrochronology** (the terms "dendroclimatology" and "dendroecology" are also used) is the taking of cores of materials from trees in order to be able to count and measure the thickness of the layers, which appear as "rings" in the cores (coring does not hurt live trees, a fact with ethical implications to which we return in Chapter 14). **Pollen analysis** is another example of a **biotic proxy**. Cores are also taken from ice sheets and glaciers; trapped gases can be retrieved from these cores at various locations, which indicates age. This provides data on the past gaseous composition of the atmosphere. Measurements from sediments in oceans and lakes can also serve as **geological proxies**.

The age of soils and some other organic materials is often determined by **radiocarbon dating**, a technique in which the beta particle emission rate from radioactive $^{14}CO_2$ (carbon dioxide containing carbon-14 or radiocarbon) is measured. Radiocarbon dating was an important methodological breakthrough for various "paleosciences," including physical geography and geology, paleontology, and archeology. It is based on the fact that the small proportion of atmospheric CO_2 that contains ^{14}C instead of the much more prevalent and nonradioactive ^{12}C becomes incorporated into living creatures at an equilibrium concentration. When an organism dies, its ^{14}C will continue to radioactively decay, emitting beta particles. Because the organism no longer takes in new ^{14}C from the atmosphere, its rate of beta emission continuously decreases over time. The decreasing rate of emission with decreasing ^{14}C, as compared to the concentration of ^{14}C that was at a maximum and equilibrium state upon the death of the organism, provides a basis for measuring the age of a material.

Turning to research on the hydrosphere, we find that geographers are interested in water in its various forms, especially liquid and frozen water; water vapor may be included but is typically seen as being more in the purview of climatologists and atmospheric geographers. Geographers make physical measurements of water temperature, pH, the concentration of various salts and other organic and inorganic compounds, dissolved gasses, the speed and direction of water movement in both the X-Y horizontal dimensions and the Z vertical dimension (**hydrometry** is the measurement of water flow in river and stream channels), and more. For example,

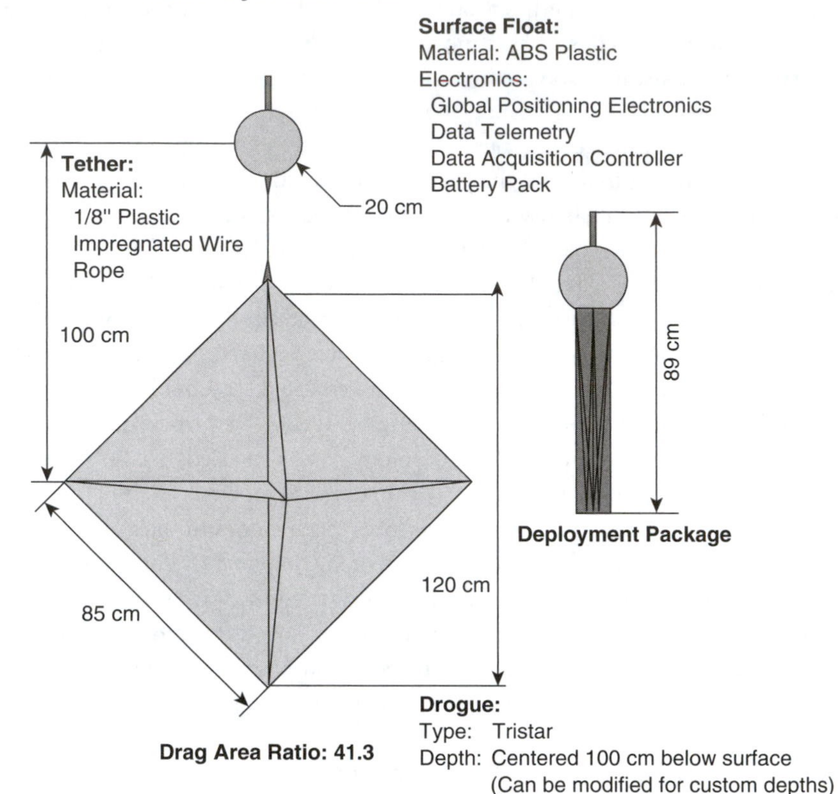

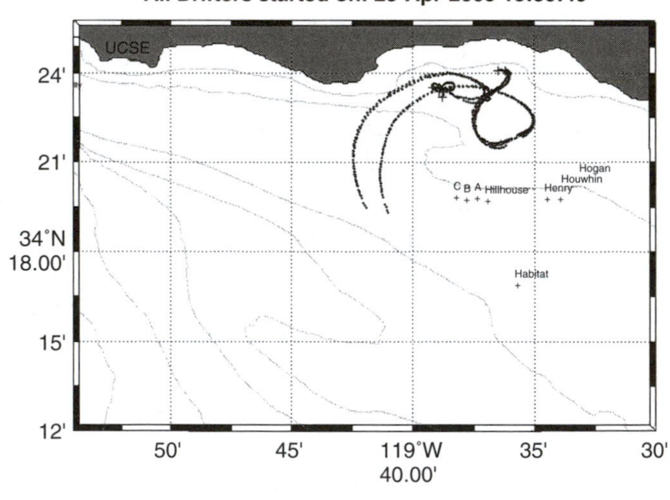

Figure 4.2 Use of drifters to measure near-shore ocean current flow patterns off the coast of Santa Barbara, California. The tracks in the lower image are records of the movements of drifters like that in the Microstar, sent by GPS receivers attached to it. In an attempt to calibrate a new method of measuring currents, the drifter tracks are compared to current measurements from a land-based radar. (Graphic by Carter Ohlmann[3]. Reprinted with permission.)

geographers and oceanographers are interested in the physical properties of ocean currents. Currents are important because they distribute energy, nutrients, and pollutants around the planet. They are driven by atmospheric winds, which pile up surface water into aquatic "ridges"; gravity generates flows in the resulting uneven surface. The Coriolis force deflects these currents clockwise or counterclockwise in the northern and southern hemispheres, respectively. Surface water is fairly warm and well mixed to a depth of a few tens or hundreds of meters; below that, at least in tropical and temperate latitudes, is the **thermocline**. The thermocline is a layer of water within much of the oceans (and some lakes) in which temperature decreases rapidly within distances of something like a few hundred meters. Particularly in the oceans, the thermocline is a relatively stable barrier to interaction between thin surface waters that are warm and well mixed, and the bulk of the deeper, colder ocean.

Scientists traditionally measure surface currents with some sort of physical **drifters** that are dropped into the ocean and tracked. In the past, the drifters would have been tracked visually or by inference when they were retrieved from the water. GPS (Chapter 12) transmitters automatically track the drifters in Figure 4.2 and send near-real-time positional information about themselves to computers on land. The researchers who did this work[3] are using the drifter tracks in Figure 4.2 to calibrate high-frequency radar as a much more efficient technique for measuring ocean currents.

Finally, geographers who study the spatial and temporal distribution of living plants and animals are known as biogeographers.[4] Of course, they measure various physical properties of plants and animals, such as their height, width, color, total mass, and so on, but simple counting is probably the major form of physical measurement that biogeographers employ; human geographers count people too, commonly via the explicit report measure of a census. In some cases, physical measurements of animals, including counts, depend on first netting or trapping the organisms (with ethical implications—see Chapter 14). In other cases, animal populations can be measured by direct observation in the field or by inferences from evidence such as the presence of "scat" (fecal excrement).

Vegetation need not be collected as physical materials to be counted or measured, of course. Geographers often travel into the field to observe, count, and otherwise measure properties of plants. However, the presence of plant species and the hypothetical aggregation complexes of species known as **vegetation assemblages** are very often inferred from local conditions, especially altitude, latitude, slope, and aspect. These factors in turn affect insolation, precipitation, temperature, soil

---

[3]Ohlmann, J. C., White, P. F., Sybrandy, A. L., & Niller, P. P. (2005). GPS–cellular drifter technology for coastal ocean observing systems. *Journal of Atmospheric and Oceanic Technology, 22*, 1381–1388.

[4]Ecologists are also interested in the distributions of living organisms, although they tend to focus more on smaller spatial, temporal, and thematic scales—thematic scale in this case referring to the taxonomic level of the organism.

composition, and wind. This inferential approach to the presence of vegetation is significantly fallible, however. The presence of plant species depends not only on these physical factors of energy, water, and nutrients, but on competition and predation from other plants and animals, and the "geographic opportunity" provided by the past activities of animals, winds, currents, and, increasingly over the last few centuries, humans.

Physical Measurements in Human Geography

Physical measurements are the major type of data that physical geographers collect, but human geographers also collect them. Usually such physical measures on humans are observations of the residues of past activities, the effects of past behaviors by people on their physical surroundings. An important characteristic of physical traces is that individuals and groups do not create them with the intention of producing data. They generally don't know that one day, the traces of their acts will be used as data. Physical traces provide sources of data without the measurement process changing a person's feelings, thoughts, or behaviors. Traces do not reflect any intentional or unintentional effort by those who have left them to produce something to cater to a researcher or a research project, or for that matter, to thwart a researcher or research project. An important shortcoming of some other data collection techniques used with humans, particularly the explicit reports discussed in Chapter 6, is that they require people to provide data knowingly and intentionally to researchers. Under these conditions, reactance often results. **Reactance** is when people's behavior (including their expressed opinions, feelings, and so on) changes because they are aware of being measured or observed.[5] As we discuss further in Chapters 6 and 11, such "researcher-case artifacts" are a major threat to the validity of results obtained via techniques such as surveys or interviews. In contrast, physical-trace measurements are an excellent example of **nonreactive measures** in human geography, along with particular versions of behavioral observations and archival data we consider in Chapter 5. Nonreactive approaches to measurement are quite valuable to human geographers, as they are to other social and behavioral scientists.

---

[5]An analogy is sometimes drawn between behavioral reactance and Heisenberg's "uncertainty principle" from physics. That principle states that both the momentum and the location of a quantum particle cannot be observed simultaneously because observing one requires exposing the particle to energy (such as light photons) that necessarily changes the other. Therefore, by analogy, people speak of something like a "behavioral uncertainty principle": *observing people necessarily changes their behavior because of reactance.* This is a poor analogy, however. Observing or measuring people does not necessarily change their behavior (we return to this in Chapters 5 and 11). Heisenberg's principle only applies to the extremely tiny quantum level in the first place. It does *not* mean we are unable to simultaneously observe the momentum and location of a macroscopic entity like a basketball, at least within the limits of precision that are generally relevant. Otherwise, some people would not be able to catch and shoot the ball so well.

Physical traces are not intentionally created as data, or as artifacts to be measured as a basis for data, but they are often created intentionally. The intentionality with which physical traces are created is an important characteristic that varies across different types. Sometimes people create traces quite intentionally. A family intentionally places certain plants, lawn statues, children's swing sets, unused automobiles, and so on in their front yards. A farmer intentionally plants sorghum and not corn (maize). A transportation planning board puts traffic lights and stop signs at some intersections and not others. In many cases, however, people create traces accidentally or unintentionally. Bare spots in a yard indicate areas of high usage, agricultural chemicals wash into watersheds, and accumulations of paper waste in a recycling bin behind the offices of the transportation planning offices suggest something about the activities of the people who created them.

All physical traces can be understood as a result of one of three actions: accretions, deletions, and modifications. An **accretion** is an intentional or unintentional addition, deposition, or accumulation. Examples of accretions include graffiti painted on city walls or slag piles left by mining operations. A **deletion** is an intentional or unintentional removal, erasure, or obliteration. Examples of deletions include aquifers emptied by intensive agriculture or broken windows in dilapidated buildings. Finally, a **modification** is an intentional or unintentional change, alteration, or conversion. Examples of modifications include a vacant city lot used as a baseball field or a river's course that has been shifted by levee construction.

Physical traces can be grouped according to four types of functions or purposes. The first is **byproduct of use**. This is the only one of the four types that does not in itself reflect function, but results, usually unintentionally, from what people have done in the environment for other reasons. The bare spots left in a yard as a byproduct of activity there is a good example. The second is **adaptation for use**. These are intentional changes to the environment that are meant to make some place more functional to its users. For example, children often make openings in fences in order to create direct routes in the environment. The third type of physical trace is the **display of self**, which is meant to express one's self or group identity. The way families paint their houses and design their front yards provides a good example. The fourth type is the **public message**. These are traces meant to communicate to others, whether officially like a road sign or unofficially like the graffiti we mentioned above that serves to communicate gang territory boundaries at the same time it expresses group identity.

Before we conclude this chapter, we note that there are some other types of physical measurements that human geographers could potentially use besides physical traces of activities. They can measure the physical properties of people themselves, although this is somewhat unusual as of yet in geography. However, we are likely to see a future increase by geographers in the measurement of physiological and neurological states and processes of humans. For example, body weight and height can be measured as indices of malnutrition (including *over*nutrition). Pulse and blood pressure can be measured as reflections of stress or anxiety. Various types of brain scans such as **functional magnetic resonance imaging** (fMRI) could tell us about people's brain activities while they think or reason about geographic topics or problems.

Review Questions

Physical Measurements in Physical Geography

- What are physical materials? How is it misleading to call them "samples?"
- What are physical models? Suggest some ways they may not be completely adequate as tools for studying particular research questions?

Geodetic Measurement

- What is geodesy, and what are some basic geodetic properties that are measured?
- What do theodolites measure? What are some other tools for geodetic measurement?

Physical Measurements of Earth Systems

- What are the four earth systems typically said to describe the physical and biological world? What are some problems with conceiving of the earth in terms of separate systems like these?
- For each of the following areas of physical geography, what are some basic variables of interest, and how are some of them measured: geomorphology, soils, climatology, oceanography and hydrology, biogeography?
- What are proxy measures in physical geography, and how does radiocarbon dating in particular work?

Physical Measurements in Human Geography

- What are physical traces of human activity, and how can they be used in the study of human geography? What are examples of physical traces that are accretions, deletions, and modifications?
- What is reactance, and what are nonreactive measures? Why are nonreactive measures valuable to geographers?
- What are the prospects for using physiological measures of humans as data in geography?

Key Terms

accretion: action that creates physical traces via intentional or unintentional addition, deposition, or accumulation

adaptation for use: type of physical trace that results from people's efforts to make some setting more functional

aerosol: tiny solid or liquid particles dispersed uniformly within the gaseous atmosphere

albedo: the reflectance of solar radiation by a surface material and structure; radiation not reflected is absorbed

altitude: geodetic information about the third dimension (Z) of location on or above the earth surface; sometimes restricted to atmospheric height above the earth surface, in contrast to elevation

anthrosphere: portion of the biosphere including human habitation and activity

aspect: the X-Y orientation of a slope face, typically expressed as a heading in cardinal directions

atmosphere: earth system that includes the envelope of gasses and other materials surrounding the terrestrial surface

barometer: tool to measure pressure in the atmosphere

barometric heighting: technique for measuring large feature elevations, based on atmospheric pressure differences at different elevations

biosphere: earth system that includes the living organisms

biotic proxy: any proxy measure based on the analysis of materials that were at one time part of living organisms; in some cases, such as tree coring, they may still be living

boundary layer: the lower portion of a fluid layer over a solid layer whose motion is influenced by the frictional influence of the underlying solid surface; both the bottoms of the atmosphere and of the oceans have boundary layers

byproduct of use: type of physical trace that results, usually unintentionally, from what people have done for other reasons

climate: relatively long-term pattern of atmospheric conditions such as temperature, wind, and precipitation; average weather in an area

clinometer: geodetic tool for measuring feature heights consisting of a sighting device connected to a gravitational horizontal or vertical index and calibrated by a spirit bubble or plumb bob

cross-sectional profile: the shape of a riverbed or streambed going across from one bank to the other, perpendicular to the direction of flow

cryosphere: frozen portion of the hydrosphere, sometimes considered a separate earth system

deletion: action that creates physical traces via intentional or unintentional removal, erasure, or obliteration

dendrochronology: proxy measure in which cores of material are taken from trees in order to count and measure the thickness of the layers, which appear as "rings" in the cores

denudation: any process that degrades landforms, including physical and chemical weathering, mass movement, and erosion

digital elevation model (DEM): digital representation of the elevations, and thus the morphology, of a portion of the earth surface

display of self: type of physical trace that results from people's expressions of their self or group identity

Doppler radar: radar (*ra*dio *d*etecting *a*nd *r*anging) that measures the velocity of water droplets in the atmosphere by analyzing reflected high-frequency electromagnetic radiation, thereby measuring the velocity of atmospheric movement

drifters: floating objects dropped into the ocean and tracked in order to measure currents

earth systems: sets of interrelated components of the natural earth in which both the components and the system as a whole take on various states as a result of processes operating on them; they are commonly organized into the four systems of the lithosphere, atmosphere, hydrosphere, and biosphere

echo sounder: tool used to measure water depths that are very deep, even too great for a sounding line

environmental cabinet: device for exerting physical control over variables like temperature, pressure, and humidity in a physical model

elevation: geodetic information about the third dimension (Z) of location on or above the earth surface; sometimes restricted to land surface height above sea level, in contrast to altitude

fixed: geodetic determination of the absolute and relative locations of points on the two-dimensional (X-Y) earth surface

functional magnetic resonance imaging (fMRI): type of brain scan used to measure human brain activity during mental processing; increasingly popular in cognitive and neurosciences, and likely to be increasingly applied to mental processing involving spatial, geographic, and environmental information

geodetic measurement: measuring spatial properties of the earth and features on it

geological proxy: any proxy measure based on the analysis of materials that were not at one time part of living organisms

geomorphology: the description and explanation of the shape or form of earth surface landforms

ground surveying: geodetic measurement of feature location and other spatial properties directly in the field

hydrographic surveying: techniques for measuring cross-sectional or longitudinal profiles in streams and rivers

hydrologic cycle: movement and transformation of water throughout the earth systems

hydrometry: the measurement of water flow in river and stream channels

hydrosphere: earth system that includes water bodies, both fresh and saline; sometimes it is considered to include only liquid water, whereas at other times it is considered to include water in all three forms of liquid, gas, and solid

hygrometer: tool to measure relative humidity in the atmosphere

insolation: amount of solar energy in a given time period, or its intensity at a given moment, incident on the terrestrial surface of the earth or, sometimes, on an elevated surface such as the top of the atmosphere or the tops of tree canopies; from "*in*tercepted *sol*ar radi*ation*"

lithosphere (geosphere): earth system that includes the terrestrial earth surface crust, including bedrock, and surface rock and soil

longitudinal profile: the shape of a riverbed or streambed going parallel to the direction of flow

modification: action that creates physical traces via intentional or unintentional change, alteration, or conversion

morphology: the study of forms; in geography it refers specifically to the measurement of landforms at all measurement levels from nominal to ratio

nonreactive measures: measures of people's behavior or activity that do not induce reactance, typically because they do not make people aware they are being measured or studied

orogenesis: mountain formation events, including tectonic plate movement and volcanism

paleoclimatology: study of past climates

pH: the acidity or alkalinity of a substance

physical materials: samples collected by physical geographers in the field that must be measured, typically in a lab, in order to produce scientific data; examples include soil, rock, water, gases, and plants

physical model: model (see Chapter 2) expressed in physical form; essentially a physical or material simulation of a portion of reality

physical trace: type of physical measurement in which "residues" left behind by human behavior or activity (or that of other animals) are recorded

planimetric information: geodetic information about two-dimensional (X-Y) location on the earth surface

pollen analysis: proxy measure in which microscopic pollen grains in sediments are analyzed

proxy measure: measurement, usually physical, that is based on a present trace of past conditions, particularly climate conditions

public message: type of physical trace that results from people's efforts to communicate to others, either officially or unofficially

radiocarbon dating: technique for dating organic materials based on their emission of beta particles from radioactive ^{14}C (carbon-14); important to various paleosciences

rain gauge: tool to measure the amount of rainfall at some location

ranging pole: tool used to measure water depths that are not too great (up to a few meters)

reactance: when people's behavior, including their expressed opinions, feelings, and so on, changes because they are aware of being measured or observed; although it does not necessarily occur in all situations, it is an important reason to use nonreactive measures

sieve analysis: one of several methods used to determine soil texture

soil: the dynamic layer of natural material on the earth's crustal surface composed of fine particles of minerals and organic matter (humus)

soil horizon: recognizable layer within the soil, more or less parallel to the surface, that is differentiated from the materials above and below it because of different soil formation processes

soil orders: categories of soil types

soil profile: cross-section of soil layers from the surface down to the bedrock

soil texture: the relative mixture in a given soil of the three particle sizes of sand, silt, and clay

sounding line: tool used to measure water depths that are more than a few meters, too great for a ranging pole; also called a lead line

stereoscopic: imagery that reveals information about depth or elevation by representing a surface from two perspectives, one offset a little from the other

theodolite: geodetic tool for measuring angles in horizontal and vertical planes by allowing directions to precisely sighted targets to be measured against internal protractors

thermocline: layer of water within much of the oceans (and some lakes) in which temperature decreases rapidly within distances of something like a few hundred meters

thermometer: tool to measure temperature in a solid, liquid, or gas

traverse: line or path segment whose distance and direction is known; used to fix locations

triangulation: fixing location by means of measuring directions

trilateration: fixing location by means of measuring distances

vegetation assemblage: hypothetical aggregation complex of plant species, often inferred from local physical conditions such as soil, slope, and precipitation

weather: relatively short-term pattern of atmospheric conditions such as temperature, wind, and precipitation

Bibliography

Goudie, A. S. (Ed.) (1990). *Geomorphological techniques* (2nd ed.). London: Unwin Hyman.

Reid, I. (2003). Making observations and measurements in the field: An overview. In N. Clifford & G. Valentine (Eds.), *Key methods in geography* (pp. 209–222). Thousand Oaks, CA: Sage.

Ritchie, W., Wood, M., Wright, R., & Tait, D. (1988). *Surveying and mapping for field scientists.* Harlow, U.K. Longman.

Souch, C. (2003). Getting information about the past: Paleo and historical data sources. In N. Clifford & G. Valentine (Eds.), *Key methods in geography* (pp. 195–208). Thousand Oaks, CA: Sage.

Thomas, D. S. G., & Goudie, A. S. (Eds.) (2000). *The dictionary of physical geography* (3rd ed.). Oxford: Blackwell.

Zeisel, J. (1984). *Inquiry by design: Tools for environment-behavior research.* Cambridge, U.K. Cambridge University Press.

CHAPTER 5

Behavioral Observations and Archives

Learning Objectives:

- How are behavioral observations and archives used to collect data in geography?
- How are behavioral observations and archives examples of nonreactive measurements, and how are they not?
- How is scientific behavioral observation different than everyday behavioral observation?
- What is "coding" open-ended records, and how are coding systems developed?
- How are validity and reliability established for coded data?

This chapter covers the two types of data collection known as behavioral observation and archives, used primarily by human geographers. Theories in human geography often involve ideas about human activity in space and place—what people do, what they have done, or what they will do. This includes the behavior of both individuals and various groups, such as families, communities, corporations, bureaucracies, and so on. A straightforward way to collect data on people's current or ongoing activities is to observe and record those activities, which is the technique of behavioral observation. If we want data about people's past activities, we could use archives. Of course, we cannot directly collect data about what people will do in the future, but we can at least use some kind of interview or survey to ask them about their intentions. Chapter 6 covers such explicit reports. We group behavioral observation and archives together in this chapter primarily out of convenience. Besides the fact that both types of data collection are used mostly in human geography (with some applications in physical geography), they frequently share another important property, along with the physical trace measures discussed in Chapter 4: They can both produce data without requiring people to intentionally and knowingly supply information to a researcher. That is, they are often good examples of the nonreactive measures we introduced in Chapter 4, at least when they are collected in such a way that people are unaware they are producing the raw material for data.

Behavioral Observations

Behavior is overt, potentially perceptible action or activity by people or other animals. It is *not* thoughts, internal feelings, purposes, or motivations. That is, behavior is what we do, not why we do it. Behavior is nearly always goal-directed, however, and some behavioral scientists don't consider aimless movement to be behavior. Behavioral observation is a type of data collection in which scientists watch and listen to behaviors of individuals or groups. The behaviors are coded into data during observation, or more commonly, **records** of the behavior are somehow made for later coding into data. Of course, we all observe human behavior all the time, including our own. However, scientific observation of behavior is a bit different. It aims to apply the typical values of scientific practice, such as systematicity and objectivity, to observing behavior. For example, scientific observation should be based on a planned strategy. At a first pass, it should produce data on behavior rather than inferences about the meaning or intention of the behavior. It should include a complete record of the setting and its inhabitants. It should involve an attempt to overcome subjectivity in observation, such as when an observer is more or less sensitive to the occurrence of particular actions because of his or her personal interests.

We noted above that behavioral observation often produces data nonreactively. The degree to which scientific behavioral observation is actually nonreactive varies quite a bit, of course. That depends mostly on whether people are aware they are being observed and recorded. Observers or recording devices can be hidden so the data collection is completely nonreactive. Or the observer or device may be in plain view, but time is allowed to pass before behavior is actually recorded so that those being observed become adapted to the observation and revert to their "normal" behaviors. The most extreme form of this may be **participant observation** in a particular setting, wherein the researcher essentially joins an ongoing setting or subcultural group by becoming a member (yes, this raises ethical questions that we return to in Chapter 14). Finally, we should repeat what we noted in footnote 5 of Chapter 4, that reactance—when people change their responses because they know they are being measured or observed—does not *necessarily* happen even when people are aware they are being observed (we wonder if performers or celebrities only become themselves when they are being observed).

There are a variety of ways to observe behavior. You might simply write **anecdotal records** about what you observe, although this typically lacks the systematicity we look for in scientific observation. Alternatively, you could fill out checklists or rating scales of activities you observe (see Chapter 6). A detailed and comprehensive record of what an individual or group does over some time period is called a **specimen record** (or "running record").[1] Such a record is not usually very useful to a researcher; it is typically best to focus on particular behaviors or particular events. That is, you will probably want to observe less than all possible behaviors at all possible times. This calls for a form of sampling.

You can sample observations whenever you have the time and inclination, but that approach is also lacking in the systematicity that scientists favor. There are two main approaches to **formal observation schedules** that systematically sample from some population of possible behaviors that could be measured. The first is **time sampling**. To do time sampling, choose specified uniform time intervals occurring at specified times, such as one hour every day at 5:00 p.m. Observe and record the behaviors of the people you are studying during those intervals. We discuss sampling designs in detail in Chapter 8, but for now, consider that either randomly or systematically chosen time intervals could be best. Time sampling is appropriate for frequently occurring behaviors, but clearly issues of temporal scale in choosing the size and frequency of intervals are important. The second formal approach to behavioral sampling is **event sampling**, appropriate for situations or behaviors that are rare and would not be expected to necessarily occur in any given time interval. To do event sampling, choose an event to observe, such as a hurricane. Observe and record the behaviors of the people you are studying whenever this event occurs during the duration of the study (one could also sample events during the study duration). Issues of concern when doing event sampling include defining what the relevant event is and establishing that the observation period of the study is adequately representative of some larger population of events. In fact, the concepts of time and event sampling are really quite broadly relevant to many types of geographic data collection used to study processes that occur over time. All geographers must think about sampling from time intervals or events whenever they use data that depend on temporally recurring phenomena.

Behavioral observations can be made and coded live and *in situ*, with an observer or set of observers who are present at the place and time of the behaviors of interest, and trained to code what they see and hear as data. Usually this is not the method of choice, however. It is difficult or impossible to always be there when you want to observe, and to be able to code reliably on the spot in real time. Instead, some type of media is used to make records of behavior, usually to be coded later in order to produce interpretable data (we discuss coding below). There are a variety of media that can be used to record behaviors. Each has strong and weak points, typically with respect to the nature of the record they create and their cost and efficiency. Possibilities include still photographs, audio recordings, video recordings, or some combination thereof. Base maps on sheets of paper or computer tablets can

[1]A classic example of a specimen record comes from Barker, R. G., & Wright, H. F. (1951). *One boy's day*. New York: Harper. The authors and their assistants closely followed a child from the time he woke up in the morning until he went to bed in the evening, taking detailed notes on what he did, where he went, who he was with, and so on. Yes, he brushed his teeth before he went to school.

be used to record the locations of activities over time. There are various automatic recording technologies, such as radio or GPS tracking (we discuss this satellite technology in Chapter 12 and the ethical questions about tracking people as part of research in Chapter 14). Transportation researchers use automatic **traffic recorders** that are pneumatic tubes or inductive loop detectors laid across roads and highways. Depending on their specific design, they can automatically generate a count of vehicles or even record their speed of travel. Devices placed in tiles on the floor can record where people step in public buildings. Of course, odometers in vehicles or connected to bicycles can certainly record distances traveled. Another example that is becoming increasingly popular is to automatically record what pages or files people view, and for how long, on computer systems such as the Internet. These **transaction logs** may be considered examples of archives (discussed below) if they are recorded previous to the study, for other purposes.

Behavioral observation is an important tool for data collection by geographers, but it has its problems. As we mentioned above, observation records must usually be coded to become interpretable data. Below, we discuss how labor intensive and difficult it so often is to code well. Behavioral observations also have the problem that the presence of observers (or their equipment) can affect the behavior of those observed. This is not limited just to reactance. Participant observers, to take a case in point, typically become part of a social situation and have an influence on it like anyone would, even if the people being observed have no idea the participant observer is a researcher recording data. Observation also suffers from the subjective and selective nature of perception. It is hard to train observers or coders to overcome their strong tendency to interpret the world meaningfully, to see what they expect or what makes sense to their prior worldviews. And both observers and recording devices literally have *points of view*—they always observe or record from some locations and positions, and not others. As a fan of sports knows, even with extremely expensive recording efforts (think of American football's Super Bowl), some behaviors just don't get recorded or observed. It really doesn't matter how many cameras and how much slow- or fast-motion technology you have at your disposal, you will still miss something. It seems that the ideal observational tool and procedure does not exist, and probably never will: a clear, continuous, and precise record of all the behaviors of all the actors, taken as if you were appropriately near or far simultaneously from all perspectives, without any trace of you being there at all!

Before we turn to archives as data, we note that besides observing humans, researchers observe the behavior of nonhuman animals.[2] Behavioral observation is

[2]Reactance and other forms of interactional artifacts (Chapter 11) can be relevant to those who study animal behavior, just as they are to those who study human behavior. An example we enjoy is the famous case of "Clever Hans," the numerically adept horse. During the early 20th century, Hans and his owner traveled around Europe, demonstrating Hans's amazing ability to tap out answers to arithmetic problems and the like posed by his owner. In fact, careful observations over several months revealed that Hans was responding to the subtle body language of his interrogator. The interrogator would lean forward slightly as Hans began tapping and then straighten up when he reached the correct final tap. We still think that's a pretty darn clever horse, even if he couldn't add.

a well-developed approach to data collection by researchers who study animal behavior, such as zoologists and psychologists. Some biogeographers do this too, but it is unusual given the geographer's typical interest in the numbers and locations of animals more than in their micro-behaviors.

Archives

In Chapter 3, we introduced archives as a type of data collected by geographers. Archives are existing records that were not collected for the purpose of a particular geographer's research. More often than not, they were not collected for any research purpose at all, but for various storage, journalistic, or record-keeping purposes. In some cases, archival records were created and kept for idiosyncratic and personal reasons—why does a person save letters and photographs? As such, archives are, by definition, clear examples of secondary data (introduced in Chapter 3). Geographers use financial records, birth and death records, marriage records, law enforcement data, hospital and clinic data, newspaper stories (in any medium), industry and business records, museum records, historical documents, diaries, letters, government records, movies, literature, advertisements (again, in any medium), voting results, tax and cadastral records, retail and wholesale data, and more.

Archives tend to provide a very nonreactive source of data. As secondary sources, they do not reflect any reactance to the measurement activities of a particular researcher. They are sometimes based on procedures in which people were asked to supply information intentionally to someone other than a researcher. Thus, archives can potentially suffer from types of biases that occur when people attempt to create particular impressions with the data they provide instead of responding truthfully. We can't think of a better example than using tax records as data; although we assume our readers are honest citizens like the two of us, a researcher would obviously be gullible to assume that all responses on tax forms were strictly truthful and sincere.

We use the term "archives" to refer specifically to records expressed in symbolic form—typically words or numbers but also images. That is, archives are records that indirectly refer to entities or processes, rather than records that directly express entities or processes. Physical materials (discussed in Chapter 4) that have been stored for some time are not archives in this sense, even if they are sources of secondary data insofar as they were not stored for the purpose of a particular study or researcher. Physical materials such as rock samples or stuffed birds are often stored in museum collections, for example. Of course, researchers use such collections as physical materials for their studies; they come to the museum and make physical measurements on the entities in the collection. We prefer to think of using collections in this way as physical measurement on secondary materials rather than as archives.

Let's consider an example of the use of archives. Early in the history of geographic thought, people's written (and drawn) records of their experiences traveling to foreign lands provided an informal source of information about the peoples and places they had visited. More recently, travel journals and diaries have been more systematically mined as an archival source of data to study the motivations and experiences of long-distance international migrants. And diaries or journals can provide information about many other phenomena too. Comments about rainfall in the journals or logbooks of explorers and expedition leaders have been analyzed as a proxy measure of historical weather patterns, for example.[3] This example shows that physical geographers as well as human geographers can use archives, even relatively unstructured and qualitative archives like diaries. Physical geographers and other earth scientists have in fact used a variety of archives as sources of data, such as ancient texts or inscriptions, governmental and private records, shipping records, and business and commercial records.

Like behavioral observations and responses to open-ended explicit reports (Chapter 6), archives must often be coded in order to produce usable data. Archives are records of activities, events, or characteristics that often cannot be directly interpreted by scientists; they must be processed first. For example, an explorer's journals probably do not give daily rainfall reports in inches or centimeters. This information has to be determined by classifying what was written in the daily entries in terms of categories of more or less precipitation, or none at all. That is, the researcher has to code the journals to extract interpretable data from them. We turn to the subject of coding next.

Coding Open-Ended Records

Both behavioral observation and archives often produce relatively open-ended records of phenomena that must be processed further in order to generate systematic data for analysis and interpretation. A video of pedestrians is a record of their behavior, but it is not in itself data. Likewise, travel journals kept by migrants are not data until their contents are "digested" for meaning. Other types of data collection similarly produce these open-ended records, notably some of the approaches to explicit reports we consider in Chapter 6. An audio recording of a park ranger talking about the way human visitors interact with the flora and fauna of the park is not scientific evidence until it can be systematically interpreted. This interpretation is essentially a process of categorizing the records or their parts. It is more than just the process of physical measurement discussed in Chapter 4. Unlike most of the physical measurements, open-ended records typically consist of words, pictures, or intentional acts that have meaning—they are semiotic or symbolic artifacts created by an

[3]Michaelsen and Larsen discuss the use of archival sources to estimate historical precipitation. They analyzed the journals of John C. Fremont and others in order to reconstruct Southern California precipitation during the 1840s. Michaelsen, J. C., & Larson, D. O. (1991). Historical documentary evidence for Southern California precipitation variability. In M. R. Rose & P. E. Wigand (Eds.), *Proceedings of the Southern California Climate Symposium: Trends and extremes of the past 2000 years* (pp. 93–117). Technical Report No. 11. Los Angeles: Natural History Museum of Los Angeles County.

entity with agency (as discussed in Chapter 3). As such, these records must be *interpreted* in order to extract their basic meaning. Not all human behaviors are semiotic in this way, of course. But even observations of the most ingenuous and straightforward acts usually require extra work to turn them into generalizable scientific data, just as the physical materials of Chapter 4 must be measured to produce data.

The process of turning open-ended records into data is called **coding**. When the records are semiotic artifacts such as verbal or graphical expressions, the process is often called **content analysis**. It consists of two parts, segmentation and classification. **Segmentation** is breaking the records into appropriate units. What counts as "appropriate" is primarily a matter of conceptual and theoretical consideration. What are you interested in? What type of entities is your theory concerned with? Are they words, phrases, sentences, paragraphs, or entire documents? Or perhaps they are images, objects, or parts thereof. In the case of behavioral observation, you might need to know about specific or average weekly acts (such as trips). These considerations are essentially about the analysis scale of your units (Chapter 2). There are some important implications of whether you use narrow or broad units. As in all areas of geographic research, it is optimal if your analysis scale matches the scale of the phenomenon you are studying. As a rule, it is more difficult to generalize with narrow units. What do I know about your travel behavior when I observe that you took one left-hand turn? Broad units may be too coarse or insensitive. What have I missed about your travel behavior if all I know is the total number of miles you travel each month? It is important to recognize that narrow units can usually be aggregated into broader units, but not vice versa.

Once you have determined the segmentation appropriate for your research and explicitly described the process by which you will create the segments, you need to decide how to classify your segments. **Classification** is grouping segments into abstracted categories—assigning segments to the categories—that capture aspects of the content or meaning of the records, aspects that are relevant to the theoretical and conceptual purpose of your research. Classification is virtually always an act of reduction insofar as a certain number of segments are translated into a smaller number of class members, usually many fewer. In rare circumstances, classification could lead to an expansion of data, if each segment is classed into several different categories.

We emphasize that developing a coding system is usually very time-consuming and difficult work. If you find otherwise, it might be because you have very circumscribed records that can be transformed rather straightforwardly into data, which essentially means your records are not very "open-ended" to begin with or you are working with a set of records that contains little variability. We believe that, in many cases, quick and easy coding is a sign that you have developed a poor coding system that will produce data of doubtful value. No matter how easy you find the process of developing your coding system, however, you must make sure to document the final system you use in detail in a **coding manual**. That will prove invaluable in training coders, in describing your procedures when you tell other people about your research, and in allowing people to evaluate or replicate your work.

Categories for coding come in many diverse forms. Their most important characteristic is that they effectively and efficiently capture aspects of the records relevant to your theoretical interests. Generally, categories should be specifically and operationally defined; they should deal precisely with your type of records and your research issues, and they should make it as easy as possible for coders to do their job well. Categories must be **exhaustive** by providing a coding option for every possible segment a coder will encounter in the records. More often than not, categories should also be **mutually exclusive**, with each segment going into exactly one category, although there are definitely situations where it makes senses to allow multiple categories for each segment or even multiple category systems for each segment (our example coding problem below demonstrates both of these situations). It's great if you can devise a coding system where every segment fits into a meaningful category, but usually it helps to include a "miscellaneous" or "other" category for eccentric segments; have coders write brief explanations for each such segment they code as "other." The nature of your records and research question obviously determines the proper number of categories to include in your system. As a general rule, however, you will usually find that three to 10 categories do the job well—more than that may be unwieldy to use or confusing to interpret.

What about the meaning of your categories? That should depend on prior conceptual and theoretical knowledge you and the rest of the research community in your problem domain have accumulated. It should depend on what specific research question or hypothesis you want to address. If you are studying homeowners' beliefs about the likelihood of a wildfire in their neighborhood, you'd better make sure to create one or more categories that concern statements about wildfires happening. Perhaps in a fantasy world of prior omniscience, this conceptual/theoretical or **top-down approach** would be sufficient to dictate your category design. We don't live in that world, however, so one nearly always combines this top-down approach with an after-the-fact **bottom-up approach** to developing categories in which you examine records to get ideas for what categories are needed. In fact, it is not uncommon for categories to be defined almost exclusively by the bottom-up approach of developing categories from the records themselves, as part of the ongoing process of actually coding. In the end, all records must be uniformly coded according to the final category system, of course. Perhaps most often, the top-down and bottom-up approaches are combined in an iterative process. That's part of the reason that developing good coding systems is such hard work.

Example Coding

A detailed example will help illustrate the coding of open-ended records. Our example is based on open-ended explicit reports (Chapter 6) generated as part of some research on how people look at and remember information from topographic contour maps and natural landscapes.[4] The studies examined how perception and memory vary as a function of the amount of training people have had in the use of

[4]The study was conducted by the first author and his colleagues: Montello, D. R., Sullivan, C. N., & Pick, H. L. (1994). Recall memory for topographic maps and natural terrain: Effects of experience and task performance. *Cartographica, 31*, 18–36.

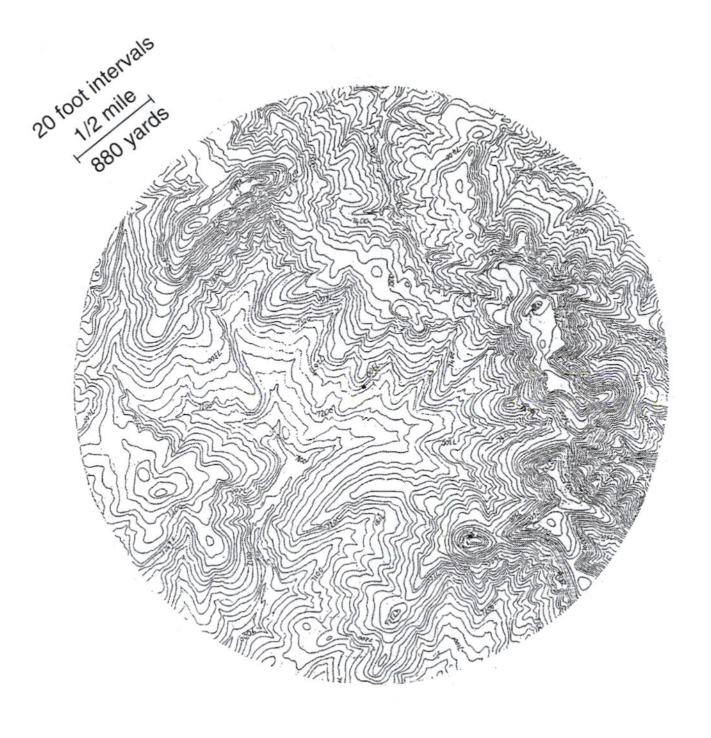

S# 2 Map D

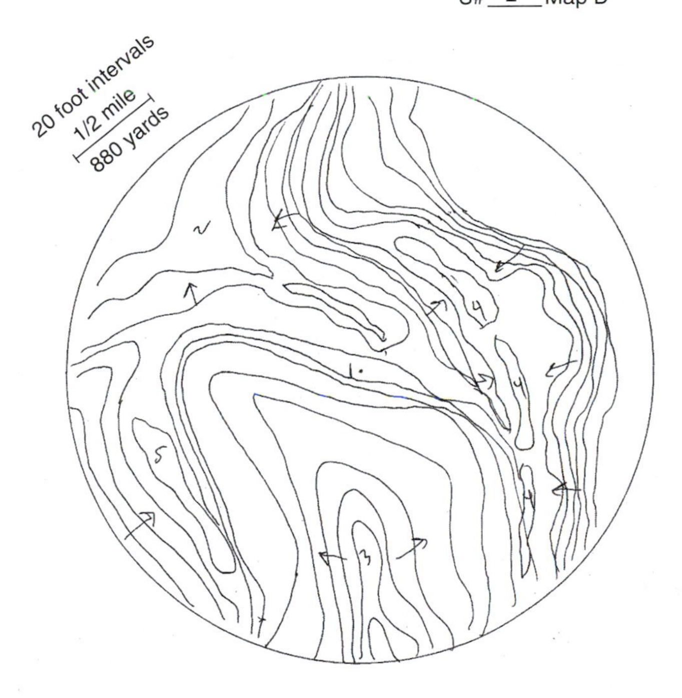

Figure 5.1 Example of open-ended coding based on a study of memory for topographic maps[4]; (a) topographic map portion D, (b) sketch map drawn from memory of Map D by an example research subject.

topographic maps, and as a function of whether the people are required to perform a task matching the map with the corresponding landscape. In one of the studies, research subjects viewed and recalled portions of topographic maps; half of the subjects subsequently matched a heading direction to one of three photographs of landscape scenes they would see if they were standing at the center location represented on the map. Subjects recalled the map portions by verbally describing them into an audio recorder. Subjects also sketched the map while describing it, explicitly labeling features with numbers, in order to help disambiguate the parts of the map being described.

Figure 5.1 shows the topographic map portion referred to as "Map D" (it happens to be from central New Mexico near Albuquerque). It also shows the sketch map of D drawn by one subject (labeled subject 2). Here is a transcription of what this subject recalled about Map D:

> Ok, #1, I'm standing on top of a hill that's right above me. And it's connected to a hill that's on the top left on the circle. Which means pretty much it's the predominant mountain of the map. Umm, below me there's a ravine that comes up from the bottom. Ok, that's #2. #3 is the ravine that comes up and goes to the left. But it's still below where I'm standing. Umm, to the top then right, well more to the right of me #3, no #4, are a bunch of little mountains or little peaks that form like a half circle surrounding the ravine and me where I'm standing, and their slopes are pretty steep to the right. #5, there's a little mountain to the right. I mean to the left and to the bottom, and it's not as high or as steep as #2 or #4. (Experimenter: Anything else you remember?) That's about it.

Box 5.1 shows the coding sheet used to classify elements of the verbal recalls. The researchers were interested in the nature of the features subjects would use to discretize the landscape, and how subjects would describe spatial properties of those features and spatial relations among them. The first step in coding this record was to segment the text into discrete features. This was relatively easy because subjects were requested to explicitly label each feature with a number. In a few instances, however, subjects did not employ the labeling system correctly, and thus features identified by coders did not match the subject's numerical labels. In our example, five features were described:

1–hill (single maximum, also self-location)

2–hill (single maximum)

3–ravine (single minimum)

4–bunch of mountains (multielement maximum)

5–mountain (single maximum)

Then spatial terms were coded. These were considerably more difficult to code than the features, because discriminations among general classes of terms could be quite subtle, and because redundant coding needed to be avoided (redundancy was very common in these transcripts). In our example, 21 spatial terms were mentioned nonredundantly. They are tallied and explained in Box 5.1.

Coding Reliability and Validity

Now that you have turned your records into *coded* records, you have data. As scientists, you want to know whether your coded values really measure what you think or claim they do. After all, other scientists will certainly want to know this when they read your work. Given the complex and sometimes vague nature of your records and your coding system, can you blame a person for being skeptical that your data, your evidence, really reflect the constructs you claim to be measuring? We're not saying that you intentionally miscoded or misrepresented your data, although that would definitely be a problem. We're talking about a system of coding that is too difficult, too underspecified, or too vague to produce adequate measurement. That is, you will want to establish, as much as possible, that your data-coding system is "reliable" and "valid." And you will want to be able to convince others of this.

Chapter 11 is all about reliability and validity. For now, we'll simply define reliability as the *consistency* of measurement; validity, specifically "construct" validity, is the *truthfulness* of measurement. The construct validity of coded records is largely determined by interpreting the apparent meaning of a detailed and transparently documented coding system, and by the statistical relationships of coded data to other types of data that are already understood to be valid. You can establish coding reliability by having two or more coders redundantly and independently code subsets of your records. These independent codes of the same records will largely agree if the coding is reliable. In Chapter 11, we label this "inter-rater reliability." Agreement among coders is usually greater when dealing with specific, concrete, and narrow categories. You and I can agree pretty reliably that a magazine advertisement included a tree in it; we would agree less reliably that the ad was trying to convince consumers that driving a certain brand of automobile would stimulate their "sense of adventure." But high reliability is pretty important for scientific data, as we discuss more in Chapter 11. So another reason that developing a coding system is hard work is that you need to put in considerable effort to refine and improve your coding system and its documentation so that you can achieve satisfactory coding reliability. You can do this by performing intermediate reliability checks on your coding system as you develop it. Have two coders code a small subset of your records with your system as it is, then compare and discuss disagreements. Make changes in the coding system to avoid the disagreements. Repeat until you get reliability high enough; we consider what that is in Chapter 11. The topographic map recalls in our example above were eventually coded with about 80% agreement on feature types and 70% on spatial terms.

To the degree that you want to take a scientific approach to your research, it is a good idea to establish coding reliability and include that in your research write-up. Particularly if your research is meant to be confirmatory rather than exploratory,

Box 5.1 Coding Sheet and Coded Terms for Example of Open-Ended Coding

Topographic Map Memory Study[4]

Coder <u>SAS</u> Date of Coding <u>Nov 11, '91</u> Subject# <u>2</u> Map <u>D</u>

1. Number of Features <u>5</u>
2. Number of Feature Mistakes <u>0</u>
   A feature is wrong, not present, or located incorrectly relative to other features (ordinally).
3. Terrain Feature Types (nouns or noun phrases):
   Code each feature by letter according to how it is first mentioned. Do not code redundantly.

	Single	Multielement
up (maximum)	1, 2, 5	4
down (minimum)	3	
flat		
slope		
mixed/combined		
water		
other		
self-location (feature identified as subject's location; may be redundant)	1	
indeterminate		

4. Spatial Terms (words or phrases):
   Count nonredundant spatial terms not directly prompted by experimenter. *Metric* is quantitative, whether precise or approximate. *Topological* is nonquantitative, expressing order, connection, adjacency, containment, etc. *Ego-dependent* is a spatial term that depends for its meaning on the person's location (on map or in lab). Cardinal directions are typically ego-independent unless person is explicitly mentioned (e.g., "west of me") and topological unless qualified by term like "precisely" or "directly." Do not code "here," "there," "then," "next," or "in this area."

	<u>Ego-independent</u>	<u>Ego-dependent</u>
<u>feature-locational</u>		
metric distance		/
metric direction		
topological/nonmetric	//	///// ///// /

feature-intrinsic properties			
size or number	///	elevation	
slope		metric	
metric	/	topological	/
topological	/	shape	/

Spatial terms marked with gray highlighting:

Ok, #1, I'm standing on top of[a] a hill that's right above me[b]. And it's connected to[c] a hill that's on the top left[d] on the circle. Which means pretty much it's the predominant[e] mountain of the map. Umm, below me[f] there's a ravine that comes up from the bottom[g]. Ok, that's #2. #3 is the ravine that comes up and goes to the left[h]. But it's still below where I'm standing. Umm, to the top[i] then right[j], well more to the right of me #3, no #4, are a bunch of little[k] mountains or little peaks that form like a half circle[l] surrounding[m] the ravine and me[n] where I'm standing, and their slopes are pretty steep[o] to the right[p]. #5, there's a little[q] mountain to the right. I mean to the left[r] and to the bottom[s], and it's not as high[t] or as steep[u] as #2 or #4. (Experimenter: Anything else you remember?) That's about it.

a-topological, ego-dependent
b-metric distance, ego-dependent
c-topological, ego-independent
d-topological, ego-dependent
e-size or number
f-topological, ego-dependent
g-topological, ego-dependent
h-topological, ego-dependent
i-topological, ego-dependent
j-topological, ego-dependent
k-size or number
l-shape
m-topological, ego-independent
n-topological, ego-dependent
o-metric slope
p-topological, ego-dependent
q-size or number
r-topological, ego-dependent
s-topological, ego-dependent
t-topological elevation
u-topological slope

we recommend subjecting a subset of your records to redundant coding by two independent coders. If you can show that inter-rater reliability is satisfactorily high, it will help convince reviewers of your research that you are measuring something with quality and dependability. And once you achieve a suitable level of reliability in your coding system, you need to make sure to rigorously train coders in its use. Also, check your data periodically throughout the coding process to make sure you maintain high coding reliability.

Since our example study was originally conducted in the 1980s, several commercial software packages have become available to facilitate the coding of open-ended records—so-called qualitative data analysis software such as *ATLAS.ti* and *NUD*IST*. These packages do help; they automatically update category definitions and coded units whenever you make a change to the coding system, for example. But don't misinterpret the promise of these packages—they cannot do the hard work of actually interpreting what someone's verbal utterances mean (their semantics). To a large extent, that is, the hardest part of coding cannot be automated. That stems from the fundamental nature of extracting and classifying the semantics underlying words. In most studies, you want to code what verbal text *means*, not what it literally *says*. But the relation of words to meaning is a highly complex problem that is anything but straightforward; it still has not been completely solved by scientists who study language and cognition. The problem is that words relate to meaning according to a "many-to-many mapping" that is typically disambiguated by the context of the situation, of the recent conversation, of the people involved, and so on. A single word means different things, often many different things, and the different meanings can be quite unrelated. In our coding example above, sometimes subjects used the word "below" to mean at a lower elevation in the landscape, and sometimes they used it to mean lower on the page of the map. In many languages, English being a prime example, a single concept can be expressed by several different sets of words, often many different sets of words. The concept of "urban area" may be captured by words like city, town, municipality, central city, village, hamlet, urbanized area, suburb, metropolis, polity, burg, conurbation, metropolitan area, and megalopolis—and that's not even counting technical census terms like "standard metropolitan statistical area" or "urban cluster" (see Chapter 6). These facts about the way all natural languages (as opposed to formal languages such as mathematics) express meaning make coding a difficult task that is hard to perform reliably. They also ensure that there will always be work for scholars of law and of religion, in case any of you were worried about that.

Review Questions

Behavioral Observation

- How does scientific behavioral observation differ from the everyday behavioral observation that we all do?
- What are some different ways to record behavior for future coding and analysis?
- What are the formal observation schedules of time and event sampling?
- What are some strengths and weaknesses of behavioral observation as a technique for collecting data?

Archives

- What are some examples of archival data sources in geography? How do archives differ from physical materials (Chapter 4)?
- To what extent are archives examples of nonreactive measures?
- What are some strengths and weaknesses of archives as a source of data?

Coding Open-Ended Records

- What types of data sources in geography require open-ended coding and why?
- What are the two parts of coding called segmentation and classification?
- What are properties of a good coding system, and how does one go about developing such a system?
- How does one establish the reliability of a coding system?

Key Terms

anecdotal records: nonsystematic form of behavioral observation

behavior: overt, potentially perceptible action or activity by people or other animals

bottom-up approach: generating coding categories by reasoning from an examination of the records that are to be coded

classification: part of the process of coding open-ended records in which segments are assigned to categories that capture aspects of the content or meaning of the records

coding: process of turning open-ended records into data by classifying them, or segmenting them and classifying their parts

coding manual: detailed documentation of your coding system for open-ended records

content analysis: term for coding when the records are semiotic artifacts such as verbal or graphical expressions

event sampling: specific formal observation schedule in which events are recorded at any time they occur during the course of data collection; appropriate for rare events

exhaustive: necessary property of coding systems in which there is a coding option for every possible segment in the records

formal observation schedules: plan for systematically sampling events from a population of events distributed over time

mutually exclusive: property of coding systems in which there is one and only coding option for each segment in the records; usually but not always desirable

participant observation: technique of behavioral observation in which a researcher becomes a member of an ongoing setting or group in order to observe it

records: representations of structures and processes, including human activity, encoded in words, pictures, or sounds; they must be coded or otherwise measured in order to produce scientific data

segmentation: part of the process of coding by breaking records into appropriate units

specimen record: detailed and comprehensive record of what an individual or group does over some time period; also called a running record

time sampling: specific formal observation schedule in which events are recorded during specified time intervals occurring at specified times

top-down approach: generating coding categories by reasoning from prior conceptual or theoretical knowledge

traffic recorders: detectors laid across highways that automatically record the passage of vehicles

transaction logs: records of actions a user takes while interacting with a database or other computer system

Bibliography

Boehm, A. E., & Weinberg, R. A. (1987). *The classroom observer: Developing observation skills in early childhood settings* (2nd ed.). New York: Teachers College Press.

Krippendorff, K. (2004). *Content analysis: An introduction to its methodology.* Thousand Oaks, CA: Sage.

Webb, E. J., Campbell, D. T., Schwartz, R. D., & Sechrest, L. (2000). *Unobtrusive measures* (rev. ed.). Sage classics series #2. Thousand Oaks, CA: Sage.

CHAPTER 6

Explicit Reports

Surveys, Interviews, and Tests

Learning Objectives:

- What are general properties and specific types of explicit reports?
- What are options for formatting explicit-report items?
- What are options for administering explicit reports, and what are some consequences of the different options?
- What are some ways to design and generate explicit-report instruments?
- What are characteristics of the U.S. census and the data it produces?
- What are basic limitations of explicit reports as a type of research data?

One of the most flexible and popular types of data collection in human geography is the explicit report. As we described in Chapter 3, explicit reports are measures of beliefs people have about all sorts of things, including themselves or other people, places or events, activities or objects. Explicit reports, including surveys, interviews, and tests, can request many different types of beliefs: behaviors, knowledge, opinions, attitudes, expectations, intentions, experiences, and demographic characteristics. The defining trait of such measures is their **explicitness**. People know they are providing information to a researcher when they are surveyed, for instance, and the responses they provide are based on opinions or beliefs they can consciously access, that is, of which they are aware. Often, explicit reports request responses that cannot readily be judged as being right or wrong; that is, the responses are personal opinions or preferences that cannot be compared to any objective standard of reality, although they can be characterized as common or unusual, and they can be related to other variables such as demographics. As we

Table 6.1 Major Types of Explicit Report Instruments

1. Surveys, questionnaires
2. Interviews
3. Sociometric ratings
4. Activity diaries, logs
5. Contingent valuation
6. Focus groups
7. Protocol analyses
8. Tests

described in Chapter 3, when the explicit responses *can* be assessed for correctness, and that is of major interest to the researcher, we call the reports tests. That is, tests are used to study knowledge rather than opinion.

Major types of explicit reports are listed in Table 6.1. The first is the **survey**, or questionnaire. Surveys require respondents to answer questions about their opinions, attitudes, or preferences. They can also ask questions about activities or demographics. Even though some of these have answers that are potentially true or false, that is not of primary interest to researchers—researchers typically accept them uncritically as true (we discuss skepticism about the veracity of responses at the end of the chapter). We use the term "survey" to refer to reports that are administered and responded to in written format. In contrast, **interviews** collect the same types of information as surveys but are administered and responded to orally.[1]

Three additional types of explicit reports are really subtypes of surveys. One is the **sociometric rating.** These are opinions or beliefs expressed by members of small groups, such as families or carpool groups, about each of the other members of the group. Such ratings are particularly valuable for studying small-group structures and dynamics. They have an interesting analytic property that results from all members rating, and being rated by, each of the other members. This allows a person's average ratings to be separated into a component due to how he or she rates others, a component due to how others rate him or her, and a component that is unique to the particular "rater-rated" dyad.

A second subtype of survey is the **activity diary** or log. Activity diaries require respondents to record what they do on a regular basis, typically on a daily basis. Geographers are often especially interested in *where* people carry out activities,

[1]Our distinction between surveys as written and interviews as oral is not universally held. For example, some authors refer to all explicit reports that do not focus on factually correct or incorrect information as surveys (or "self-report" measures), reserving the term "interview" for those surveys administered orally by a researcher, and the term "questionnaire" for those self-administered in writing. What is important is not so much what they are called, but that written and oral administration can produce somewhat different responses in different situations, and their costs certainly differ.

such as where they shop or which road they take. Transportation surveys are a good example. Activity diaries could be considered behavioral observations on one's self, but because they require people to explicitly report on their activities, we include them here rather than with the behavioral observations of Chapter 5. Similarly, activity diaries as explicit reports should not be confused with the everyday, common use of the term "diary." As the term is commonly used, diaries are confidential records or personal letters that people write to themselves to record their experiences, activities, thoughts, and feelings. Such diaries are nonreactive measures, of course, because their authors do not expect that researchers will use their diaries as data. We cited traditional diaries as a source of archival data in Chapter 5.

A third subtype of survey is **contingent valuation**. This type of report requires people to rate or rank how much they value something, typically something that may have great subjective value but is very difficult to assign value to "objectively." Contingent valuation is commonly used in studies of how people value particular landscapes or environmental actions, for example, preserving open space or maintaining quiet parks. It has been applied in other contexts too, such as studies of mass transit use. Ratings of value are expressed directly in dollars or indirectly in terms of units of something else that the respondent values, such as hours of time or miles from family members.

Two more types of explicit reports are really subtypes of interviews. **Focus groups** are unstructured (discussed below) interviews carried out with small groups of respondents, with as few as two or three members, or as many as 30 or more (7–20 is a common range). Typically, focus groups discuss a particular topic and are led by a facilitator or moderator. The most notable characteristic of focus groups is that they involve a simultaneous interview with a group of people, so various **group dynamics** develop. Of course, members make comments that jog other members' memories. But the comments also influence what other members think or are willing to say; nonverbal cues do this too. For both conscious and unconscious reasons, people will often change what they say and how they say it in order to make impressions on others they know are observing. These various social effects are amply documented in a vast literature in social psychology and sociology on persuasion, conformity, and self-presentation. Furthermore, the outcome of a focus group depends on the particular people who make up the group; for example, outspoken group members can monopolize a focus-group discussion. Transcripts or notes of focus groups are usually not carefully coded (Chapter 5) or analyzed quantitatively; they should not be unless extensive steps are taken to deal with their limitations. In most cases, focus-group results are only very useful during exploratory phases of research.

A second subtype of interviews is **protocol analysis**. This is an open-ended (discussed below) interview in which people "think aloud" about the contents of their conscious mind while reasoning about some problem or issue. It attempts to identify what people are thinking about during some mental task, including what they are looking at, listening to, imagining, inferring, or retrieving from long-term memory. For example, a researcher could ask people to talk about their thoughts while navigating a geographic information database system, a study that could help make the system's design more user-friendly. Protocol analysis is not about general or average thoughts, or about justifications or rationalizations. Done properly, it is about the *products* of thought at different stages, not the *processes* of thought. As we consider at the end of the chapter, mental processes are typically unavailable to conscious access, and in fact, so are many mental products. Many uses of protocol analysis ask people to talk about (or point, draw, and so on) what they are thinking or attending to while they reason; this is called "concurrent" protocol analysis. Alternatively, people may talk about their thinking *after* they finish reasoning, perhaps because the talking would interfere with the primary task; this is called "retrospective" protocol analysis and may be relatively immediate or delayed. Whether the protocol is concurrent or retrospective, it is typically quite valuable for the researcher to perform a detailed analysis of the primary task the person is working on as an important way to help interpret the protocol.

Finally, as we have defined them, **tests** require participants to respond to questions that can be assessed as right or wrong. Researchers administering tests are specifically interested in the factual correctness or incorrectness of responses to their tests. A geographer might use tests, for example, to study a person's knowledge of population demographics or economic development (of course, if the geographer is using the test to evaluate your performance in a class, we call it *teaching* rather than research). Tests are typically scored for accuracy, including number or size of errors, in order to generate the data. Sometimes people are timed while taking tests, and their response times are recorded to produce data. Response times can be theoretically interesting, especially when used comparatively, because they reflect how long it takes someone to remember or reason about something. For example, if a GIS user takes longer to solve a problem with one interface than with another, that is evidence for the relative superiority of the faster interface, other things being equal.

Format of Explicit Reports

All types of explicit reports ask people to respond to questions, statements, pictures, or other stimuli. A specific survey, interview, or test is called an **instrument**; the specific questions or statements that make it up are called **items**. We have distinguished between explicit reports expressed in written versus oral form, but in both cases, reports are most often expressed verbally, in words. However, responses to explicit reports can be expressed in numbers, gestures (such as pointing), graphics, or manipulable objects. In the case of manipulable objects, respondents can be asked to mark lines, draw pictures, sketch on maps, or construct physical models. In this section, we discuss some alternatives for the formats of these items.

Perhaps the most important aspect of item format is the distinction between closed-ended and open-ended items, one important way that many people distinguish between quantitative and qualitative research methods (as we discussed in Chapter 3). **Closed-ended items** provide a finite number of specific response options for respondents to choose as answers, typically a modest number. They provide responses in a fixed-response format. **Open-ended items** do not provide specific response options for respondents. They provide responses in an unstructured or free-response format. As compared to closed-ended items, open-ended items allow responses that were not anticipated by the researcher. Also, open-ended items allow responses of any length, assuming enough time or space is allowed, and responses that are meaningful to respondents in their own words. However, as we explained in Chapter 5, open-ended responses are not data in themselves—they are records of (typically verbal) behavior. If they are to serve as anything more than impressionistic suggestions, they must be coded (we explained how to code in Chapter 5). Interestingly, some research has shown that closed-ended items often prompt respondents to choose response alternatives that they would not spontaneously make to open-ended items.[2]

Another distinction concerns the way the items are formatted for different respondents. **Standardized items** are presented in a predetermined and consistent format to all respondents. This consistency may be in the wording of the items, their response formats, the order of items, how they are administered, or any combination thereof. In one or more of these ways, **nonstandardized items** are not consistent across respondents. The most common examples of nonstandardized items are explicit reports that contain **follow-up questions**, the specific wording of which depends on responses to earlier items. Follow-up questions are of two types. Some follow-up questions have a so-called **branching format** in which questions vary "automatically" as a function of responses to earlier questions. For example, an item can be presented as appropriate depending on the answer a respondent gives to a previous demographic question, such as ethnicity or income; questions about retirement might follow only if people give their age as over 60. The second type of follow-up question is even more nonstandardized. These are **free-format** follow-up questions that attempt to clarify and dig deeper into the meaning of responses. Their specific wording is determined in real time by an interviewer who reacts spontaneously to a previous answer.

Table 6.2 lists common types of closed-ended items. Closed-ended items should always provide exhaustive alternatives to respondents and should usually provide mutually exclusive alternatives too, so that the respondent can comfortably pick one and only one alternative. In some contexts, however, it may be perfectly reasonable to allow respondents to pick more than one alternative; this is typically true with adjective checklists, for example. Whenever a respondent might plausibly want to answer something you have not provided as an alternative, include an "other ___" response; this makes a closed-ended item somewhat open-ended. Although some researchers prefer to force respondents to express a belief about something, we believe respondents should be allowed to choose "don't know" or "no opinion" in most cases. When analyzing these alternatives, be careful that you separate "don't know" or "no opinion" responses from the rest as appropriate, such as when you calculate proportions or averages.

---

[2]It is not clear whether this means that closed-ended alternatives remind people of things they really do believe, or that they put ideas into people's heads they would not otherwise have.

Table 6.2 Major Types of Closed-Ended Items

1. **Rating scales**. Respondents provide a number or mark a line to indicate the amount or extent of something, including the degree of belief they have in something.
2. **Forced-choice alternatives**. Respondents pick one alternative from a list of choices, such as "yes-no," "true-false," or "multiple choice."
3. **Ranking of alternatives**. Respondents put two or more alternatives in order from most to least, or least to most.
4. **Adjective or activity checklist**. Respondents mark each alternative that applies from a list of adjectives or activities.
5. **Paired or triadic comparisons**. Respondents specifically compare entities, two in the case of paired comparisons, and three in the case of triadic comparisons. These comparisons can be ordinal ("Which two of the three neighborhoods are most similar?") or metric ("What is the distance from the courthouse to the visitor center?").
6. **Sorting task**. Respondents sort phrases, pictures, or other stimuli into groups based, typically, on similarity.

Closed-ended items like rating scales occasionally produce patterns of responses that suggest a respondent was not answering sincerely. These are called **response sets**. They come in various forms. **Social desirability** is the tendency for respondents to give answers they think the researcher, or some imagined group of peers or superiors, want to hear. This may be reflected in a set of answers that are "just too good to be true," which we admit may not be easy to detect. Response sets are often reflected in forms of consistent responding, such as yea-saying, nay-saying, extremism, or moderatism. For example, a respondent might only use the highest value on a rating scale. To deal with this, it is sometimes recommended to alternate the "sense" of items so that a response to one side would not always reflect agreement with the same view. This probably does induce respondents not to respond reflexively, but it also has the real potential to confuse them. As another example, someone responding to a survey of "environmentally responsible behaviors" might make dubious claims about always recycling (every candy wrapper), never leaving lights on, sharing rides everywhere, and taking showers in cool water. Including instructions that urge respondents to use several values on the scale could help here. In any event, you have every right (even a *responsibility* of good data analysis) to remove participants from the data set when you detect response sets. However, this must be based on some rule that is objectively describable other than that someone gave responses counteracting your hypothesis; you will describe this rule in the Results section of the research manuscript (Chapter 13). Of course, insincere responses are not always possible to spot as some pattern that sticks out. Such responses are hopefully (and we think probably) rare, but they undoubtedly happen. When they do, they typically add "noise" to the data, which lowers reliability (Chapter 11).

Rating Scales

Let's consider rating scales in a little more detail. As we define them in Table 6.2, rating scales ask respondents to provide a number or mark a line to indicate the amount or extent of something, including the degree of belief they have in something. Several types of rating scales are shown in Table 6.3. We use the term **generic rating scale** to refer to any rating scale that does not have one of the more specific formats and purposes of the other types. The **semantic differential** requires respondents to rate a set of entities (objects, events, places) on each of several adjectives. Semantic differential scales are designed to measure attributes of connotative, rather than denotative, meaning—not what an entity literally means but what it suggests or implies. **Likert scales** require respondents to express their degree of agreement or disagreement with a series of statements. **Paired comparison ratings** require respondents to compare entities (as described in Table 6.2) with the use of a quantitative rating scale.

There are a couple of general design issues that apply to all types of rating scales. First is the question of how many scale options to provide respondents. In most situations, the answer is five to nine options, preferably toward the low side when orally administering an instrument. The rationale behind recommending five to nine options is that fewer options provide less measurement resolution than people can validly make, while more options provide too much resolution. Although one may be tempted, for example, to ask respondents to rate on a 100-point scale (people are used to "100-point thinking" in our culture), evidence suggests that people cannot validly make discriminations at that fine a level.[3] Although respondents might claim to distinguish a preference of 70 from one of 75, never mind 72 or 73, they probably cannot do so in a way that validly reflects their feelings. It would be spurious precision, as we defined in Chapter 2. When working with young children, one should use a scale with two to five values; a charming series of happy, neutral, and unhappy faces is sometimes used. Whatever the specific number of scale options one uses, it is best to use an odd number in order to leave room for a midpoint of neutrality, when that is appropriate. And as for other closed-ended items, you might or might not find it a good idea to offer respondents the opportunity to choose "don't know" or "no opinion" as response options (these are not shown in Table 6.3).

Finally, we want to come back to the issue we touched on in Chapter 2 concerning the measurement level of data obtained from rating scales. As we noted in Chapter 2, several authors of statistical and methods textbooks in geography (and in other disciplines) advise treating rating-scale data as ordinal. That is, the interval between "3" and "4" should not be assumed to equal the interval between "4" and "5." Following this advice has the fairly substantial implication that only nonparametric ordinal statistics (see Chapter 9) should be applied to rating-scale data:

---

[3]These limits on human discrimination are discussed in a famous paper that also deals with limits to working memory capacity: Miller, G. A. (1956). The magical number seven, plus or minus two: Some limits on our capacity for processing information. *Psychological Review, 63*, 81–97.

Table 6.3 Types of Rating Scales

Generic Rating Scale

For each place, circle a number from 1 to 9 to rate how much you would like to take a vacation there in the next year:

not at all	1	2	3	4	5	6	7	8	9	very much

______ Disney World, Orlando, Florida

______ Yellowstone National Park, Wyoming

______ Las Vegas, Nevada . . .

Semantic Differential

Rate each city on each scale by marking a number from 1 to 7:

big	1	2	3	4	5	6	7	small
hot	1	2	3	4	5	6	7	cold
safe	1	2	3	4	5	6	7	dangerous . . .

Likert Scale

Rate your degree of agreement or disagreement with each statement by circling one phrase:

strongly disagree	disagree	neither agree nor disagree	agree	strongly agree
1	2	3	4	5

"Individual people are more responsible for energy conservation than is the government."
"The U.S. should reduce its dependence on foreign oil."
"The federal government should increase funding to develop efficient solar energy."

Paired Comparison Ratings

For each pair of cities, rate how similar the two cities are to each other by marking a number from 1 to 7:

completely different	1	2	3	4	5	6	7	exactly the same

______ London, U.K. – Bangkok, Thailand

______ Lima, Peru – Moscow, Russia

______ Cairo, Egypt – New York, U.S.A. . . .

no means, no standard deviations, no analysis of variance (ANOVA), no Pearson correlations or least-squares regressions (it also has implications for graphing and mapping—see Chapter 10). Contrary to this, several other textbook authors, mostly outside of geography, consider it acceptable to treat rating-scale data as metric. In fact, rating-scale data have been treated as interval level in thousands of published studies by top researchers in psychology, sociology, and other fields (rarely would it constitute ratio-level data, as "0" on a liking scale from –3 to +3 is an indifferent preference, not "no" preference at all). These researchers consider this treatment to be both justifiable, accurately reflecting the underlying nature of the attitudinal and preference constructs being measured, and functional, allowing the most powerful and flexible statistical analyses possible.

We endorse the view that rating-scale data generally can be treated at the interval level, particularly when the scale is an attempt to measure an underlying construct that can reasonably be considered continuous, such as preference or degree of confidence. We certainly agree that people do not treat equally spaced scale values as *exactly* equally spaced differences in the property being measured. However, we also believe, and some evidence supports, that people do treat equally spaced scale values as *approximately* equally spaced differences, especially when the scale presents visually and/or numerically equally spaced values to the rater (for that reason, we do not recommend the use of wordy verbal labels for each scale value that might not be interpreted as evenly spaced concepts). That is, we believe that rating scales as people understand and use them are not precise interval scales, but that they are much more than merely ordinal. If they were just ordinal, the underlying difference between "3" and "4" could as easily be 10 times the difference between "4" and "5" as it could be one-tenth the difference—and this is untenable. Instead, we see intervals in rating scales as being understood and used by people somewhat vaguely and with some variation across times, individuals, and even positions in the scale. This conception of a rating scale would be captured visually by an animation in which intervals between scale values were a little fuzzy and quivering. Given this conception, we consider it justified treating rating-scale data as interval rather than ordinal, accepting that we thereby introduce much greater power and flexibility into our research at the cost of a little conceptual or empirical error. If you, or your advisor or editor, don't accept this position, feel free to treat the rating scale data as ordinal. At worst, you might fail to uncover some interesting patterns in your data.

The Administration of Explicit Reports

Besides their format, there are a variety of other options about how explicit reports are administered. They may be self-administered or researcher-administered (of course, people do not generally *interview* themselves). They may be administered individually or in groups. The medium of administration could be in person (face-to-face), through the mail, over the telephone, or on the Internet (more below on Internet research). Interviews are often done with the help of computer programs that display the questions and accept typed-in answers. Also, interviews may be recorded, either audio or video. A variety of considerations determine the best way to administer explicit reports in a given research situation:

1. *Cost.* The cost of doing research with explicit reports, whether expressed in terms of money, time, or effort, can vary greatly as a function of the way they are administered. A case in point is oral interviewing versus written surveying. Interviewing is usually much more costly; even if interviewers are not paid, they must be selected, trained, and supervised. But the cost of sending out mail surveys, including the cost of postage, is not trivial. Internet surveys have become an attractive alternative because of their low cost.

2. *Number and nature of items.* Usually, surveys can contain more items, and more complex items, than interviews. This is because of the extra demands placed on memory by listening and responding orally, as compared to reading and writing. Complex items or complex response formats are difficult or impossible to administer orally, although there are techniques for orally administering response options more effectively. Even more than complexity, the various administration options can differ greatly in their effectiveness at getting honest and forthcoming answers to sensitive questions. Such questions include not only such obviously personal matters as sexual behavior and substance use, but personal finances, political beliefs, and environmentally responsible behaviors. The degree of anonymity or confidentiality a respondent perceives varies quite a bit across administration options, and that definitely affects the honesty of responses (anonymity and confidentiality are relevant to the ethical treatment of human research subjects, as we discuss in Chapter 14). Again, there are various techniques that can be used to increase respondents' forthcomingness, including clever ways to ask questions that don't require respondents to provide a personally identifiable answer that could even possibly be linked to them.

3. *Response rate.* Response rate is an important consideration in choosing an administration method. More people throw mailed surveys away, considering them junk mail, than refuse to talk on the phone, although frustration with incessant "sales calls" has reduced the response rate to phone solicitation over the last decade. People apparently ignore e-mail solicitations at an even higher rate than they do paper mail solicitations. In-person requests for participation are the most effective in terms of response rate, but of course, they are costly.

4. *Potential for follow-up.* Administration options differ a great deal with respect to the possibility they allow for follow-up questions that depend on the nature of responses to earlier items. This can be among the most important considerations in choosing among options. Here we are talking mostly about free-format follow-up questions that attempt to clarify and dig deeper into the meaning of responses. That can be done effectively only with oral administration, whether in person, on the telephone, or during Internet exchanges (e-mails can be sent to follow up on responses). In contrast, branching-format questions that "automatically" pose follow-up questions tailored to particular responses to earlier questions are possible with any and all administration options. Until artificial intelligence becomes much more advanced, however, it is generally impossible for a computer program

to interpret the meaning of a response and then ask a follow-up in order to clarify part of that meaning. A sentient human being who is thinking about the response as it is given is required.

5. ***Nature of respondents.*** This is a fairly obvious, common-sense consideration. But common sense is sometimes surprisingly rare, and in any event, no consideration is obvious if one fails to think of it. Different types of respondents differ in their ability to handle different administration methods, different types of questions, and so on. Respondents differ in ways that have implications for administration options. Relevant ways include age, sex, formal or informal education, social status or position, language spoken, disabilities (sensory, motor, cognitive), and so on. Different cultural and subcultural groups of people have been exposed to different norms that can have large effects on the viability of different administration options. Depending on the sex of the respondent, an interviewer of a particular sex can be quite inappropriate in some cultures. Revealing some kinds of information to a researcher who is not actually present may be strange to the point of not working.

6. ***Possible interviewer artifacts.*** In Chapter 11, we discuss potential threats to the validity of research that might arise because of the appearance, demographic characteristics, or personal style of researchers. These "researcher artifacts" are an issue only when the researcher is in the presence of the respondent when he or she responds (not quite—the respondents' expectation that different kinds of people will look at their responses could have an impact). For example, male respondents may not answer questions about their "sense of direction" in the same way when speaking to female interviewers as they do to male interviewers; female respondents may not either. The wording of items may reflect bias on the part of the researcher, which may in turn bias responses; the nonverbal and "paraverbal" cues (body language, intonation) used by an interviewer when orally administering items, including follow-up questions, may bias or lead responses. In addition, some methods of administering or scoring explicit reports allow or require the researcher to interpret what respondents said or did in order to turn it into recorded data. This can obviously open the door to the potential of biased interpretation on the part of the researcher. We discuss these researcher artifacts more in Chapter 11, including some approaches to minimizing them.

Using the Internet to Collect Explicit Reports

Within the last decade, the possibility of using the Internet, including the World-Wide Web, to carry out research with explicit reports has become a reality. The major benefit of this approach, whether e-mailing surveys or having people respond to a Web form, is its efficiency and low cost. This benefit is potentially huge. Researchers have recently been using the Web to solicit and record responses by thousands or tens of thousands of people, at extremely low costs. What's more, data entered on Web surveys, including forced-choice options and open-ended text, are automatically stored in digital form, and to varying degrees, can be analyzed automatically as well. Web surveys and tests can be automatically modified as they are being administered to take account of earlier answers—the branching-format follow-up questions we discussed above.

But Internet research is not without risks to the quality of the data. There has been considerable concern that samples of respondents obtained over the Internet, often people who choose to go to a particular Web site and fill out its survey, are unusual in some ways and do not represent some larger population very accurately. In other words, there is concern about the presence of a sampling bias in Internet research. In fact, empirical research[4] has demonstrated fairly convincingly that Internet samples are not perfectly representative of the general population. However, they are better than college student samples in this respect, being more heterogeneous on most variables, including sex, race and ethnicity, socioeconomic status, and age. In particular, Internet samples are superior with respect to regional diversity. In addition, assessments of personality and psychological adjustment suggest that Internet samples these days are not filled with "geeky antisocial nerds." As much as Internet users were once an unusual subset of the population, finding people who never use the Internet will become increasingly difficult in the future.

There are other potential difficulties with Internet research. Perhaps respondents participate repeatedly, either because they don't know that it is inappropriate or they are protégés of the Daley school of survey research ("respond early and often"). However, research suggests that this does not occur that commonly, and in any case, there are techniques for identifying repeat responders and omitting them. Some research indicates that participants in Internet research drop out at higher rates, but this is not much of a problem, given the high numbers of people such research can contact in the first place and the fact that nonparticipation bias is apparently no greater with this type of research than with more traditional types. Researchers are sometimes concerned that Internet respondents will take studies less seriously and provide more frivolous responses, but most research indicates that the results of Internet studies are similar to traditional studies on comparable topics. In addition, there are potential ethical difficulties (Chapter 14) with Internet research that are getting quite a bit of attention, including the potential for loss of privacy and the fact that Internet researchers may not always be able to ensure that respondents can provide genuine informed consent in order to participate.

Finally, the Internet provides quite a bit more to researchers than just an efficient way to administer surveys and tests. The Internet in its various guises has led to the emergence of a host of new social and behavioral phenomena that are of interest in their own right to researchers, including geographers. Such new phenomena are the subject of ongoing and future research, and include the diffusion of innovations via

---

[4]A good example is Gosling, S. D., Vazire, S., Srivastava, S., & John, O. P. (2004). Should we trust Web-based studies? A comparative analysis of six preconceptions about Internet questionnaires. *American Psychologist, 59,* 93–104.

the Internet, social interaction in digital worlds, and the emergence and structure of online communities.

Designing and Generating Explicit Instruments

Having considered various structures for report items, we now turn to the question of how to create the items themselves. How do we get ideas for items? The answer is pretty much the same as the answer we gave in Chapter 2 as to how to generate research ideas—any way that works! Intuition and prior knowledge, whether educated or naive, is a basic starting point. Existing literature will provide many ideas, including specific items and even entire instruments that already exist. Even if you don't find an instrument in the literature you can use in your research, you should make sure that whatever you come up with to use is compatible with whatever wisdom is to be found in that literature. For example, make sure your items deal with the concepts the literature says are important in a particular domain. A good source of ideas for generating items is to conduct one or more open-ended, unstructured interviews with single respondents or focus groups.

However you generate items, the way you construct them is extremely important, because that will greatly influence how respondents understand them and come up with responses to them. That has major implications for the reliability of our instruments, their construct validity, or both (see Chapter 11). Avoid confusing, biased, and ambiguous wording by using clear and unambiguous language, understood consistently by all respondents. For example, if you ask respondents to state whether their "income level is low, medium, or high," you have the problem that different people interpret low or high incomes very differently. One should steer clear of items that involve double negatives; a classic example is the multiple-choice question that inspires respondents to pull out their hair because it asks "which is not true about. . .?" and then provides "none of the above" as an option. Items need to be **unidimensional**—they need to ask only one thing. An item that asks "do you favor preserving farmland and forests?" won't do, because a person could easily favor one landcover differently than the other. Avoid biased and emotionally charged wording, as such items can be leading or create unfortunate reactions in respondents, like quitting the data collection session. Who says no to this: "Do you favor preserving pristine and beautiful wilderness areas?" Finally, design the visual appearance of surveys so they are easy to understand and use (apply graphical principles like those discussed in Chapter 10), and so they communicate a serious attitude on the part of you, the researcher.

What about the length of explicit report instruments? They obviously must be long enough so you can obtain the information you need, but they should not be longer than that. This is harder than it sounds; many researchers, especially when they are inexperienced, readily fall into the attitude that "well, while we're at it, we might as well ask them that too." Unnecessary questions create an unwieldy and unfocused instrument that can bore or overtax respondents. Excessive length can also decrease the response rate and lower the thoughtfulness of responses.

And given that administering, coding, and analyzing explicit reports is usually very laborious and cognitively challenging, brevity and conciseness are surely blessings.

The order of items in an explicit-report instrument is often important. Sometimes items must follow other items because they depend on information in those earlier items or in the respondent's answer to earlier items. But even when item order is not dictated, it may influence responses because of the context created by what previous questions induced a person to think about or feel, because of fatigue, and so on. Respondents are typically much more comfortable with the format of rating scales by the end of an instrument than on the first item, for example. These are called **order effects**. They come in many potential forms, including some that are due to the absolute position of an item within a sequence of items, and others that are due to the specific context created by the immediately preceding item or items (in which case they may be called **context effects**). There are several approaches to dealing with order effects. One can randomize the order of items for each respondent, which is relatively easy to do with computer-administered instruments, or at least use several different random orders; of course, which respondent gets which order must be decided randomly. Alternatively, if one has a small number of items, one can give each possible order to equal numbers of respondents. When this approach, known as **counterbalancing**, is feasible, it is optimal because it perfectly balances all possible order effects.[5] Sometimes items may group logically into subsets; the item order within each subset may be fixed for all respondents, but the order of the subsets may be counterbalanced or randomized.

We conclude this section by offering a strategy for creating explicit report instruments. We emphasize that creating instruments is very hard work and really cannot be done well on the first try, even by experienced "research geniuses." We stress this point because we have seen so many inexperienced researchers jump right into research with explicit reports, generating items, designing the instrument, and beginning final data collection within a period of one or two weeks. That just doesn't work, but we sure wish we had a euro for every time an inexperienced researcher proceeded in that fashion, only to bemoan later some ambiguity or mistake that prevented them from drawing the conclusions or answering the questions they set out to. We recommend you follow a cycle that we call "GPM . . . PMT." That stands for "generate, pretest, modify . . . pretest, modify, test." In other words, first use any combination of the methods we discussed above to generate items you think will tap into your constructs of interest. Then pretest your first draft; such a preliminary study designed to practice and evaluate any component of your primary data collection is called a **pilot study**. Do this by showing it to your friends, colleagues, or office mates. Administer it to a small sample of participants from a research pool or class of students. At some point, you should definitely pretest your instrument with a small sample of the type of respondents you will eventually use for real data collection.

---

[5]As the number of items grows, the number of possible orders can quickly become too large to do this. There are elaborate strategies for generating less than all possible orders, such as the *Latin square* technique, that nevertheless deal more systematically with various potential order effects than does complete randomization.

To pretest, don't just administer your instrument, but ask your pilot respondents how they interpret items, instructions, and so on, and whether they are confused about any part of the instrument or its administration. In other words, carry out a pretest that is open-ended and unstructured. The next step is to make modifications to the instrument in response to feedback you get from the pretest. Then pretest it and modify it again. You should sometimes repeat this sequence of pretesting and modifying several times, especially with large, complex, or expensive studies. That's why we put the ellipsis in GPM . . . PMT. When you feel that your instrument is clear, adequately focused, and of the proper length—but only then—you can proceed with the actual testing phase. By the way, make sure to pretest the instrument with whoever is going to administer it, too (research assistants, interviewers).

The Census: An Important Secondary Source of Explicit Report Data for Geographers

A **census** is a count of the number of people in a country and an assessment of their characteristics, such as family structure, economic activities, and so on. They are carried out by national governments within a fairly short time period, so that the population does not change too much from the start to the end of data collection. Censuses provide an important source of explicit-report data for geographers. It is a secondary source, as census data are not collected for the specific purpose of a particular research project. The answers to census questions vary both spatially and temporally. Many governments find the answers to these questions to be sufficiently important to expend substantial resources in an attempt to answer them. Historically, governments have wanted to answer some of these questions for taxation and military purposes. With the advent of democratic governments, censuses have been required to fairly apportion representatives to administrative regions, for example, U.S. states. Increasingly, census data are used for planning, public health, and many scientific purposes.

Governments of ancient Egypt, Babylonia, China, India, and Rome conducted censuses at irregular time intervals; systematic and regular census enumerations did not really start until the middle of the 18th century. The U.S. Constitution (Article 1) mandates a census of the population in order to apportion taxes and congressional representatives. Although we focus on the *U.S.* census in this chapter, many other countries conduct censuses of various kinds; in fact, the United Nations has encouraged several countries to conduct regular censuses. The United States has conducted a census of its population every 10 years since 1790. The actual questions used in the census are approved by Congress for each enumeration; consequently, the questions asked have varied over the years. Substantial changes have occurred for questions associated with numbers of slaves owned, fertility, foreign-born status, mental and physical capability of household members, income, housing characteristics, and racial and ethnic identity. These changes introduce some difficulties for longitudinal studies (Chapter 7). At the same time, they are also a fascinating documentation of changing social and political values over the course of U.S. history.

The U.S. census is designed as a complete enumeration of the population; this precludes the use of statistical sampling for some critical population information. The question of complete enumeration versus statistical sampling has recently become a significant political issue.[6] From 1790 to 1930, all census questions were asked of all appropriate persons. Over the years, Congress added questions to the census, the country's population grew, and conducting a census with so many questions became logistically and financially difficult. Consequently, in 1940 the **Census Bureau** began to administer a small set of basic questions, often referred to as the **short form**, to the entire population and a much larger set of questions, the **long form**, to a fraction of the population (about one in six). The short-form items concern basic demographic and housing information: name, age, sex, race, year of birth, marital status, relationship to head of household, Spanish/Hispanic/Latino origin, and a few questions regarding housing characteristics. (It is interesting that for the first time, the 2000 census allowed individuals to indicate more than one racial category; in fact, fewer than 3% of respondents did.) It is important to us as geographers that demographic and housing characteristics are spatially located ("geo-referenced" in Chapter 12). The fact that the surveys are sent to specific addresses, not people, means that spatial location is implicitly measured.

The long form in the 2000 census had 26 additional population questions and 20 additional housing questions. Notable among these were questions about income, employment, migration, commuting and transportation behavior, education, birth location, military service, cost and financing of housing, and so on. Administering the long form presents some methodological problems, particularly **nonresponse bias**, discussed in Chapter 8 as "nonparticipation bias." The Census Bureau goes to great effort to ensure that long forms are filled out, including employing "nonresponse enumerators" who actually visit the addresses of forms that were not returned. Because of these difficulties and the coarse temporal resolution of the census (once a decade), the Census Bureau implemented a new continuous census in 1996: the **American Community Survey** (**ACS**). The ACS measures approximately 3 million people a year, about 1% of the entire population, with a survey that is essentially the long form. It is anticipated that the 2010 census will abandon the use of the long form altogether and send the short form exclusively. The ACS will replace the information obtained by the long form and increase its temporal resolution.

---

[6]The so-called **sampling debate** concerns this question of complete enumeration versus sampling and statistical estimation. The U.S. constitution calls for a "census," which implies a complete count. Nonetheless, as we discuss in this chapter, it is well established that some people are undercounted, particularly those who are poor, homeless, and so on. In the last couple of decades, several social scientists and statisticians have recognized that high-quality sampling and statistical estimation would very likely improve the accuracy of counts of underrepresented groups, and for much less money. Aside from the merits of the two arguments, the question of a complete count versus sampling and estimation has clear political implications insofar as liberals and conservatives believe they would on average benefit or suffer, respectively, from more accurate counts of groups such as the urban poor. See Wright, T. (1998). Sampling and Census 2000: The concepts. *American Scientist, 86,* 245–253.

Another important question regarding a census is "Who should be counted?" One approach is referred to as a **de facto census** of the population. A de facto census attempts to count who was in a given area at a given time. One way to think of this is to imagine dropping a giant cage over a particular area. Count and gather data from everyone that you "captured," and you have a de facto census of that area. A de facto census is not concerned whether people are tourists, students, illegal immigrants, and the like, only whether they are there. In contrast, a **de jure census** attempts to count people who legally belong to a particular area; that is, it does not count tourists, illegal immigrants, and so on. The U.S. census is neither purely de facto nor de jure; it uses a more vague definition of who it is attempting to enumerate—people at their "usual residence." This approach to the census is considered appropriate because of the fundamental and constitutionally mandated purpose of the U.S. census to demarcate congressional districts. This leads to some interesting issues for geographers who use the census as a source of research data. For example, in Las Vegas, Nevada, the census does not count the continuous and extremely large population of tourists. In the New England states, particularly Boston, the census may or may not count the students from around the country and world who attend the many colleges and universities there. Residents of prisons, who are not allowed to vote, are nonetheless counted (and rather accurately!) by the census. People who have immigrated into the U.S. illegally similarly do not vote but are counted (less accurately, to be sure). Likewise, some areas of the country have a much higher proportion of people under 18 years of age who are counted but do not vote. Congressional districts are drawn to contain equal populations of people counted in the census, regardless of whether or not they are allowed to vote. Thus, the question of "Who is counted?" raises the additional questions of "Who is not counted that should be?" and "Who may have been counted more than once?"

It is generally accepted that the U.S. census suffers from some degree of an **undercount** of the population that it is attempting to enumerate; this is a threat to the "construct validity" we define in Chapter 11. Actually, the net undercount results from both a larger undercount and an **overcount** that is somewhat smaller. It is likely that people who have immigrated illegally, people without permanent residences, and people of lower socioeconomic status living in inner cities are undercounted; one piece of evidence supporting the idea that the latter are undercounted is a comparison of military conscription data to census counts. An overcount results when students are counted twice, once by their parents at their "permanent" home location and once at their college or university residence. "Snowbirds" or other people who seasonally travel to another region of the country may be counted more than once. The Census Bureau goes to great effort to prevent both overcounting and undercounting; nonetheless, it is recognized that there is a measurable amount of both. It is also likely that the Census Bureau is making progress in diminishing the magnitude of the undercount. It's kind of intriguing to think about how the Census Bureau can know what it didn't measure. We won't discuss this issue in detail here, other than to note that measuring **coverage error** (overcount and undercount) is mostly accomplished by methods known as "demographic analysis" and "dual-system estimation."

What Do U.S. Census Data Look Like and How Can You Obtain Some?

The U.S. census is a substantial and sincere attempt to characterize the entire U.S. population demographically, geographically, and economically. The 2000 census counted over 281 million people in over 100 million households. In 1790, the census was likely recorded with feather pens on crude paper. The year 1950 marked the first use of digital computers to process and record census information. Since 1980, census data have been stored in geographic information systems (GISs, see Chapter 12). Initially they were stored with **DIME files** (Dual Independent Map Encoding files); now they are stored with **TIGER files** (Topologically Integrated Geographic Encoding and Referencing files). Storing census data in a GIS enables easier access, analysis, and mapping and visualization of the information. Privacy concerns are an important issue for the Census Bureau, and the actual completed survey forms are not made available to the public for 70 years. Also, census data are aggregated to spatial units designed to ensure the privacy of individual respondents. U.S. census data can be obtained from numerous public and private sources for a surprisingly wide range of costs, from free to rather expensive. A good start is the Web site of the Census Bureau at http://www.census.gov.

Census data are provided at a range of spatial resolutions. The spatial hierarchy of major census units is as follows (see Figures 6.1 and 6.2): entire United States; census regions (four); census divisions (nine); states (50 and the District of Columbia, plus nonstate U.S. territories like Puerto Rico and Guam); counties (basic administrative and legal subdivisions of states); county subdivisions (such as minor civil divisions, towns, townships); census places (incorporated cities and unincorporated census-designated places); **census tracts**[7] (small, relatively permanent subdivisions of a county, typically containing about 5,000 people); **census block groups** (clusters of blocks within census tracts); and **census blocks** (the smallest geographic unit of the census, typically bounded on all sides by streets, railroad tracks, streams, lakes, and so on). The boundaries of these units change somewhat from census to census. When conducting longitudinal studies, you typically find that the smaller your geographic unit of analysis, the more likely its boundaries will be different from one year's census to another; that is, blocks are very likely to be different, regions are very unlikely, and so on.

In addition, the census bureau classifies urban areas into a fairly confusing alphabet soup of acronyms. This is largely because cities are increasingly difficult to define and delimit precisely; their "commutersheds" have become so large that a hierarchy of categories is required to characterize the urban-rural distinction in any realistic way. In 1950, the Census Bureau used county-level data to come up with the concept and definition of a **standard metropolitan statistical area (SMSA)** as counties with a core city of at least 50,000 residents and a population density of at least 1,000 persons

[7]Census tracts are the smallest units at which long-form data are made available, for privacy reasons.

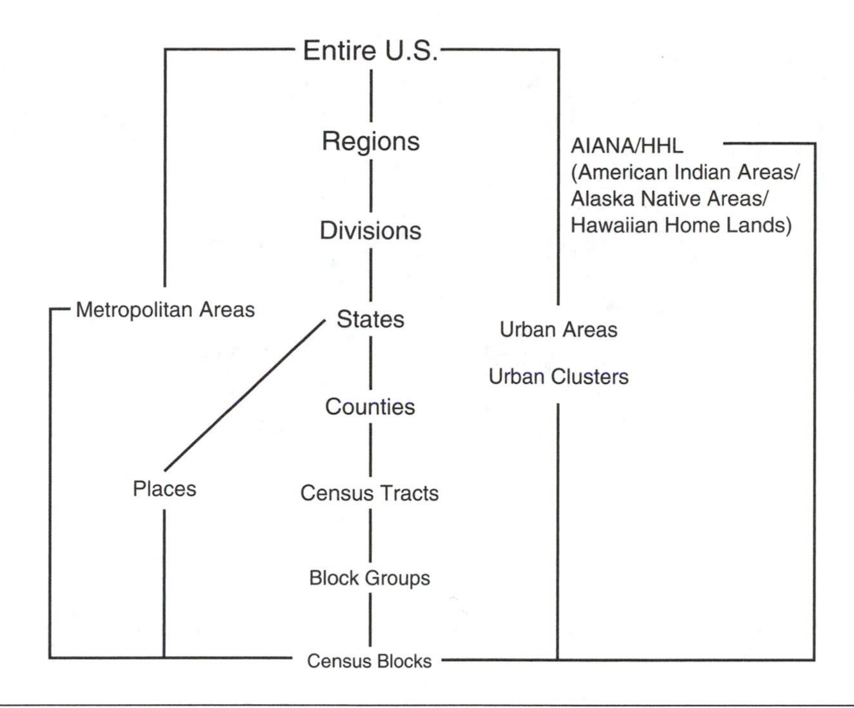

Figure 6.1 Some major geographical units used by the 2000 U.S. census (adapted from the U.S. Census Bureau Web site: http://www.census.gov)

per square mile. The SMSA was supposed to be "an area containing a recognized population nucleus and adjacent communities that have a high degree of integration with that nucleus." Subsequently, the Bureau created MSAs (metropolitan statistical areas), CMSAs (consolidated metropolitan statistical areas), PMSAs (primary metropolitan statistical areas), and CBSAs (core-based statistical areas). The building blocks of these defined urban areas are counties. The classifications have been criticized for being confusing and irrelevant to issues of population and urbanism; nonetheless, the county is the finest spatial resolution at which myriad kinds of data are regularly and systematically collected. This is a cogent reason the Census Bureau continues to define many urban areas by using counties as the **minimal mapping unit.** Nonetheless, counties do vary dramatically in their areal extent; for example, San Bernardino County, California, is much larger than the whole state of Rhode Island. Also, many studies demand a finer spatial resolution than counties provide. In fact, many businesses obtain, archive, and analyze data at much finer spatial resolutions than the county level, for example, at the ZIP-code level.

The Census Bureau has tried to keep up with the times by developing finer-resolution definitions for *urban* that are not based on county-level data. This has been possible largely because of the power of computerized GISs. For the 2000 census, the Census Bureau created a more flexible definition of urbanized areas, defining **urban areas** (UAs) and **urban clusters** (UCs). According to the Census Bureau, an urban area consists of a "densely settled core of census block groups and census

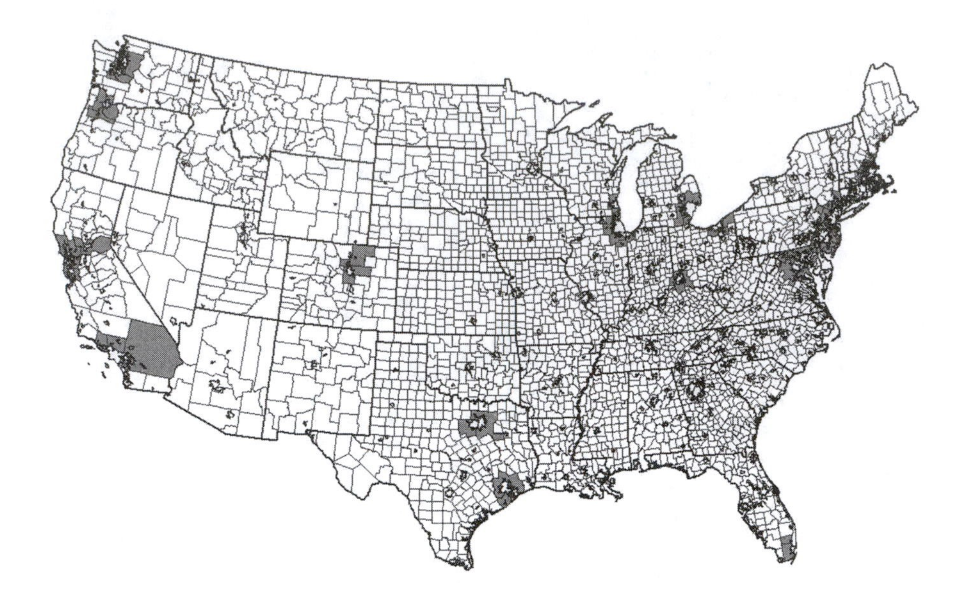

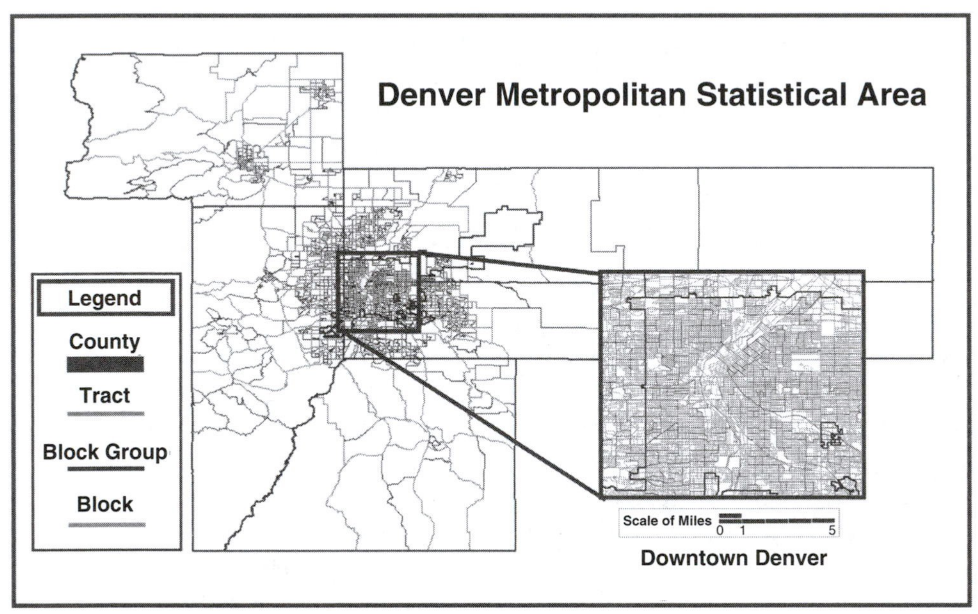

Figure 6.2 Maps showing U.S. census geography for the conterminous United States: (a) States and counties, with urban areas (UAs) and primary metropolitan statistical areas (PMSAs) shaded. (b) Counties, tracts, and block groups for most of the Denver metropolitan statistical area. Inset shows block-level detail for downtown Denver.

blocks that meet minimum population density requirements, along with adjacent densely settled surrounding census blocks that together encompass a population of at least 50,000 people, at least 35,000 of whom live in an area that is not part of a military installation." An urban cluster is defined in the same way but is smaller; it requires a minimum of 2,500 and a maximum of 50,000 residents. Given such efforts on the part of the Census Bureau, we believe geographers can now accurately characterize human population patterns much more precisely than ever before. Although it is difficult to link definitions of urban and rural to many types of data that are gathered only at the county level, the finer spatial resolution manifest in the newest definitions of urbanized areas does improve the validity and detail of many studies, such as those on urban sprawl, commutersheds, and exurbia. In fact, all the data gathered at the block and block group can be linked to these new definitions of urban. We appreciate that the Census Bureau maintained their county-level characterizations of urban/metropolitan areas at the same time they created a more precise definition of urban from the finest-resolution data available to them.

Limitations of Explicit Reports

We conclude this chapter by stepping back a little, taking a more critical look at the use and meaning of explicit reports in geographic research. First, we return to the observation that explicit reports are quite popular in human geography. They are so popular probably because they strike geographic researchers as so direct—if you want to know where people vacation, and why, just ask them. If you want to understand the meaning of place to people, just ask them. As we have described, these measures are called explicit for the very reason that people *know* they are responding to a request for information by a researcher, and they respond only with what they think of explicitly. As we noted in Chapter 3, this explicitness both increases and limits the value of explicit reports as types of data. We have considered the value of explicitness; we now consider some of its limitations.

To believe that all explicit report data are completely true and accurate would be naive. There are a variety of psychological, social, and cultural processes that influence memory for personally held beliefs and feelings, and their communication to others. We have already referred to some of the more straightforward of these processes, such as that respondents may give socially desirable answers. Or respondents may do the opposite—attempt to thwart or damage the efforts of the researcher. A malicious, impish, or plain antisocial respondent (yes, we have met them) may lie or attempt to deceive the researcher. Finally, we have already discussed the possibility that interviewers can produce distorted responses, because of either their personal characteristics or the way they administer or interpret report instruments.

The limits of memory are clearly important. Both its accuracy and completeness should be questioned. Of course people forget. But even more, psychological research has shown that memory is an elaborative process, a *construction* of what happened as much as a reconstruction or simple retrieval. The emotionality of events can influence this, and not always in the way many people believe (namely, that intensely emotional events are recalled more completely and realistically). Both the recency of the time period asked about and the amount of aggregation of information required by a memory question influence the accuracy of responses. For example, a common type of report item asks people to say how often they do something. The accuracy of responses varies as a function of the time frame over which people are asked to aggregate their activities. Keeping the time frame short and recent leads to more accurate reporting. "How many times do you take the bus each week?" is better than "How many times do you take the bus each month?"

There are some subtler and, we think, more interesting limits to the utility of report data than just the fallibility of memory. The cognitive science of language is a core concern here—how language encodes and expresses one's mental contents; we touched on this in Chapter 5, and we return to it in Chapter 12, where we discuss implications of the "ontology" of geographic features for geographic information systems. This is a complicated issue with a long history of scholarly and scientific debate that has by no means been resolved in the various disciplines that address it (such as anthropology, linguistics, psychology, and artificial intelligence). But we can make a few fairly noncontroversial observations about language, meaning, and mind. The first is that mind does not equal language, and not all meaning and experience can be expressed verbally. Language does express meaning, of course, but in a surprisingly complex and unobvious way; luckily, we don't need to understand how language works in order to use it effectively. Words label categories of knowledge or belief, but despite lots of attention by smart people, we still do not know exactly how. For example, language expresses meaning in a profoundly contextual way—the context being provided by the current situation, past situations, cultural experience, and more. Translation across languages is therefore impressively difficult, and it still has not been completely automated. It is no surprise that altering the wording of items in explicit-report instruments can produce different responses, even when they seem formally equivalent. "Do you think the federal government should forbid corporations . . ." will not mean exactly the opposite of "Do you think the federal government should allow corporations . . ." At the least, this suggests that researchers should interpret responses to explicit-report items in a relative way; for example, comparisons of responses by subgroups exposed to the same wording can be interpreted, but absolute levels of responses to a particular wording perhaps should not be.

Furthermore, as we have noted, explicit reports require respondents to access their mental states (beliefs, emotions) and externalize them for the researcher. But in fact much mental processing goes on subconsciously or unconsciously; Freud may have overstated our interest in sex and death (then again . . .), but he was apparently right about the great portion of mind that is outside of conscious awareness.[8] A person can't provide an explicit answer to something about which he or she is unaware, and that's apparently quite a lot. For example, we can access what we are thinking about but often not the process by which we arrived at that thought. We can usually answer "what" questions much more faithfully than "why" or "how" questions. For example, ask somebody why he or she thinks natural scenery is so beautiful or how he or she managed to steer the car around a turn without crashing. You may get sputtering or some vacuous restatement of the question like "because it makes me feel at one with nature" or "by turning the wheel." We humans are funny creatures insofar as ignorance and chaos make us very uncomfortable.

[8]Theories and debates about conscious versus unconscious mental processing are a big part of the history of cognitive and behavioral science in the 20th century. Besides Freudian psychodynamic ideas, there is introspectionism, behaviorism, and various versions of cognitivism. Modern cognitive psychologists recognize distinctions between automatic and controlled processes, and "implicit" and "explicit" memory.

The illusion that we understand something is very comforting. So when asked to explain something, we often offer rationalizations or personal theories, all the while deceiving ourselves and the researcher that we are directly accessing the mental activities that actually explain a belief or behavior.

It should also be remembered (hopefully explicitly!) that most geographers who study attitudes, beliefs, and memories do so because of what it can tell them about behaviors—what people do and will do, and why. But there is by now extensive literature from social psychology, sociology, and political science that explicit mental states and behaviors have a surprisingly tenuous correspondence. We cannot go into this here in great detail, but the common-sense belief that we do what we do entirely because of what we believe just isn't very true. A sampling of factors that actually help explain behavior besides beliefs in the domain in question include cultural norms and expectations, habits, action opportunities and constraints, personal gain or loss, past experience, and beliefs that one's actions are public or anonymous.

Let's consider a specific example of all this. As we noted above, contingent valuation surveys have been used in an attempt to put a monetary figure on how much people would be "willing to pay" for various aspects of "nature." These surveys can be (and have been) severely criticized, however. Such a survey of 1,000 Americans might show that they would be willing to pay $28 for the continued existence of the grizzly bear, $33 for the bald eagle, $19 for the bison, and so on for other "charismatic mega-fauna." Summing the willingness to pay for enough species would eventually exceed the average household income of the respondents. Another criticism of contingent valuation surveys is their failure to account for people's lack of understanding of what they are valuing. For example, researchers are often interested in the value of "services" provided by various types of ecosystems or biomes. For example, swamps and bogs, now affectionately known as "wetlands," provide many valuable services to humans such as flood control, fish hatcheries, wastewater absorption and cleansing, and so on. Ordinary citizens rarely appreciate the true cost of paving over a wetland for a parking lot. Nonetheless, these costs often come back to haunt citizens in the form of capital losses from disastrous floods, regional economic problems associated with depleted fisheries, and capital expenditures like sewage treatment plants. In other words, a survey of ordinary citizens regarding the value of a local wetland may be a very inaccurate measure of its true value.

Our discussion of the limitations of explicit reports like contingent valuation is not meant to dissuade you from using them in research. Explicit reports are much too useful for that, even with their limitations. We do hope you will think carefully about using them and consider other approaches to data collection. Contingent valuations, for example, can be complemented with measures of what people actually spend on trips to national parks or on homes that have unspoiled views of nature, a technique known as **hedonic valuation**. We favor using explicit reports along with other types of data; any research idea is rightly more believable when it has multiple types of data supporting it, not just multiple data sets. And when you use explicit reports, exercise caution and restraint in interpreting them. For example, responses may be instructive *relative* to responses to other items or in other situations, but they may be of doubtful value in the absolute. Appreciate that explicit reports are more likely to tell you something accurate about some things than about others.

Review Questions

- What are the following types of explicit reports: surveys, interviews, sociometric ratings, activity diaries, contingent valuation, focus groups, protocol analysis, tests?

Format of Explicit Reports

- How do closed-ended and open-ended items differ, and what are some strengths and weaknesses of each?
- How do standardized and nonstandardized items differ, and what are some strengths and weaknesses of each?
- What are some specific types of closed-ended items? What are some specific formats for rating scales?
- What are arguments for and against treating rating-scale responses as metric data?

The Administration of Explicit Reports

- What are options for administering explicit reports, and what issues need to be considered when deciding how to administer them?
- What are strengths and weaknesses of using the Internet to collect explicit report data?

Designing and Generating Explicit Instruments

- How are items for explicit reports created and revised?
- What are some desirable characteristics of explicit-report items?
- How do order effects arise, and what can be done to minimize them?

The U.S. Census

- What is a population census, and how has a census historically been administered and used in the United States?
- What are strengths and weaknesses of the U.S. census as a source of data for scientific research?
- What are the various spatial units in which U.S. census data are organized?

Limitations of Explicit Reports

- How are the quality and usefulness of explicit-report data compromised by the following issues: memory, truthfulness, open disclosure, the linguistic encoding of meaning, access to cognitive processes, and the mind-behavior relationship?

Key Terms

activity diary: type of survey in which respondents record their regular activities; also known as an activity log

adjective or **activity checklist:** type of closed-ended explicit-report item in which respondents mark each alternative that applies from a list of adjectives or activities

American Community Survey (ACS): survey newly created by the Census Bureau in 1996 to be administered yearly to about 1% of the U.S. population; essentially replaces the long form

branching format: follow-up questions that are prestructured to occur whenever a particular response is given to a previous item

census: a count of human population size and characteristics carried out by national governments within a fairly short time period; important secondary source of data for geographers

census block group: geographical unit of the U.S. census consisting of clusters of blocks within census tracts

census block: smallest geographical unit of the U.S. census, consisting of areas typically bounded on all sides by streets, railroad tracks, streams, lakes, and so on

Census Bureau: bureaucracy charged with designing, administering, and distributing the U.S. census

census tract: geographical unit of the U.S. census consisting of small, relatively permanent subdivisions of a county, typically containing about 5,000 people

closed-ended item: item on explicit reports that provides a finite number of specific response options for respondents to choose as answers

context effect: potential influence on responses to particular items on an explicit report that arise because of the context created by items administered previously; a relative order effect

contingent valuation: type of survey in which respondents rate or rank how much they subjectively value something

counterbalancing: a way to deal with order effects by administering all possible orders to equal numbers of respondents; not feasible when certain item orders are required or with large numbers of items

coverage error: generic term for both overcount and undercount on the census

de facto census: an approach to conducting a census in which people are counted according to where they happen to be when they are counted

de jure census: an approach to conducting a census in which people are counted according to where their permanent legal residence is when they are counted

DIME file: Dual Independent Map Encoding file, the standard way to organize U.S. census data within geographic databases since 1980; recently supplanted by TIGER files

explicitness: property of measures in human geography that require people to be aware they are responding in order to provide data to a researcher, and the responses depend on beliefs of which the person is aware

focus group: type of unstructured interview in which respondents in a small group participate in discussions of particular topics

follow-up question: common example of nonstandardized explicit-report item whose specific wording depends on responses to earlier items

forced-choice item: type of closed-ended explicit-report item in which respondents pick one alternative from a list of choices, such as "yes-no," "true-false," or "multiple choice"

free format: follow-up questions that are not prestructured but are formed more or less spontaneously on the spot by an interviewer

generic rating scale: general type of rating scale not designed for a specific purpose or in a specific format such as semantic differentials, Likert scales, and so on

group dynamics: social influences that unfold over time within focus groups and affect what people say and how they say it

hedonic valuation: a measure of how much people value things based on how much they spend to acquire it, used as a complement to contingent valuation

instrument: generic term for a specific example of an explicit report

interview: survey administered and responded to in oral form

items: generic term for questions, tasks, or other units on explicit reports

Likert scale: rating scale on which respondents express their degree of agreement or disagreement with a series of statements

long form: the much larger set of questions concerning more detailed demographic, economic, mobility, and housing information on the U.S. census, administered to a sample of about one-sixth of the population

minimal mapping unit: the smallest spatial unit used to define, measure, and display an areal concept; the U.S. Census Bureau uses counties as the minimal unit for defining cities as "statistical areas"

nonresponse bias: problem of inference in interpreting census data created by the nonrandom subset of the population that does not respond to the Census Bureau's request for information; type of nonparticipation bias, discussed in Chapter 8

nonstandardized items: administering explicit-report items in a way that is neither completely predetermined nor consistent to all respondents

open-ended item: item on explicit report that does not provide a set of specific response options for respondents to choose as answers, but allows them to provide idiosyncratic responses

order effect: potential influence on responses to particular items on an explicit report that arise because of the order, either absolute or relative, in which the item was administered

overcount: threat to the validity of a census arising from people being counted who should not be counted, or people being counted more than once

paired (or **triadic) comparison:** type of closed-ended explicit-report item in which respondents directly compare entities, two or three at a time, either ordinally or metrically

paired comparison rating: rating scale on which respondents compare entities with the use of a metric rating scale; a metric paired comparison

pilot study: preliminary study designed to practice and evaluate any component of your approach to primary data collection of any type

protocol analysis: type of unstructured and open-ended interview in which respondents "think aloud" about their reasoning during some problem or issue

ranking of alternatives: type of closed-ended explicit-report item in which respondents put two or more alternatives in order from most to least, or least to most

rating scale: type of closed-ended explicit-report item in which respondents provide a number or mark a line to indicate the amount or extent of something, including the degree of belief they have in something

response set: pattern of responses to explicit items that suggest a person was not sincere in his or her responses

sampling debate: controversy over whether the U.S. census should continue to attempt a complete enumeration of the population *or* use sampling and statistical estimation to achieve a more accurate count, potentially

semantic differential: rating scale on which respondents rate a set of entities on each of several adjectives, as a means of expressing attributes of connotative meaning

short form: the small set of questions concerning basic demographic and housing information on the U.S. census, administered to the entire population

social desirability: type of response set that occurs when people provide responses according to what they think is appropriate or socially positive

sociometric rating: type of survey in which members of a group express beliefs about each of the other members of the group

sorting task: type of closed-ended explicit-report item in which respondents sort phrases, pictures, or other stimuli into groups

standard metropolitan statistical area (SMSA): way the U.S. Census Board first defined cities in 1950 on the basis of county-level data; now supplemented or superseded by a variety of related concepts, including MSAs, consolidated MSAs, primary MSAs, and core-based SAs

standardized items: administering explicit-report items in a predetermined and consistent format to all respondents

survey: explicit report in which respondents answer questions about their opinions, attitudes, preferences, activities, or demographics, typically in written form; also known as a questionnaire

test: explicit report in which respondents answer questions about their knowledge, and the researcher is interested in the correctness of the answer

TIGER file: Topologically Integrated Geographic Encoding and Referencing file, currently the standard way to organize U.S. census data within geographic databases; recently supplanted DIME files

undercount: threat to the validity of a census arising from people not being counted who should be counted

unidimensional: desirable property of explicit report items that ask about only one idea rather than two or more

urban area (UA): new way the U.S. Census Board defines cities with data at a finer resolution than county level, instead defining them on the basis of densely settled block groups and blocks that encompass at least 50,000 people

urban cluster (UC): new way the U.S. Census Board defines cities with data at a finer resolution than county level; same as urban areas but smaller, requiring a minimum population of only 2,500 people and a maximum of 50,000

Bibliography

Barrett, R. E. (1994). *Using the 1990 U.S. Census for research.* Thousand Oaks, CA: Sage.

Birnbaum, M. H. (2004). Human research and data collection via the Internet. *Annual Review of Psychology, 55,* 803–832.

Ericsson, K. A., & Simon, H. A. (1993). *Protocol analysis: Verbal reports as data* (Rev. ed.). Cambridge: MIT Press.

Fowler, F. J. (2002). *Survey research methods* (3rd ed.). Thousand Oaks, CA: Sage.

Kraut, R., Olson, J., Banaji, M., Bruckman, A., Cohen, J., & Couper, M. (2004). Psychological research online: Report of Board of Scientific Affairs' Advisory Group on the Conduct of Research on the Internet. *American Psychologist, 59,* 105–117.

Stone, A. A., Turkkan, J. S., Bachrach, C. A., Jobe, J. B., Kurtzman, H. S., & Cain, V. S. (Eds.) (2000). *The science of self-report: Implications for research and practice.* Mahwah, NJ: Lawrence Erlbaum Associates.

U.S. Census Board Web site: *http://www.census.gov*

Weeks, J. R. (2005). *Population: An introduction to concepts and issues* (9th ed.). Belmont, CA: Wadsworth/Thomson Learning.

Weisberg, H. F., Krosnick, J. A., & Bowen, B. D. (1996). *An introduction to survey research, polling, and data analysis* (3rd ed.). Thousand Oaks, CA: Sage.

CHAPTER 7

Experimental and Nonexperimental Research Designs

Learning Objectives:

- What are the three forms of empirical control in scientific research (physical, assignment, and statistical), and how are they related to the distinction between experimental and nonexperimental research?
- What does the distinction between laboratory and field settings mean?
- What is the difference between within-case and between-case research designs, and what are its implications for research?
- What are developmental research designs, and how do longitudinal and cross-sectional designs differ?
- What are some implications of the distinction between single- and multiple-case designs?
- How does computational modeling compare to traditional experimental research, in terms of how variables are controlled and how data are generated and analyzed?

In any research study, variables are created or measured in ways that allow particular comparisons to be made and not others. The structure of possible comparisons in a study that results from the way variables are created or measured is the issue of **research design**. Research design largely determines which questions can be asked of data and greatly influences the validity with which those questions can be answered, as we discuss below and in Chapter 11.

Empirical Control in Research

The distinction between variables that are "created" versus those that are "measured" suggests what may be the central distinction in research design, that between **experimental** and **nonexperimental studies.** As we discuss in detail below, at least one variable is created in experimental studies, whereas variables are only measured in nonexperimental studies. To understand this central distinction, let's return to our list of the goals of a science from Chapter 1. One of those goals was explanation, which we described as the explication of causal relations among entities and events, an explication that accounts for patterns within data. Scientific researchers want to be able to explain patterns in their data, and they want to be able to do it validly. In order to increase their ability to draw valid causal conclusions, researchers exercise empirical control over the phenomena they study. **Empirical control** is any method of increasing the ability to infer causality from empirical data[1] (not to be confused with the fourth scientific goal from Chapter 1 of exercising practical or material "control" over phenomena).

Empirical control is exercised in one of three ways, or a combination thereof. The first is **physical control** (mentioned in Chapter 4). This is physically controlling the data collection situation in order to reduce the potentially distracting influence on our data patterns of factors that are not of interest to us. Some examples of physical control include isolating cases, simplifying or purifying tasks or procedures, and increasing the consistency of the research situation across data collection episodes. Although some social and behavioral scientists employ physical control in their research, it is more common in the biophysical sciences, particularly laboratory physical sciences like chemistry; that's the reason they wash chemical beakers so carefully.

The second type of control is **assignment control.** Researchers using assignment control *create* at least one of the variables in the study in order to test ideas about its effect on some other measured variable. The created variable is usually at the nominal level, so that each value or level of the created variable is a discrete experimental **condition** to which cases are exposed that are assigned to it. Researchers incorporate assignment control in their studies when they determine which cases to assign to which conditions of the study. For example, a researcher studying soil erosion could create five values of a variable called *Solvent pH,* which concerns how aqueous solutions differing in pH cause different rates of erosion in particular soil types. Values of this variable could be "pH 5," "pH 6," "pH 7," "pH 8," and "pH 9" (neutral and a little acidic or basic). The study involves assignment control because the researcher decides which soil material (the case) would be assigned to be dissolved with which solvent solution. Alternatively, assignment control takes place when the same cases are subjected to all of the conditions of the study in an order determined by the researcher, defined below as a "within-case design." For example,

[1]Empirical control often increases statistical power and precision (Chapters 8 and 9) as well.

a researcher studying GIS interface design could expose samples of GIS users to two systems that use either verbal labels or iconic symbols to represent particular GIS operations; all users would be exposed to interfaces with both symbol types but in an order controlled by the researcher. Either way of controlling the assignment of cases to a variable is known as **manipulation**, even though they do not involve physical control, as might be implied by that term. Manipulation is most often done by **random assignment** to conditions, because that is a simple and straightforward technique that leaves the manipulated variable uncorrelated with all the non-manipulated (measured) variables, on average. Variations on random assignment include **matching**, in which two cases that have some characteristic in common are assigned to contrasting conditions.

The third and final way to exercise empirical control is **statistical control**. In studies employing statistical control, the researcher measures, but does not create, variables in the study. But the researcher takes explicit steps to identify, measure, and statistically analyze any variable that he or she thinks might have an effect on the main variables of interest in the study. For example, a geographer studying land-use changes might want to focus on the possible effect of changing family structure, such as the addition of more children, on the transition from one type of agriculture to another. To examine the effect of family structure without being distracted by the possible effect of, say, earnings from off-farm employment, the geographer could make sure to measure earnings and enter them into a regression equation along with the family-structure variable. Any effect of family structure would then be over and above the effect of off-farm earnings.

Having described the three forms of empirical control (physical, assignment, statistical), we can return to the important distinction between experimental and nonexperimental research designs. The term "experiment" is often used colloquially to refer to any scientific research study. Technically, however, it has come to refer specifically to studies that involve the manipulation of one or more variables—assignment control, in other words. This technical use of the term "experiment" is reminiscent of Hume's third principle of causality we discussed in Chapter 2: *Controlling the cause will control the effect.* Nonexperimental studies may involve physical and/or statistical control, but they do not involve manipulated variables. We absolutely do not mean to imply that nonexperimental studies are necessarily less "scientific," however. In fact, geographic researchers in many topic areas, including many physical geographers, rarely or never conduct experimental studies. That is simply because some constructs cannot easily be manipulated, if at all. It is impossible for a researcher to assign mountain chains to levels of a variable expressing orientation to dominant wind patterns; it is similarly impossible to assign cities to different base ratios of economic activity. At the end of this chapter, we consider an approach to research design that emerged in the 20th century as a very promising technique to expand the domain of cases and constructs that can be studied experimentally—computational modeling.

So all true experiments have one or more manipulated variables and one or more nonmanipulated variables. The manipulated variable is "created" by the researcher, who may accurately be called the "experimenter." The distribution of values of the manipulated variable across cases is independent of what the cases do or the properties they have, so it is called an **independent variable** (IV or "factor"). The nonmanipulated variable is only measured by the researcher. Its distribution of values across cases depends at least partially on what the cases do or the properties they have, so it is called a **dependent variable** (DV). The IV is the potential *causal* variable of interest in an experiment, and the DV is the potential *effect* variable of interest. Strictly speaking, it is misleading to refer to variables in nonexperimental studies as independent or dependent variables (in a sense, all are dependent), even though one or more variables in such a study is often considered a potential cause and others are considered potential effects. It is more correct in such a study to refer instead to "predictor" and "criterion" variables, as we will in Chapter 9 when we discuss *X* and *Y* variables in regression analyses.

The experimental-nonexperimental distinction is so important to research design because of its considerable implications for our ability to infer causality from empirical data. Experimental manipulation makes it logically much easier to establish that the IV is the variable responsible for patterns of variation in the DV, as opposed to some other causal variable being responsible. These "alternative" causal variables may not even have been thought about by the researcher, let alone measured or manipulated. To understand this rather fascinating aspect of scientific logic, consider that all empirical attempts to establish causality in science involve finding correlations between putative cause variables and effect variables. That is, variable *A* can be a cause of variable *B* only if the presence of *A* is always or usually accompanied by the presence of *B*, and the absence of *A* is always or usually accompanied by the absence of *B*.[2] However, even though finding patterns of correlation between variables is necessary to establish causality, it is not sufficient. That's because two variables can be correlated without one directly causing the other, as Figure 7.1 shows. Perhaps a third variable might directly cause both *A* and *B*, which in turn causes *A* and *B* to be correlated. This "third variable" could be a single variable *C* or a larger set of variables of any size or complexity. Even if the correlation between *A* and *B* is due to direct causality, the presence of the correlation itself does not establish whether *A* caused *B* or *B* caused *A*. Thus, the well-known and important adage that "correlation is not causality" is more correctly stated this way: *Correlation is causality, but the specific pattern of that causality is ambiguous.*

So manipulation makes it logically more likely that the variable we think is the cause of correlations in our data really is the cause, although it does not absolutely

[2]Philosophers such as Francis Bacon in the 17th century and J. S. Mill in the 19th century referred to the method of establishing causality via manipulation as the "Method of Concomitant Variation."

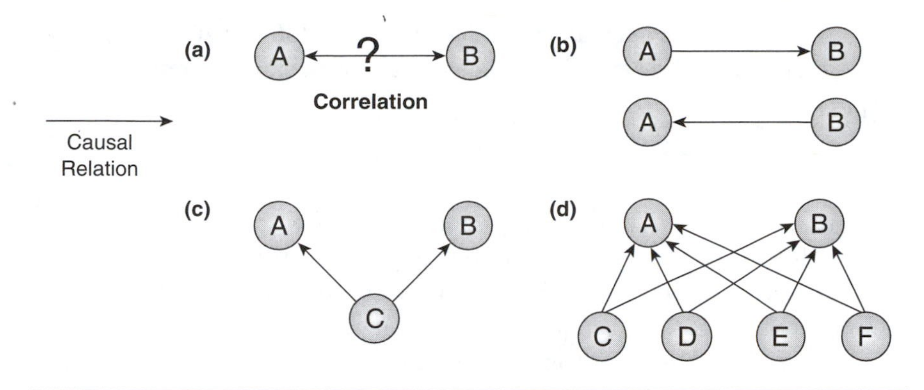

Figure 7.1 Alternative causal patterns that can explain an empirically observed correlation between variables *A* and *B*, shown in (a). The correlation may come about because *A* causes *B* or *B* causes *A* (b), because a third variable *C* causes both *A* and *B* (c), or because a set of any number of third variables that are interrelated to any degree of complexity causes both *A* and *B* (d). Without additional information, the underlying pattern of causal interrelations that explains the observed correlation is completely ambiguous.

ensure this. When one variable consistently changes value in response to changes in another variable that we, the researchers, have brought about, it becomes very unlikely—too much of a coincidence—that some other third variable could really be the cause of those changes. In nonexperimental studies, the chance that some third variable, or **confound**, is the actual cause of correlations in our data is usually quite substantial. In some specific research designs, as we discuss more below, the risk of a confound causing the correlations can be very high. In that case, the observed correlation between the two variables of interest is not the result of a direct causal relation between them; instead, the apparent direct causality is **spurious.** An example of attributing spurious causality to a correlation would be mistaking a proxy measure of past climate for the ultimate cause of the climate (see Chapter 4).

In the minds of many scientists, the likely existence of a confound may be the single most damning flaw in a research study because it sheds such doubt on the validity of causal conclusions (we learn in Chapter 11 that the validity of causal conclusions is called "internal validity"). So if you are interested in drawing causal conclusions when you are planning a study, thinking about possible confounds is critical. And the existence of confounds in other researchers' studies is one of the central issues you should consider when critically interpreting them, as when you review for a journal. It is vital to note, however, that all possible confounds are *not* necessarily important threats to the validity of causal conclusions. To be an important threat, both of the following circumstances must hold:

(a) The potential confounding variable must vary differentially across values (conditions, in an experiment) of the causal variable of interest.

(b) Even if the potential confound does vary in this way, it must also be related to the effect variable of interest.

In fact, every study in geography, experimental or otherwise, even those done very well, involves any number of potential confounds. But potential is not necessarily actual. For example, in our study of soil erosion described above, one could argue that it was temperature rather than the solvent pH that caused differential erosion. This critique is invalid unless (a) the temperature of the solvent solution varied sufficiently across pH conditions *and* (b) solvent temperature actually does correlate with the rate of soil dissolution. In this example, (b) is generally true but (a) would probably not have been true—the researcher would likely have used physical control to make sure the solvents were applied at the same temperature. Our general point is that when you critique the research of others, it behooves you to express concern for potential confounds only if they satisfy circumstances (a) and (b).

This is a good place to discuss a concept most people have heard of but may misunderstand a little—the concept of the **control group** (control condition). This is one or more experimental conditions that are added to an experimental design in order to allow particular comparisons that are logically relevant to the conclusions of an experiment. Many people think of the control group as a special experimental condition in which a certain "treatment" manipulation is withheld from the cases. For instance, most people know that medical researchers cannot establish the effectiveness of a new drug unless its use is compared to a condition where it is not used. This is correct, but we think it is an overly restrictive way to think about control groups. We don't use the term much ourselves because we know that all experiments require the necessary conditions to support whatever comparisons you want to make. The control group is really no different than any other condition in this respect. As a matter of fact, the comparisons you want to make in an experiment may or may not even involve some type of "empty" control condition. Medical researchers learned decades ago that a necessary comparison condition for drug trials is the "placebo" condition (an inert pill, for example), not the "no-pill" condition.

We're also in good position now to make sure we clearly understand the distinction between "random sampling" and "random assignment." They really don't have much to do with each other. Random sampling, as we discuss in Chapter 8, refers to selecting cases from a sampling frame to be in a sample. It influences the validity of the generalizations we draw from samples to sampling frames, including generalizations about statistical patterns that have nothing directly to do with causality. Random assignment refers to determining which cases in an experiment to expose to which condition or sequence of conditions. It influences the validity of causal conclusions, as we discussed at length above. In other words, the only similarity between random sampling and random assignment is their randomness. It's not only possible to have either one without the other, it's actually quite common. Researchers who work in disciplines or topical areas that stress one type of validity tend not to stress the other type. For example, political pollsters put a lot of energy into successful random sampling, but they usually do not assign cases to conditions at all because their studies are not experiments.

Laboratory vs. Field (Naturalistic) Settings

In Chapter 4, we mentioned a distinction between lab and field settings for conducting research studies. To many people, an "experiment" conjures up an image of a white-coated scientist working away in a sterile and technical-looking room known as a laboratory. Aside from the fact that even those geographers who *do* conduct experiments in labs usually don't wear white lab coats, this stereotype is misleading in a more substantial way. It conflates the distinction between experimental and nonexperimental designs with the distinction between laboratory and field settings. A **laboratory**, or lab, is a specially designed setting, nearly always an interior room in a building, that allows researchers to exert physical control while conducting studies on their phenomenon of interest. Both physical and human geographers collect data in laboratories. Temperature, humidity, and chemical contamination can be controlled; so can sounds, sights, and other distractions. The **field** setting, in contrast, is also a setting where geographers conduct studies on their phenomenon of interest, but it is the setting in which the phenomenon normally occurs. Field settings are essentially "naturalistic" settings, by which we mean they are data collection locales that are the places where phenomena of interest go on as they normally do. Given the interests of most geographers, data collection in the field is more common than in the lab.[3] Field settings are usually outdoors, but they need not literally be *fields* (although they are to many biogeographers and agricultural geographers!). Again, both physical and human geographers conduct studies in the field—at forests, seashores, drainage basins, glaciers, cloud formations, and eroded mountains; also at rural villages, central business districts, front yards, manufacturing plants, legislative sessions, and national parks.

We discuss the concept of validity at length in Chapter 11. The issue of experimental versus nonexperimental designs has implications for the validity of causal conclusions, as we discussed above. The issue of laboratory versus field settings has implications for the validity of conclusions we make about how our research findings generalize to other settings, cases, measures, and so on. In Chapter 11, we learn that this is the issue of "ecological validity," which is the natural verisimilitude of the research setting. Our point here is that the issues of experimental versus nonexperimental and lab versus field are mostly independent from one another. A researcher can do a lab experiment, a lab study, a field experiment, or a field study. Perhaps field experiments are even superior in some research domains. For example, the surveys we discussed in Chapter 6 are very common and flexible ways of collecting

---

[3]Most geographers think of field research as requiring the researcher to actually travel to the setting where data are collected. So even though satellites collect remotely sensed data on phenomena in their naturally occurring setting, most geographers would not consider studies using such data to be field research because the researcher sits in an office at a computer terminal in order to acquire data.

data in nonexperimental research, but they are used in experimental research too; the experimenter may decide which respondent gets which version of the survey. Therefore, it is not really correct to contrast "survey research" with experimental research. For that matter, researchers may administer surveys and other explicit reports in a controlled lab setting *or* in a field setting.

Basic Research Designs

We can now turn to an overview of some of the basic alternatives of research design, the choices we have available for how we design variables and set up measurements to allow us to make certain comparisons. As we noted above, our design choices go a long way toward determining which questions we can ask of data and the nature of the truth we can expect from our answers.

First, we should make sure we are clear about the distinction between variables and **levels of variables.** As we defined them in Chapter 2, variables are the attributes or properties of cases that vary, either across cases at the same time, within cases over time, or both. A study must have one or more variables. A study with only one variable is rather unusual, as it is only able to consider the distribution of that single variable across cases. Studies generally have at least two variables, allowing for the investigation of relationships, and they usually have more than two. In any event, a basic question of research design concerns how many variables to measure and/or manipulate. This still leaves the question of how many values each variable will take in a study. We know that two is a minimum—it wouldn't be a *variable* otherwise. The different values a variable takes (or is designed to take by an experimenter) are different levels of the variable. In experiments, they are conditions; with any discrete variable, these levels might be called "groups" or "classes" (for example, groups of two types of drainage basins). With nonmanipulated variables, this question of levels is in part a question of measurement resolution (precision).

A second basic issue concerns whether levels of the variables differ between cases or within cases. Above, we contrasted experimental designs where each case gets exposed to one and only one condition of the independent variable with designs where each case gets exposed to all conditions of the independent variable. (Intermediate designs in which cases are exposed to more than one but less than all conditions are possible too, but they lead to complexities of analysis and interpretation and should usually be avoided.) This distinction holds for nonexperimental designs, too; a study of biota in limestone caves versus biota in sandstone caves is different than a study of biota in the same limestone caves over time. Whether experimental or nonexperimental, the first type of design is a **between-case design**. Cases are at different levels of an independent or predictor variable, whether placed there by an experimenter or not, so that analyzing these variables involves comparing data *between* different cases. The second type of design is a **within-case design**, also known as a repeated-measures design. Cases are at all levels of a variable at different times, again in an order determined by an experimenter or not, so that analyzing the variable involves comparing data *within* the same cases.

Both classes of designs have their strengths and weaknesses. Generally, within-case designs

(a) are more efficient, as fewer cases are necessary to get the same amount of data;

(b) lead to higher precision of estimation and power of hypothesis testing (Chapter 9), even given the same amount of data, because random noise associated with irrelevant variations among cases is reduced (measurements of a city or mountain made at one time are usually more similar to measurements of that city or mountain made at another time than they are to measurements of another city or mountain at the same time); and

(c) reduce confounds because comparison groups are more nearly equated, for the same reason as in the previous point.

These are certainly enormous benefits, so enormous that you should always do within-case designs if you can. Unfortunately, you cannot. Within-case designs have some severe limitations. They always employ multiple measurements of the same cases and, in experimental designs, multiple exposures to different conditions of the independent variable (IV). Therefore, they often run the risk of changing the cases somehow as the study goes on, including producing order effects, such as those discussed in Chapter 6 that result from the administration of multiple items in explicit reports (fatigue, practice effects, carryover effects, and so on). These can often but not always be addressed by randomizing or counterbalancing the orders of conditions (see Chapter 6). But within-case designs have a much more serious problem than order effects, really an ultimate limitation. Often they simply cannot be done. Many variables fundamentally reflect inherent properties of different cases that can hardly if at all be made to exist within the same case over time. A city in Europe cannot become a city in Africa. Saltwater fish cannot become freshwater fish. Residents of Chinese descent cannot become residents of Korean descent. What's more, measurement sometimes destroys cases, which absolutely precludes a within-case design. A study of insect migration that involves capturing and killing insects is one example. It is an example that suggests that ethical considerations too can preclude the use of within-case designs. We return to this in Chapter 14, and we leave it to you here to entertain additional ethically relevant scenarios of this type.

Specific Research Designs

Table 7.1 presents several specific research designs, both nonexperimental and experimental. As we pointed out above, research designs such as these vary in the validity of causal conclusions they support. Some designs are so poor in this respect that they rarely if ever deserve to be considered for use. Others are relatively weak in this respect but deserve consideration because they are less costly to carry out, in all senses of cost. And of course, we made it clear above that experimental designs are simply impossible in some research areas.

Table 7.1 Assorted Research Designs*

Nonexperimental Designs							
One-Group, Single Measurement	O_1						
One-Group, Multiple Measurement	O_1	O_1	O_1	O_1			
One-Group, Posttest-Only	E	O_1					
One-Group, Pretest-Posttest	O_1	E	O_1				
One-Group, Multiple Pretest-Posttest	O_1	O_1	E	O_1	O_1	O_1	O_1
Two-Group Nonmanipulated, Single Measurement	O_1 O_2						
Experimental Designs							
One-Group Manipulated Within	M_1	O_1	M_2	O_1			
Two-Group Manipulated Within	M_1	O_1	M_2	O_1			
	M_2	O_2	M_1	O_2			
Two-Group Manipulated Between, Posttest-Only	M_1 M_2	O_1 O_2					
Factorial Four-Group Manipulated, Posttest-Only	M_1N_1 M_1N_2	O_1 O_2					
	M_2N_1	O_3					
	M_2N_2	O_4					

*"O" is an observation or measurement, "E" is an event that naturally occurs, and "M" and "N" are applications of manipulations. Subscripts on "O" indicate which independent group of cases is observed; numerical labels on "M" and "N" indicate conditions of the manipulated factors.

Turning to nonexperimental designs first, we see that the simplest possible design consists of a single measurement on a single group of cases. This design is useful only if you want to infer the absolute level of some variable in some population—no comparison among groups of cases, no evidence of the effects of an event or manipulation, no contrast among different variables. If you repeat the measurement several times, you at least get descriptive evidence for change over time that might suggest something worth investigating with a more sophisticated design. Next, we can take the opportunity to examine the possible effect of a naturally occurring event (that is, not manipulated by the researcher), for example, a volcanic eruption or the enactment of new legislation. A **posttest-only design** is quite deficient, however ("pre-" and "post-" mean before or after the event or manipulation). A single measurement after the event leaves you wondering what the measurement would have been before the event. A **pretest-posttest design** at least gives you a baseline measurement for comparison, although the design still leaves open whether some ongoing trend or some other event, rather than your event of interest, caused the difference from pre- to post-. We get much better evidence for the effect of a naturally occurring event if we take multiple measurements over time, preferably both before *and* after the event.[4]

If we want to examine the possible effect of a variable that is intrinsically part of our cases, rather than an externally occurring event, we can conduct a between-case study by sampling from two identifiable subpopulations of cases. This allows us to identify relationships between group membership and the measured variable, but as we know from Figure 7.1, that alone leaves causality quite ambiguous. We show only two groups in Table 7.1, but of course any number of groups can be sampled. Finally, note that all designs, whether nonexperimental or experimental, can involve measurement of more than one observed variable. This has the obvious benefits not only of increasing the number of variables you can make conclusions about but allowing comparisons of patterns across variables that may shed light on whether your event or manipulation is the actual cause of changes to any of your measured variables.

Turning to experimental designs, we see that the simplest within-case design exposes a single group of cases to one condition of a factor and a measurement, and then exposes them to a different condition and another measurement. This design is considerably strengthened if we assign our cases into two groups, typically randomly, by exposing each group to one of the two possible orders of the two conditions of the factor. That way, we'll have evidence about the possible effects of being exposed first to one of the two conditions before being exposed to the other. If we want (or need) to conduct a between-case experiment, our simplest option is to assign our cases into two groups, exposing each to different conditions of the manipulated factor, and then measuring them. You can design something a little more complex by manipulating two (or more) factors. In the simplest possible such **factorial design**, the cases would be assigned into one of four conditions formed by crossing the two conditions of one factor against the two conditions of the other (this is a "2 × 2 design"). For example, we could conduct an experiment on the effectiveness of cartographic "in-vehicle navigation systems" in automobiles as a function of their orientation flexibility and information content. Starting with a sample of 120 drivers, we could assign 60 each to receive map displays on a dashboard computer oriented in one of two ways: a fixed orientation with north up or a variable orientation that turns as the car turns so that forward stays up. Within each group of 60, we could assign 30 to receive maps containing labeled landmark features, whereas the other 30 could receive maps without labeled landmarks. The

---

[4]Designs that incorporate a large number of repeated measurements over time, to investigate either the effect of some event or just a normally occurring developmental trend, are called **time-series designs**. These are fairly common in geography, but their statistical analysis is special in ways rather analogous to the analysis of data distributed over space (Chapter 9). The presence of different patterns of **temporal autocorrelation**, wherein values of variables are more or less similar as a function of when they were measured, must be evaluated in the data. For example, many variables show evidence of "seasonality" by taking on characteristics patterns of high or low values at different times of the year (including temperature, precipitation, burglaries, and tourist activity).

value of factorial designs goes well beyond simply being able to test multiple factors all at once. Such designs also allow for the investigation of factorial **interactions**, wherein the effects of one factor depend on the condition of the other factor; more complex interactions involving more than two factors are also possible. In our example, one possible interaction pattern would be if the labeled landmarks proved useful to drivers when the map kept a fixed north-up orientation but not when the map turned as the car turned.

There are an unlimited number of possible research designs that go beyond the basic ones we introduce here. An important class of designs is called **quasi-experiments**, which are studies without manipulated variables that nevertheless attempt to establish causal relations by applying systematic statistical control via more complex research designs. As we suggested above, the number of variables in a study, both manipulated and measured, can be increased *ad infinitum.* The number of levels, or the measurement resolution, of these variables can also be increased to any level you might want. However, not only do research designs of increasing complexity cost more and more (including requiring more cases and more measurements), they also potentially lead to interpretative difficulties that eventually exceed the powers of our limited human minds.

Developmental Designs (Change over Time)

Geographers and other natural and social scientists are often interested in processes of change over time. How do beaches accrete and erode over time? How do patterns of migration, both human and nonhuman, shift over time? Systematic (nonrandom) processes of change like these are examples of "development." There are many specific examples of systems in human and physical geography that develop, and many processes responsible for the course of those developments. The planet's solid surface develops, cultures develop, atmospheric composition develops, cities develop, individual people develop, and ecosystems develop. Broadly speaking, "evolution" is another term for development. Understood in this broad sense, Charles Darwin's ideas make up just one of many theories of "evolution," although of course it is an especially noteworthy theory about the genotypic and phenotypic development of species over generations.

Developmental designs are studies designed to conduct research on developmental processes. There are two basic approaches to the design of developmental research studies. In the first approach, two or more groups of cases, each at different "ages" or levels of development, are compared at the same time. For example, different forests at different levels of succession or the economies of countries at different stages of economic maturity are compared. This is the **cross-sectional** or "synchronic" approach. Its biggest advantage is its relative ease and efficiency. However, such a design provides no direct evidence on development, only the indirect evidence of comparing the product of development as realized in two or more static groups of cases at two or more developmental ages. These groups of cases of the same age—that is, "born" at the same time—are called **cohorts.** Cross-sectional designs always compare two or more cohorts of cases. A specific cohort sometimes has characteristics that result from the particular generation of which it is a part, rather than characteristics that have always described or will always describe cases at that age. Such **cohort effects**, wherein properties of cases originating at the same time are mistaken for properties of cases of a certain level of development, are threats to the internal validity (Chapter 11) of cross-sectional designs.

The second approach to developmental studies is the **longitudinal** or "diachronic" design. A group of cases at one level of development is compared to itself over time. By definition, such a design requires repeated measurements. For example, the geographic knowledge of children can be studied by testing a group of children when they are five years of age, the following year when they are six, and a third year when they are seven. As another example, the location and size of a streambed can be compared over a 10-year period by taking measurements every year at the same locations of the same stream. The biggest advantage of longitudinal designs, over and above the other benefits of within-case designs discussed above, is that they provide direct evidence on development—the cases actually change over time during the course of the study. Unfortunately, they are typically quite time-consuming and expensive. Such designs also suffer from various forms of **attrition** or "mortality," the loss of cases during the course of the study; researchers and research support can also quit or "pass away" during the course of the study. Attrition is especially problematic to internal validity when different types of cases are systematically more likely to drop out of the study than others, a condition known as "differential attrition." For example, a study of spatial behavior over time in which tracking devices are attached to people's cars might suffer from the fact that people going to particular types of places rather than others might be more likely to quit the study before it is finished (we return to this in Chapter 14, but please use your imagination for now). Longitudinal designs can also suffer from **history effects**, idiosyncratic events or conditions that hold during the particular time period of the study. Finally, many longitudinal studies that are conducted over several years or, especially, several decades potentially suffer from **instrumentation**, which is a change over time in the way measurements are made. A classic example is the way questions on the U.S. census are modified over the decades, as we discussed in Chapter 6.

A hybrid approach to developmental designs is known as a **sequential design**. There are many specific variants of the sequential design, but they all combine aspects of cross-sectional and longitudinal designs. Two or more groups of cases differing in their level of development at one time are sampled, as in cross-sectional designs, but these groups are measured repeatedly over time, as in longitudinal designs. The sequential design provides some of the advantages of each approach while mostly avoiding some of their threats to validity, such as cohort and history effects.

A final word about developmental designs concerns temporal scale. As we pointed out in Chapter 2, temporal scale is important to geographers, as is spatial scale. In developmental designs, both time and temporality are critically important with respect to phenomenon scale and analysis scale. What should the duration of the study be, and at what time should data collection begin and end? What should the interval between measurements be, which, when combined with the duration of the study, determines the total number of measurements to be made? What is the temporal form of the developmental process being studied within the time frame of the study? Is it a straight line going up or down? Is it "monotonic" but nonlinear (see Chapter 9), with most of the change occurring near the beginning of a study's time frame? Is it a "U-shaped function" that comes down and then goes up, or an "inverted U-shaped function" that goes the other way? These questions must be addressed when designing and conducting developmental research.

Single-Case and Multiple-Case Designs

Much of what we have said about research studies above and in previous chapters has explicitly or implicitly assumed that a study involves observations of several cases rather than just one. This is not necessarily true. There are research designs, both experimental and nonexperimental, that involve just a single case, such as a single location in a river or a single neighborhood in a city. A **single-case experiment** is a repeated-measures design with one case. This is a particularly effective design when one is interested in demonstrating the practical ability to change the value of a single case on some variable at will. For example, you could measure a case at one time, then expose it to some condition and measure it again. That would not be a very strong design because any change in the measured variable could result from some other event that occurs at about the same time or some preexisting trend that was already ongoing. An improvement would be the **reversal design** ("return to baseline"), wherein, after the second measurement, you would remove the applied condition and measure a third time. More complex designs are possible.

However, *non*experimental single-case designs are much more common in geography. A **case study** is an intensive and comprehensive descriptive study of a single case. Case studies can involve any number of data collection types, including any mixture of quantitative and qualitative approaches, although in human geography, they virtually always involve an emphasis on qualitative approaches. Case studies can provide a very rich, holistic, and wide-ranging description of a single case, but they are at best *suggestive* about causality rather than definitive. To some researchers, therefore, they are considered most useful in the early, exploratory stage of a research program, or as adjuncts to multiple-case designs that help make general conclusions more concrete and personal (much the way so many newspaper stories these days are introduced by an anecdote about a single person or other entity).

As compared to multiple-case designs, single-case designs are efficient (only one case to obtain and measure) and provide a rich, more complete picture of the characteristics of a meaningful unit. But multiple-case designs have some very definite and important strengths as compared to single-case designs. Using multiple cases obviously gives you much more ability to explore how findings generalize to various types of cases. It goes a long way toward helping you to avoid spurious conclusions drawn from idiosyncratic cases you might have happened to choose in a single-case design. And given the statistical nature of most phenomena in geography (see Chapter 9), only the multiple-case design allows you to see the effect of a single variable when many actually play a causal role in your system of interest; it allows you to find a "signal" in a background of "noise."

These benefits of multiple-case designs really matter a lot when researchers accept the goals and characteristic philosophical values of a scientific approach to research. We went over these in Chapter 1 and introduced the distinction between **nomothetic** and **idiographic** approaches to knowledge. A nomothetic approach attempts to be general over cases, times, places, and so on; nomothetic approaches attempt for a "lawlike" understanding or at least an understanding based on probabilistic general truths. An idiographic approach attempts to be specific to particular cases at particular times and places; idiographic approaches strive for a potentially "idiosyncratic" understanding. We noted in Chapter 1 that scientists strongly prefer a general understanding. An approach that is exclusively idiographic is therefore fundamentally nonscientific, whatever truth it may achieve. Don't overinterpret this; scientists do not stubbornly refuse all nongeneral or conditional truths just to maintain some status as "real scientists." In fact, a completely nomothetic, general understanding is apparently not possible—it does not fit the facts of reality as we can appreciate them. A search for "*The* Law of Nature," a single unifying force, has reputedly occupied physics at its highest conceptual levels for some time, but it seems safe to conclude that no such "über law" will ever characterize any area of research in geography, nor in many other natural or social sciences. Reality seems to lie somewhere between extremely nomothetic and extremely idiographic approaches. In truth, different research approaches tend to take positions somewhere on a continuum between the two extremes. Many hybrid and intermediate approaches attempt to take advantage of strengths of both, for example, a case-study design replicated on several cases. To recognize that some truths generally hold, for example, in certain climates and not others, or with certain cultures and not others, is quite acceptable as a scientific conclusion.

Computational Modeling

In Chapters 2 and 3, we introduced and defined models as simplified representations of portions of reality that include sets of interrelated hypotheses expressing theories about structures and processes of systems of interest in the world, including causal processes. We learned in those chapters that models are pervasive in all areas of geography and in other sciences, especially considering that they come in several forms, including conceptual, physical, graphical, and computational forms. Our interest here is in computational models, defined in Chapter 3 as models of theoretical structures and processes expressed in mathematical form (more precisely, in a **formal language**). Computational models are typically instantiated as sets of equations and other logical/mathematical operations expressed in a computer program or set of programs and attendant databases. However, there is nothing about computational models that intrinsically requires computer representation other than the practical difficulty or impossibility of doing them manually. There are many specific applications of computational modeling in geography. In physical geography, there are models of climate, hydrology, geomorphology, ecology, and other phenomena. In human geography, there are models of population growth, travel and migration, spatial decision-making, retail and manufacturing location, agricultural and urban land use, cultural diffusion, epidemiology, and other phenomena.

Like all models, computational models are necessarily *simplified* representations of reality. That is, they intentionally ignore or distort aspects of the true richness of reality by incorporating **simplifying assumptions**. In fact, in their initial versions, computational models are often blatantly unrealistic in their simplicity. For example, many models in geography start out making the assumption that their natural or human process takes place on a uniform **isotropic plain**. This is a flat, featureless, and unbounded landscape on which all forms of spatial movement and interaction can occur equally easily in any direction. Such an idea may inspire you to wonder what planet these models are supposed to reflect—what about roads, rivers, mountains, and all the other features that serve as barriers to interaction in certain directions and facilitators of interaction in other directions? Another common assumption in geographic models is that of **economic rationality**. This is the idea that economic agents (individuals, families, corporations) will act exclusively to maximize profit and will be informed in this by complete and accurate knowledge of everything that has implications for profitability. How could such blatantly false assumptions be useful to geographers and other scientists?

The answer is that simplification, even extreme simplification, is necessary in scientific work. It reflects the scientific preference for parsimony we discussed in Chapter 1. Furthermore, it reflects our cognitive limitations as humans—we need to simplify in order to understand complex reality. It might sound a little contradictory, but people must simplify (a form of intentional distortion) if they are to understand anything. The categorical thinking that universally characterizes humans is a manifestation of this cognitive need for simplification. Reality in all its rich complexity, with its contextual factors, probabilistic nature, heterogeneous existence across a huge range of spatial and temporal scales, multivariate and reciprocal causality, emergent patterns, time-lagged and distance-lagged influences, and more, is just too much for our puny little minds to grasp fully and completely. But those who make the scientific bargain we discussed in Chapter 1 accept the compromise of simplicity. They want to know *something* partially and imperfectly rather than *nothing* completely and perfectly. At least they believe, as a matter of faith or personality, that a systematic but imperfect search for truth is better than an irrational and haphazard search, a search based on nothing but invention and fantasy, or no search at all. This aspect of scientific research reminds us a little of the tale about the person who was looking in a parking lot for his lost car keys. His friend asked him why he was searching there, considering he had lost his keys on the other side of the lot. "The light's better over here," he replied. Scientists use methods such as simplifying assumptions to shed light on understanding the world; sometimes the methods are bright but do not shine directly on the most relevant portion of reality. Nonetheless, they allow the scientists to see.

In Chapter 3, we listed computational modeling as one of the basic types of data collection in geography. Unlike the other types of data collection, in which portions of reality are directly measured, portions of reality are *simulated* by models. The **model outputs** (usually numerical) are treated conceptually as if they were measurements. These "simulated data" are typically compared to standard empirical measurements made on the portion of reality to which the model refers, as a way to evaluate the fit of the model to reality (we discuss this more below). However, sometimes models are created and thought about as if they were creations of new realities rather than simulations of existing realities. It is somewhat arguable whether a person who models in order to create new realities is really doing science. You could say they are not because they are not trying to "create and evaluate knowledge about reality" (Chapter 1). Instead, they are doing something like product invention or artifact creation. On the other hand, "reality" certainly includes the intentional products of human activity (buildings, wheat fields), so maybe creating and studying models that do not correspond to the reality that already exists is just the "science of the artificial."[5]

We discuss computational modeling in this chapter on research design because it can also be thought of as an alternative to traditional experimental designs. As we explained above in this chapter, researchers conduct experimental studies when they manipulate variables—when they control the assignment of cases to levels of independent variables. Computational modeling is such an important 20th-century development because it provides a way to conduct experimental studies when they would otherwise be very difficult or even impossible to conduct. Modeling gives a form of empirical access to events or systems that no longer exist or do not yet exist, that are very rare, or that operate over spatial and temporal scales that are too large to bring into the laboratory. Models afford a form of assignment control when it would otherwise be impossible or ethically unacceptable to manipulate variables. For example, people have often suggested solving particular environmental problems, such as species invasions or the greenhouse effect, by introducing particular chemical or biological agents into the environment; many such introductions are ethically dubious to carry out in reality (see Chapter 14).

Even when traditional empirical approaches *are* viable, modeling has a definite advantage over traditional empirical approaches: the superior ability of models to represent much more of a whole system with many of its complexly interacting parts. Numerous scientific insights of the 20th century involved the realization that causal relations in many systems are not simply binary and unidirectional links, like one pool ball smacking into another. Causality is frequently multivariate, involving many entities or states interacting simultaneously or with complex patterns of **temporally** or **spatially lagged** influences. Causality is frequently **reciprocal**, with one entity or state having causal influence on another that returns an influence back on the first. These complex causal relations are difficulty or impossible to study with more traditional means of conceptualizing and empirically evaluating theories.

For example, consider Figure 7.2. It depicts a conceptual model of the earth's systems, showing influences on climate and its change over timescales of decades to centuries.[6] Climate processes operate at multiple spatial and temporal scales,

[5]Echoing the title and theme of the well-known book: Simon, H. A. (1969). *The sciences of the artificial.* Cambridge: MIT Press.

including scales too large to represent physically in the lab (although processes that operate over very large or small, or slow or fast, scales often do not need to be explicitly represented in such models). Furthermore, climate reflects complexly interacting states and influences of the atmosphere, ocean, land, ice and snow, and terrestrial and marine biota—including humans and their activities. This is reflected by the climate researchers' use of the term **forcing** to refer to a causal factor in computational models, analogous to an independent variable in true experiments. The term forcing, however, connotes the fact that causal variables in complex systems, such as climate, are inputs of matter and energy that exist within the complicated network of interacting multivariate causality that is the system. The actual observable effect of such a cause often depends on **feedback** influences from other parts of the system. In the case of **positive feedback**, the effect of a forcing can be magnified considerably; for example, increasing water evaporation on earth can lead to increasing surface temperature (water vapor is a major greenhouse gas), which in turn can lead to even more evaporation and water vapor, and so on. In the case of **negative feedback**, a forcing can be dampened so that its influence is negligible; an example is the fact that increasing water in the atmosphere from increasing surface temperature can lead to increasing cloud formation, which in turn would *decrease* insolation on the earth surface and cool it down. In many cases of feedback, changes in a forcing do not necessarily result in much change to outcome variables until the forcing changes reach particular thresholds. Characterizing the magnitude and pattern of changes to outcomes as a result of changes to forcing factors is the subject of **sensitivity analyses**. Finally, we should not forget that climate change requires modeling insofar as most people would find it unethical to carry out experiments to manipulate forcings that they believed might actually change climate in the real world.

Researchers typically distinguish between models whose parameters are based on physical laws (law-based or **numerical models**) and models whose parameters are based on estimates made from data (**empirical models**). Empirical modeling is common in both physical and human geography, but numerical modeling is much more common in physical geography (some population models do incorporate simple population-growth laws). The laws instantiated in numerical models in geography and other earth sciences come from physics, chemistry, biology, and so on. The model in our climate example would utilize the laws of thermodynamics, Newton's laws of motion, and so on. In Chapter 2, we described the difference between probabilistic and deterministic processes, in the specific context of causal processes. Empirical models are always probabilistic or stochastic in nature, because the data upon which they are constructed are understood to derive from samples rather than populations, and because empirical measurement is understood to contain error or noise. Numerical models may be deterministic or stochastic. **Deterministic numerical models** incorporate mathematical expressions of laws to derive parameters without any random input or noise. Given an initial state and set of boundary conditions, a well-constructed deterministic model gives a single precise output. **Stochastic numerical models** ("Monte Carlo models") incorporate a simulation of random processes in them. The randomness is typically supplied by

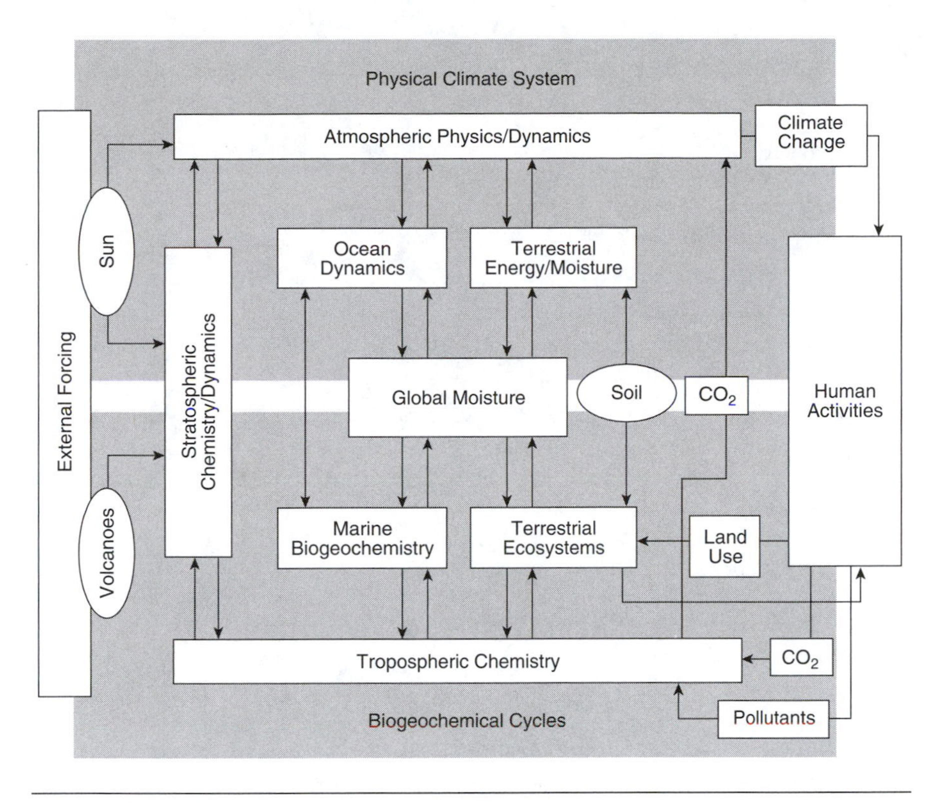

Figure 7.2 Conceptual model of earth's systems depicting influences on climate and its change over timescales of decades to centuries.[6]

a random number generator function available in whatever computer software is used to implement the model in the computer (see Box 8.1). Each "run" of a stochastic model will give a somewhat different output for the model. Usually, researchers who use stochastic models focus on patterns that emerge over many runs of the model. A fascinating scientific insight of the 20th century is that such stochastic models can be "chaotic"—their outputs can vary widely as a function of small changes to input values.

Steps of Computational Modeling

We describe the basic logic of conducting computational modeling as a series of steps in Table 7.2. The table presents an idealized view of computational modeling; depending on the purpose and topical domain of the model, these steps are certainly applied in somewhat different ways and are not fully realized in all instances. First, you must identify and express the conceptual underpinnings of your model.

[6]Adapted from Earth System Science (1986). *A program for global change.* Washington, DC: National Aeronautics and Space Administration.

Table 7.2 Steps of Computational Modeling

1. Create conceptual model
   - derive it from existing theories and concepts, and any other sources
   - visualize it in appropriate graphic form
2. Create computational model
   - identify relevant parameters for structures and processes, their boundary conditions, and so on
   - base it on scientific laws (numerical models) and/or empirical data (empirical models)
   - express model in formal language (mathematical equations, computer operations)
   - write computer program, programming from scratch or using appropriate commercial modeling software
3. Run the computer program
   - produce output as form of "data collection"
   - debug as necessary
4. Compare model output to empirically obtained data
   - base data on measurements of past or current events or states
   - evaluate model both qualitatively and quantitatively
   - interpret discrepancies between model output and empirical observations by refining the model, returning to Step 1, 2, or 3
5. Accept, use, and communicate model
   - eventually, accept model and use to forecast future, explain phenomena, and so on
   - communicate model and its uncertainties to researchers and other interested parties

All computational models have a conceptual model underneath them, at least implicitly; it is highly desirable that you make the conceptual model explicit at this stage. You can derive the entities and their interrelating processes that you consider relevant to include in the model from existing theories and concepts, or any other source of ideas; often enough, the entire conceptual model already exists in published scientific literature. It is quite helpful in most modeling situations to visualize the conceptual model as a flow diagram, network structure, or some other graphic; the climate model in Figure 7.2 is an example.

Then you translate the conceptual model into computational form. You must identify and quantify specific **model parameters**, quantify processes that express causal influences, and so on. The identification and **calibration** of model parameters is based on applicable physical laws, prior empirical data, or both, as we described above. You calibrate a model by adjusting its parameters to fit the context of your particular modeling situation; for example, you have to tune parameters to the particular spatial and temporal scale of your model. You express the computational model formally as mathematical equations, computer operations, and so on. Then you instantiate this formal expression on a computer as a program or set of programs, either programming from scratch or using appropriate commercial modeling software.

Proceed with the "data collection" phase by running the model. Generate model output by providing inputs that are processed through the computational steps of the model. The inputs are based on measured or simulated states of the relevant parameters in the model. You must typically **debug** the model at this step, often extensively. That is, you determine that the program is computing as intended, and locate and fix any errors in it.

You then evaluate and interpret the model output. Do this by comparing "data" output from the model to empirical measurements of the actual system that is being modeled. These measurements are based on past or current events or states, if they exist at all; it is not uncommon for appropriate data to be lacking. Evaluation occurs both qualitatively and quantitatively. Qualitative evaluation involves such things as identifying whether the influence of variables operates in the right direction; for example, does increasing one variable cause another variable to increase or decrease? Quantitative evaluation typically involves statistically testing numerical discrepancies between the model output and the empirical measurements using some version of a statistical **goodness-of-fit** test.[7] Interpret discrepancies between the model output and the empirical observations you used to evaluate it. This will usually lead you to refine the model by returning to steps 1, 2, or 3. You can identify missing parameters, recalibrate parameter boundary conditions and process influences, and so on. You might conduct the sensitivity analyses we described above at this step.

Eventually, you accept the model and use it to describe complex systems, forecast the future, explain phenomena, or achieve practical control over them. One of the most intriguing aspects of model use can be communicating it to other people, including the various uncertainties involved in its outputs and interpretation. This task calls for all of the expertise and cleverness in displaying data you can muster, especially when, like climate change, your model potentially has great implications for society and at the same time is politically contentious.

[7]We cover statistical data analysis in Chapter 9. There we discuss statistical "hypothesis tests" that generate probabilities about the correspondence between sample results and possible population truths. The mathematical mechanics of goodness-of-fit tests are essentially those of hypothesis testing, but there is an interesting difference between a researcher's attitude toward the probabilities of standard hypothesis tests and those of goodness-of-fit tests. When conducting hypothesis tests, as we learn in Chapter 9, one usually wants sample data to *differ* significantly from likely outcomes of the tested null hypothesis (to be improbable if the null were true). When conducting goodness-of-fit tests, in contrast, one usually wants the model output *not* to differ significantly from the measured data.

Review Questions

- What is research design, and why is it so important to conducting and interpreting research?

Empirical Control in Research

- What are the three forms of empirical control in research?
- What distinguishes experimental from nonexperimental research, and what are the implications of this distinction for the meaning of research?
- Why is statistical correlation ambiguous with respect to causal relationships? What are confounds? What is a control group?
- What is the distinction between laboratory and field settings, and how does this distinction relate to the experimental/nonexperimental distinction?

Basic Research Designs

- What are between-case and within-case designs, and what are strengths and weaknesses of each?
- What are several specific research designs, and what are some of their strengths and weaknesses?
- What are developmental research designs? What are cross-sectional, longitudinal, and sequential developmental designs?
- What are some implications of the difference between single-case and multiple-case designs?

Computational Modeling

- Why is computational modeling so useful to geographic researchers?
- Why are simplifying assumptions incorporated into computational models?
- What distinguishes numerical models from empirical models? What distinguishes deterministic models from stochastic models?
- What are the basic steps to conducting research with computational models?

Key Terms

assignment control: form of empirical control in which researchers create at least one of the variables in the study—that is, researchers determine which level of the variable cases are exposed to

attrition: problem with longitudinal designs of losing cases during the course of the study; also called "mortality"

between-case design: research design in which different cases take on different levels of an independent or predictor variable

calibration: adjustment of model parameters to fit the context of a particular model, the specific ranges of its input values, the scale of its units of analysis, and so on; calibration may be based on applicable physical laws, empirical data, or both

case study: intensive and comprehensive descriptive study of a single case

cohort: group of cases of the same age (originating at the same time)

cohort effect: threat to the validity of cross-sectional designs wherein a property of cases that arises because they come from the same cohort is mistaken for a property of cases at a certain level of development

condition: one of the discrete levels of a manipulated variable in an experiment

confound: variable in a study that is not intended to be a causal variable but affects one of the measured variables; also called a "third variable"

control group: condition of an experiment that is meant to provide the necessary comparison of "no manipulation"; in fact, all experiments require whatever conditions are necessary to make particular comparisons

cross-sectional design: developmental research design in which two or more groups of cases, each at different "ages" or levels of development, are compared at the same time; also called "synchronic"

debug: process of determining that the program running a computational model is computing as intended, including locating and fixing errors

dependent variable (DV): measured, nonmanipulated variable in an experiment that plays the role of a potential effect

deterministic numerical model: numerical model constructed with parameters that assume no random inputs or noise; generates a single precise outcome given a single set of inputs

developmental design: research design to study systematic (nonrandom) processes of change over time in human or physical systems

economic rationality: common simplifying assumption in geographic models that economic agents act exclusively to maximize profit and are informed by complete and accurate knowledge of everything that has implications for profitability

empirical control: any method of increasing the ability to infer causality from empirical data, including physical, assignment, and statistical control

empirical model: computational model with parameters based on estimates from data; common in both human and physical geography

experimental study: research study in which assignment control (manipulation) is exercised over one or more variables

factorial design: experimental research design in which two or more independent variables (factors) are manipulated

feedback: reciprocal causal influence in a complex system

field: setting where researchers conduct studies on phenomena of interest as they "naturally" occur; usually outdoors in wilderness and inhabited settings

forcing: term for a causal factor, an input of matter or energy, in a computational model; connotes the factor's nonlinear and uncertain influence on effects in a complex system like climate

formal language: abstract mathematical or logical symbol system that precisely defines entities and processes that operate on the entities

goodness-of-fit test: statistical test to evaluate how well computational model outputs match empirically observed values

history effect: threat to the validity of longitudinal designs that arises because of idiosyncratic events or conditions that hold during the particular time period of the study

idiographic approach: specific, idiosyncratic approach to creating knowledge in research and scholarship; more characteristic of humanities than sciences

independent variable (IV): manipulated variable in an experiment that plays the role of a potential cause; also called a "factor"

instrumentation: threat to the validity of longitudinal designs that arises because of changes over time in the way measurements are made

interaction: when the effect of one IV on a DV in a factorial design is different at different levels of another IV; more complex interactions involving more than two factors are possible too

isotropic plain: common simplifying assumption in geographic models that processes operate on a flat, uniform landscape with no barriers to interaction in any direction

laboratory (lab): specially designed setting, usually a room inside a building, that allows researchers to exert physical control while conducting studies

levels of variable: the number of different values a variable can take, usually in the context of manipulated variables that take on a modestly sized number of values or "groups"

longitudinal design: developmental research design in which one group of cases is compared to itself over time as it develops; also called "diachronic"

manipulation: the assignment control of one or more variables in an experiment

matching: variation on random assignment to conditions in which two cases that have some characteristic in common are assigned to contrasting conditions in an experiment

model output: the simulated data that computational models produce after they are run

model parameters: quantitative statements in computational models that mathematically express the characteristics of factors in the model and how they interrelate and change during runs of the model

negative feedback: specific pattern of feedback wherein an increase in one forcing causes a change in another forcing that in turn causes the first forcing to decrease; dampens changes in the system

nomothetic approach: general, even lawlike approach to creating knowledge in research; more characteristic of sciences than humanities

nonexperimental study: research study in which assignment control (manipulation) is *not* exercised over any variables

numerical model: computational model with parameters based on prior scientific laws, primarily from physics and chemistry; common only in physical geography

physical control: form of empirical control in which researchers physically modify or restrict the data collection situation in order to reduce the potentially distracting influence of confounding factors that are not of interest

positive feedback: specific pattern of feedback wherein an increase in one forcing causes a change in another forcing that in turn causes the first forcing to increase more; magnifies changes in the system

posttest-only design: weak research design in which a single episode of measurement occurs after an event has taken place

pretest-posttest design: research design that is better than the posttest-only design because measurement episodes occur both before and after an event has taken place

quasi-experiment: study without manipulated variables that attempts to establish causal relations more validly by applying systematic statistical control over alternative causal variables

random assignment: assigning cases to conditions in an experiment according to a random decision rule

reciprocal causality: when two entities influence each other

research design: the structure of possible comparisons in a study that results from the way variables are created or measured

reversal design: single-case experiment wherein the case is measured before the manipulation is applied, after it is applied, and a third time after the manipulation is removed ("return to baseline")

sensitivity analysis: characterizing the magnitude and pattern of changes to outcome variables in a computational model as a result of changes to forcing factors

sequential design: developmental research design that is a hybrid of cross-sectional and longitudinal designs

simplifying assumption: aspect of the representation of reality in a model that is unrealistically simple but helps researchers understand the complexity they are modeling

single-case experiment: experiment conducted with a single case

spatial lag: influences in models, including feedbacks, that express their effect over some spatial distance (via some continuously connected mechanism that may not be evident)

spurious causality: attributing causality to a variable that correlates with another variable when it is not actually its cause

statistical control: form of empirical control in which researchers explicitly identify, measure, and statistically analyze variables they think might have an effect on the main variables of interest

stochastic numerical model ("Monte Carlo model"): numerical model constructed with parameters that assume random inputs or noise; generates different outcomes given a single set of inputs

temporal autocorrelation: nonindependence among measurements of phenomena as a function of the time of their occurrence relative to other phenomena; it is the temporal analogue of spatial autocorrelation

temporal lag: influences in models, including feedbacks, that express their effect after some time delay

time-series design: research design that incorporates a large number of repeated measurements over time in order to study developmental phenomena

within-case design: research design in which, over time, every case takes on each different level of an independent or predictor variable; also called a "repeated-measures design"

Bibliography

Bradford, M. G., & Kent, W. A. (1977). *Human geography: Theories and their applications.* Oxford: Oxford University Press.

Cook, T. D., & Campbell, D. T. (1979). *Quasi-experimentation: Design and analysis issues for field settings.* Boston: Houghton Mifflin.

Kenny, D. A. (1979). *Correlation and causality.* New York: Wiley.

Kirkby, M. J., Naden, P. S., Burt, T. P., & Butcher, D. P. (1993). *Computer simulation in physical geography* (2nd ed.). Chichester, U.K.: Wiley.

Lane, S. N. (2003). Numerical modeling in physical geography: Understanding, explanation and prediction. In N. Clifford & G. Valentine (Eds.), *Key methods in geography* (pp. 263–290). Thousand Oaks, CA: Sage.

Schneider, S. H. (1992). Introduction to climate modeling. In K. E. Trenberth (Ed.), *Climate system modelling* (pp. 3–26). Cambridge and New York: Cambridge University Press.

Shadish, W. R., Cook, T. D., & Campbell, D. T. (2002). *Experimental and quasi-experimental designs for generalized causal inference.* Boston: Houghton Mifflin.

Stake, R. E. (1995). *The art of case study research.* Thousand Oaks, CA: Sage.

CHAPTER 8

Sampling

Learning Objectives:

- What are populations, samples, and sampling frames?
- What are probability and nonprobability sampling designs?
- What are the implications of the particular sampling frame and sampling design used in a study?
- What are some implications of sampling from continuous fields instead of discrete entities?
- What should you consider when deciding on the appropriate sample size (number of cases) for a study?

In Chapter 1, we discussed scientists' preference for parsimonious explanation and, consequently, their preference for general over idiosyncratic truths. That means scientific geographers strive for truths that apply as widely as possible. But geographers run into a dilemma here. They want to be general in their conclusions, but when they conduct studies, they are necessarily limited to relatively small numbers of cases, places, times, variables, measures, and other research entities. This is often because researchers have only so much time, money, and other resources available to do their studies. But there is an even more fundamental reason for this dilemma. Even with great resources, researchers typically cannot even potentially access all the entities to which their theories and models are relevant. For example, if we test a theory of urban development, we want it to apply to all cities at all times and places, as much as is possible given existing ideas and evidence. Even if we limited our focus, as we likely would, to cities of particular size ranges or cities within particular cultural or economic systems, we would still be concerned with a very large number of cities, some that no longer exist and others that will only exist in the future. We cannot measure cities that do not exist when we conduct our study, but we want theories that apply to these cities, if possible.

As another example, we might measure microbes in soil taken at various depths. No matter how many measurements we take, we cannot access every volume of soil in a large field, let alone everywhere on the earth where that soil type exists. Nor would it be feasible to test for every possible microbe that could be in the soil. And what about soil from the past or soil that has yet to be created? Even with virtually unlimited time and effort, therefore, we cannot do a study that includes every case, place, time, variable, or measure in which we are interested.

The solution to this dilemma is sampling. **Sampling** is any way of selecting a subset from the entire set of entities of interest, called a **population**; the resulting subset is a **sample**. Because the sample is an incomplete subset, it is always necessarily smaller than the population from which it comes. Often it is much smaller. But size does not define whether a set of entities is a sample or population. That is determined by your research goals—by what you are interested in or want to generalize your research conclusions to. In basic science, researchers usually want to generalize their conclusions to very large, even hypothetically infinite, sets of entities. So even when they can access large sets of entities, the sets are typically samples. For example, researchers who study the reflective "signatures" of different earth-surface land covers detected via satellite remote sensing (Chapter 12) routinely have samples of millions of sensor readings. On the other hand, quite small sets of entities can constitute entire populations; for example, a teacher who wants to know how well the class did at learning the material taught in a particular term does not want to generalize to other students or terms or class material, so a total of two exam scores from 15 students can constitute a population of scores. Furthermore, when a researcher's goal changes, what was once considered a population may become a sample. The distinction between samples and populations can be subtle, but it is important. If we have the entire population of entities of interest, we can analyze it by simply describing it in various ways. If, as is so often the case, we have only a sample of the entities of interest, we not only describe that subset but also consider what the sample might tell us about our larger population of interest; in Chapter 9, we discuss the logic of doing this formally as part of "statistical inference."

As we suggested above, researchers sample many different types of entities, not just cases. In fact, virtually anything that plays a role in empirical research can be understood as a sample from a larger set of possible choices that could have been made. An urban geographer whose research team records graffiti while hiking along streets and alleys is sampling graffiti, cities, neighborhoods, years, streets, and alleys. Even the particular people collecting the data and the particular directions they turn their heads while looking constitute samples of some larger population of possibilities. Most discussions of sampling focus on sampling cases, as we will for most of the rest of our discussion. However, keep in mind that most of our discussion could apply to a variety of research entities other than just cases.

We also pointed out above that researchers sample because they are fundamentally limited in their ability to access all the cases they want to generalize to, but that in practice, they are usually also limited because they have only so much time, money, personnel, and so on. That is, there are practical reasons that researchers use samples instead of populations. Perhaps the act of accessing and measuring cases somehow changes or even destroys them. In some types of research, physical materials must be sampled for study because they have to be destroyed in order to be measured. Sometimes researchers must sample because the phenomenon in which they are interested may change in important ways if they take too long to access and measure their cases. For example, relationships between precipitation and snow pack depth must be measured before the season changes. So there are a variety of reasons that sampling is both conceptually and practically necessary in scientific research. But remember that when it is possible and practical to access and measure the entire population of interest instead of taking a sample, it is usually much better to do so. As we make evident below and in Chapter 9, sampling contributes a great deal of complexity and uncertainty to the research process that it would be wonderful to avoid. Unfortunately, such situations are highly rare—the stuff of research fantasy for most scientific geographers.

Sampling Frames and Sampling Designs

Having defined populations as the entire set of entities of interest, and samples as incomplete subsets of populations, we now consider how samples are obtained, and what that means for the design and interpretation of research. We must first recognize that samples are drawn from populations, but that the entire population of interest (sometimes called the "target population") is typically unavailable for sampling. Consider our example of research on urban development. We probably want to generalize beyond cities that currently exist, but of course we can only access and measure cities that do currently exist. The cities we can actually put into our sample are thus a direct subset not of the entire population of interest but of some other subset, a subset that is smaller than our population but larger than our sample. That new subset is called a **sampling frame**—the subset of the population from which cases are actually drawn to become part of the sample. In most research situations, the sampling frame is smaller than the target population, although unlike samples, sampling frames may be the same size as the target population on which they are based, namely, in situations where any member of the target population can potentially become part of the sample. Figure 8.1 shows the conceptual relationship among populations, sampling frames, and samples.

What constitutes a researcher's sampling frame depends on the way he or she identifies and accesses potential cases for measurement. For example, a researcher might identify people from a list of registered voters; he or she might solicit their participation by calling them on the phone. The sampling frame is all of the people on the list of registered voters whose phone number is correct and who are in town during the study period in order to receive your phone call. As another example, a researcher might identify stands of a particular species of cactus by driving along a dirt service road through a mountain range. The sampling frame in this example is all of the cacti visible from all of the roads that the researcher might choose to drive along. Geographers identify cases from property or tax records, industrial records, phone books, census data, maps and images, students in particular classes, people standing on certain street corners, areas that they can walk or drive or sail or fly to,

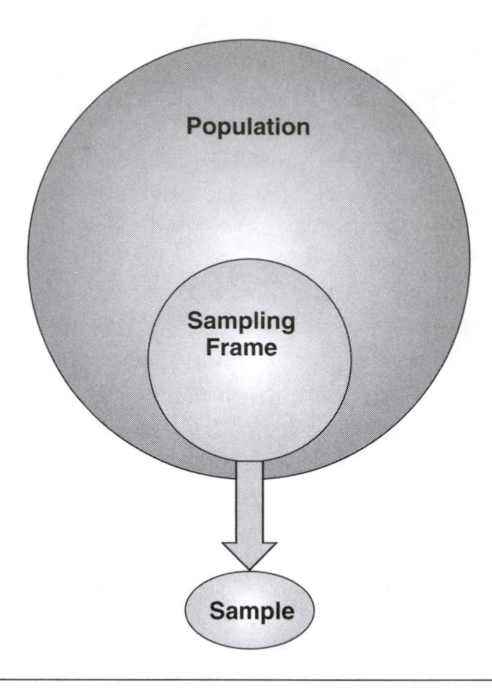

Figure 8.1 The conceptual relationship among population, sampling frame, and sample.

and so on. They solicit the participation of the cases (if the cases are entities with rights, such as individual people or companies) through face-to-face contact, on the phone, through the mail, over the Web, and more.

Once a sampling frame is obtained, the question arises as to how to select the actual sample from the frame. The procedure used to identify cases from the sampling frame to go into the sample is called the **sampling design**. There are several different sampling designs, each of which has different implications for the statistical relationship between the sample and the sampling frame. That is, sampling designs determine what we can know ahead of time about the chances that a case in the frame can get into the sample (by definition, every case in the frame has a nonzero chance of being sampled). These different designs can be grouped into two main categories: **nonprobability sampling** and **probability sampling.** In nonprobability designs, the probability of a particular case being selected is unknown—the researcher using a nonprobability design cannot say beforehand what the precise chance is of any particular case being selected to become part of the sample. Specific nonprobability designs include **convenience sampling**, in which the researcher simply accepts every case he or she can conveniently get hold of to be in the sample, up to the point where the sample is large enough. Obviously there are many specific ways this can be carried out; an example would be geographers who accept all of the remotely sensed imagery for an area they can get hold of, at whatever time and resolution it happened to be recorded. Mining pits provide a convenient opportunity to study the stratigraphy of sedimentary rock; however, it's unlikely that the location of the mining pit was chosen in order to provide the most informative data to a researcher. Another specific nonprobability design is **snowball sampling.** This is a design in which the researcher uses a case he or she has already put into the sample to find out about further cases that could be selected. A researcher studying the spatial behavior of heroin users might ask one user about other users.

In contrast, probability sampling includes any design in which the probability of each case in the sampling frame has a known probability of being selected. There are a variety of specific probability sampling designs used by geographers. The simplest and most common is **simple random sampling.** In this design, each member of the sampling frame has an equal chance of being selected into the sample. In addition, however, each equal-sized *subset* of members of the sampling frame has an equal chance of being selected into the sample (Box 8.1 discusses ways to generate random numbers for sampling and other uses). That is, selecting any particular member of the sampling frame has no effect on the probability that any other particular member will be selected. This second characteristic distinguishes simple random sampling from **systematic random sampling**, in which a first member is randomly selected, then every "nth" member after that is selected. For example, a county could be selected randomly from a list; then every 10th county after that could be selected. In such a sampling design, each county has an equal chance of being selected, because each has an equal chance of being selected first. But each subset of counties does not have an equal chance of being selected—two counties next to each other on the list cannot both be in the sample, for instance. Systematic sampling potentially poses a problem if there is some nonrandom pattern within the sampling frame list. Alphabetic ordering may contain such patterns, as when close places are similarly named. If the list is itself randomly ordered, systematic random sampling is equivalent to simple random sampling.

There are more complex probability sampling designs besides simple and systematic random sampling. In **stratified random sampling**, the sampling frame is segmented into subsets or classes called "strata," based on some relevant variable. Potentially useful stratification variables in geography could include sex, age, race or ethnicity, political party, socioeconomic status, species, soil type, season, ecosystem, altitude, and so on. Even literal "strata" in sedimentary rock could provide a basis for stratified sampling (and a decent pun, no doubt). Of course, stratification is only possible if you know the class of each potential case on the stratification variable. Assuming you do, you would then randomly select the number of cases from each stratification class that matches its proportionate size in the sampling frame.[1] If 7.8% of the households in the sampling frame include four or more children, for example, then cases would be selected to make sure the sample

---

[1]In some research studies, cases are selected equally from classes rather than proportionately, so that the number of cases from each class is equal. When this is done, the resulting sample will not represent the sampling frame very closely, but the statistical power of comparisons between stratification classes will be maximized. This sampling design makes sense if one is more interested in comparing stratification classes than in characterizing the population as a whole.

Box 8.1 Generating Random Sequences for Sampling and Other Uses

Researchers often need to create random sequences. They are required for simple random sampling and other forms of probability sampling. They are also needed to create random assignments of cases to conditions in experimental studies (Chapter 7). Stochastic numerical models (also Chapter 7), such as those used in transportation or climate simulations, usually include a simulation of random processes. How does one create these random sequences? If one needs only a few random digits, any simple physical system that generates random patterns can be used, including rolling dice, drawing shuffled cards, or flipping coins (although recent statistical research suggests the original upward face has a slightly greater chance of coming up after the toss!). To use these simple methods, assign a digit to each potential entity, making sure that the system you use generates enough different digits to cover all of your alternatives (for example, a single roll of the die produces only six possible outcomes, insufficient for ordering seven entities). If you can use the same entity more than once in a given sequence, called **selecting with replacement**, you can accept sequences in which the same digit comes up more than once; in fact, you must use such sequences that include repeating digits because you would be violating randomness otherwise. If you need sequences in which each entity is selected once and only once, called **selecting without replacement**, you should exclude a digit that repeats one you have already selected in that sequence and go to the next one.

When you require larger numbers of digits, you can use tables of random numbers (found in many statistics books). Here is a small example of such a table that we generated:

47359872271185676654332724051291678701662806568797288431 0914

29363528841075314372317841658558082282704060388836520889 8875

04302510310979592261854670839466912768993746203190959507 9313

02301823752371420223402401980777367395242374354632119643 6249

83666157955576100002100657826552475373840283402624564713 8151

When using random number tables, you can start anywhere and move in any direction as you go along; in fact, it is best to start in different places and move in different directions when using the table on different occasions. Assign digits to entities appropriately. If you have seven entities to randomize, you could use the digits 0–6, skipping any occurrences of the digits 7–9. If you have more than 10 entities to randomize, treat the digits as double digits. Given 30 entities to randomize, you could use the double digits 00–29, skipping any occurrences of the digits 30–99. Make sure that you make all choices with digit sequences of the same length. That is, don't choose some entities with a single digit and others with double digits—they are not equally probable.

Of course, in many situations, it is more convenient to use a computer program to generate the random sequences (technically they are "pseudo-random"); tables of random digits are created this way. Geographers who practice stochastic computational modeling necessarily generate random digits this way. Many programs can create random digits with various properties, including statistical, mathematical, and even spreadsheet programs. As in the less "technological" methods described above, random number programs can be programmed to sample with or without replacement, and with many other restrictions that might be appropriate in particular sampling situations. For example, researchers doing telephone surveys often use some form of "random-digit dialing" to obtain probability samples of respondents. In that situation, it is a good idea to generate only the telephone suffixes (the last four digits) randomly, restricting the area codes and prefixes to those actually in use in the region being sampled; a great many phone numbers generated entirely randomly will not be in use in a region.

contained, as nearly as possible, 7.8% households with four or more children. Compared to simple random sampling, stratified random sampling reduces sampling error (the variability of random samples from one another—see Chapter 9) because all possible samples will be the same in terms of the number of cases they contain from each stratification class. Importantly, stratification does not reduce sampling bias *on average* as compared to simple random sampling, but it does guarantee that any single sample is unbiased on the stratification variable. Stratified sampling is thus useful to ensure that small or important classes are adequately sampled. These benefits sound rather impressive, but it is important to recognize that they are only *potential* benefits. They are realized to the degree that the strata are internally homogeneous but different from the other strata. And the stratification variable must be something statistically related to the other variables that are measured as part of the research. If all ethnic groups have about the same average attitudes about oil drilling, for example, stratifying by ethnicity in a study of attitudes about energy sources will make no difference to your data analysis, although it will allow you to assure consumers of your research that you have represented all ethnic groups proportionately.

Geographers sometimes use **cluster sampling**. The geographer first chooses one or two geographic areas or features, such as cities, parks, or neighborhoods, typically out of convenience rather than randomly. Then he or she randomly selects clusters of cases, such as individual businesses or vernal pools, within each area. This saves costs, as compared to simple random sampling, but it can readily introduce bias in the sample because many characteristics are not evenly or randomly spaced out over areas (as any geographer knows!), so that cases within a single area can be rather different than those in another area. The severity of this bias is reduced if more than just one or two different areas are sampled.

Another probability sampling design often used in geographic research is **multistage area sampling**. The spatial extent of the sampling frame is divided into geographic areas or features, as in cluster sampling. A certain number of these areas, more than one or two, are randomly selected. Then each of these areas is divided into smaller areas or features; a certain number of these smaller areas are randomly selected within each of the larger areas. This can be repeated as much as needed to get down to the desired unit of analysis. For example, five U.S. states may first be selected, then five counties within each state (Louisiana and Alaska would be excluded because they don't have counties), then five census tracts within each county. Multistage area sampling is essentially spatially hierarchical stratified sampling. It has potential benefits similar to those of stratified sampling, but based on geographic areas rather than thematic stratification variables. It is an attractive sampling design from the perspective of a geographer, because it samples from different spatial areas with a spectrum of human and natural characteristics, variations across space that are of such central importance and interest to geographers. Unfortunately, the spatial areas that data come attached to may not match the spatial areas in which structures or processes of interest actually occur. The choice of the appropriate geographic units, in terms of scale and location, is an important and substantial question that we consider in some detail below and in Chapter 9.

Implications of Sampling Frames and Designs

It is important to remember that the cases that provide our data make up the sample, whereas the cases to which we ultimately want to generalize make up the population. But as Figure 8.1 suggests, the relationship of the sample to the population depends both on the relationship of the sample to the sampling frame and that of the sampling frame to the population. So it is important to consider both relationships when designing or interpreting research. For both relationships, the essential issues to consider are the same. The first issue concerns how representative the smaller set (sampling frame or sample) is of the larger set (population or sampling frame, respectively). **Representativeness** is the degree to which the smaller set resembles the larger set. A second issue concerns the question of which is the specific larger set (population or sampling frame) to which we should generalize from the smaller set (sampling frame or sample, respectively). This is the question of **generalizability**—what larger set can we validly draw conclusions about from the evidence of the smaller set? Our choice of a sampling frame determines what population our sampling frame is representative of, and to which population it generalizes. Likewise, our choice of a sampling design determines what sampling frame our sample is representative of, and to which sampling frame it generalizes.[2]

Clearly, the use of a nonprobability sampling design means the researcher cannot say with much certainty how well the sample represents the sampling frame, which means the link from the sample to the population is more uncertain.[3] Nonetheless, nonprobability sampling designs are actually quite common in

[2]Sampling frames and designs have additional implications for increasing or reducing variation due to sampling error in our data, as we discussed in the context of stratification. They can affect the precision of estimation (confidence interval width) and the power of hypothesis testing (chance of finding population effects that actually exist). We also discuss precision and power in Chapter 9.

[3]Even with perfectly conducted probability designs, researchers never know for sure how representative the sample is (except on strata or cluster variables). Such a sampling design only ensures representativeness *on average*, but researchers work with a particular sample, not the average sample. That's why probability designs make the link from sample to sampling frame (and hence to population) *more* certain rather than *perfectly* certain.

geography, and in other natural and social sciences. In some disciplines, and in some topical areas of geography, most sampling of cases is done in a nonprobability manner. Researchers who study rivers rarely sample rivers probabilistically, for example. When one remembers that many entities are sampled in research, not just cases, it becomes even clearer that various forms of nonprobability sampling are actually quite common in research. For example, little or none of even the most carefully conducted survey research asks respondents questions that are probabilistically sampled from some population of all possible ways of wording given questions.

Does this mean that a great deal of research, maybe most of it, is of little worth? In particular, is the common lack of a probabilistic sampling design a critical flaw, as many statistics and methods textbooks claim or imply? Some readers may find it surprising that we think the answer is *no*. In fact, a focus on sampling randomness and representativeness varies greatly across disciplines, subdisciplines, and research purposes. Sampling representativeness is more important for some research goals, such as the estimation of population parameters by pollsters, than others, such as testing hypotheses about causal relationships among constructs. Also, remember that the link from samples to populations always goes through sampling frames, which are typically not identical to either the sample or the population. Sampling designs tell us about the link from samples to sampling frames, but what about the equally important link from sampling frames to populations? In fact, sampling frames are nearly always based on feasibility rather than representativeness. If it were feasible, all researchers would use sampling frames that were as close as possible, if not identical, to the population to which they hope to generalize. Only the difficulty or impossibility of obtaining such a sampling frame causes researchers to use incomplete sampling frames that are often nonrepresentative of the population. For example, geographers who study migration would generally be thrilled to select from a sampling frame that included all people who are moving, who have ever moved, or who will ever move. But this is impossible, and not only because many past and future migrators are deceased or not yet born. Even many migrators currently living don't speak the language of the researcher, or they live on another continent, or they are homeless, or they are inaccessible for some other reason. Instead, such a geographer might accept the use of short-form data from two decades of the U.S. census (Chapter 6), even though that does not come close to reflecting the activities of all American migrators ever, let alone all human migrators ever. Thus, we see that even researchers who go to great lengths to administer probability sampling designs still use sampling frames that may not represent the population of interest all that well. When you realize that the population of interest in basic science so often includes past and future cases, you realize that rarely does a given sampling frame ever allow true probability sampling from that population.

We do not mean to imply that these truths about the way research is actually conducted are not problematic or have no consequences for how research should be interpreted. Either the relative rarity of true probability designs or the common use of only feasible frames, or both, is in fact a nearly ubiquitous shortcoming of research studies. The shortcomings can partially be overcome with more and better-designed research studies, but in most situations, only partially. However, as we have just considered and look at further in Chapter 11, these shortcomings are less important in some research areas than in others and are typically not fatal flaws in research. However, they should always be considered as possible weaknesses of a particular study, and such considerations constitute part of the "Discussion" section of a research paper or talk (Chapter 13).

Nonparticipation and Volunteer Biases

Above, we mentioned the need to solicit the participation of cases like individual people or companies because, of course, these cases have the ethical right to refuse participation in your research (more on research ethics in Chapter 14). There is a specific threat to the representativeness and generalizability of research involving cases like this that arises when some of the cases that are given the opportunity to participate choose not to do so. If nonparticipants are different from participants, the possibility of a **nonparticipation bias** exists. Clearly the sample may become considerably less representative of the sampling frame if the nonparticipation rate is fairly high, but only if nonparticipants differ from participants in ways relevant to your research goals. For example, research on environmental attitudes may be biased if people with "pro-business" views systematically decline to participate because they perceive the researcher and/or the research to have a "pro-environment" slant.

It is always a good idea to record the number and characteristics of nonparticipants, as much as possible. When you do this, you can compare the characteristics of nonparticipants to those of participants or the entire sampling frame. Try to find out why they refused to participate, perhaps during a more intensive follow-up. Of course, you can and should take steps ahead of time to maximize participation, not only because of potential bias but for plain old efficiency.[4] There is a great deal of available wisdom[5] on the varying effectiveness of different forms of contacting potential cases. Personal (live) appeals are nearly always most effective, but of course they are very time consuming. Research has shown that if a person publicly agrees to something, he or she is fairly likely to carry through with it. Thus, survey response rates are higher if potential respondents are first contacted and agree to participate. The use of mail or telephone reminders helps. E-mail requests and reminders to participate can be used nowadays as well, although the rate at which people ignore such requests may be higher even than it is for regular mail. Providing incentives can help too, although researchers usually don't have enough resources to offer really valuable incentives; this obviously depends on the type of potential cases in your sampling frame. Anything that can clarify and shorten a study, such as shortening a survey, can improve participation rates; so can making the study more interesting or relevant sounding to potential participants. If sensitive subject matter is the focus of the

[4]Nonparticipation can lead to the general problem of insufficient sample size, discussed in the next section. In most research situations, this is dealt with simply by making sure to solicit participation from more cases than are needed.

[5]Wisdom about the effectiveness of different types of appeals comes from a great deal of systematic research but also from the experiences of many politicians and sales people. After all, an appeal to participate in a study is essentially a sales pitch.

research, increasing the apparent anonymity of the procedure can help. Working with more carefully trained research assistants that have better "people skills" can obviously increase participation rates too.

On the flip side of nonparticipation bias is the possible sample bias that arises when cases get into the study by volunteering to participate—the so-called **volunteer bias.** This is also known as "self-selection bias" because cases get into studies by selecting themselves. Of course, many studies that involve nonprobability sampling designs are conducted on cases that have, for whatever reason, volunteered to participate. There is quite a bit of research on the characteristics of volunteers as compared to those who tend not to volunteer. For example, volunteers score higher on certain personality dimensions, such as empathy. Most of the methodological issues involving volunteer bias, as well as the solutions, are similar to those for nonparticipation bias.

Spatial Sampling From Continuous Fields

Geographers sample entities and events that are located in space and time, as do all other scientists. Unlike in some other scientific disciplines, however, the spatiality and temporality of cases is often of central importance in geography. Depending on your topical area within geography, you must carefully consider the spatial and/or temporal distribution of your cases when you create your sampling frame and design. In Chapter 5, we presented formal observation schedules for sampling events in time, specifically human behaviors in that chapter. We noted there that time and event sampling are applicable to temporal sampling in all areas of geography. Here we focus on sampling in space. We consider some of the distinctive implications that the spatial distributions of cases have for how we should sample them.

We already touched on spatial sampling in this chapter, in the section above on sampling designs. Both cluster sampling and multistage area sampling are based on obtaining discrete cases distributed in spatial clusters or regions. This is typically the situation in human geography, where cases include such discrete entities as individual people, families, neighborhoods, cities, businesses, schools, governmental institutions, provinces, countries, and so on. In physical geography, in contrast, sampling is very often done from entities such as oceans, rivers, the terrestrial surface, soil layers, and the atmosphere. The properties of these entities that interest geographers are often phenomena that are continuously distributed in space, including many of the physical properties we discussed in Chapter 4: elevation, temperature, precipitation, air pressure, salt concentration, CO_2 concentration, solar energy, and so on. These continuous distributions are known technically as "fields" in contrast to "objects" (we consider the conceptual distinction between fields and objects further in Chapter 12). When studying continuously distributed properties, geographers sample locations within the fields and measure the properties at those locations. Our focus in this section is on this spatial sampling of continuously distributed phenomena.

We noted in Chapter 2 that all measurement must necessarily have finite precision, so that actual data of any type always consist of discrete values. By the same token, we can apparently conceive of properties as being truly continuously distributed in two or three dimensions of space, but our sampling from the space necessarily has finite precision. These considerations point to the strategy geographers use to sample from continuous fields. Organize or break continuous space into discrete objects, perhaps very small and numerous objects, and then sample and measure from these objects. The plans for creating and choosing the discrete objects together constitute the sampling frame and design for sampling from a continuous field distribution. The objects may be points, lines, areas (polygons), or volumes. Of course, the first three objects are not true geometric points, lines, and areas, but rather three-dimensional entities that can be treated for many analytic purposes as if they were zero-, one-, or two-dimensional.

For example, we learned in Chapter 4 that rainfall is measured by rain gauges that can be treated as being at point locations sampled from the field of the two-dimensional land surface; obviously, the openings to rain gauges must have nonzero area or no raindrops at all could fall into them. How many gauges do we need to estimate the continuous field of precipitation well? Where should we place these rain gauges? Or consider taking samples from vegetation assemblages. A common way to do this is to create linear features called **transects**. The word "linear" suggests the approximate one-dimensionality of transects; that is, they are thin or thick "bands." They are not necessarily straight lines, although that often makes sense for sampling because straight lines will be uncorrelated with distributions that are not linearly distributed along the direction of the transect—the straight line can be a convenient way to achieve sampling randomness.[6] Once transects are created along the ground, they can be treated as individual cases by recording all instances of the plants of interest along the transect, or further sampling discretization can occur by recording measurements at sampled points along the transect.

The entire region from which your discrete features could be sampled is your sampling frame.[7] The particular locations you actually sample and measure depend on your sampling design. Just as with nonspatial sampling, discrete feature locations could be chosen with the use of either a nonprobability or probability design. For instance, often enough, locations are chosen by convenience, definitely a nonprobability design. However, if a probability design is used, there are a large variety of them that could be applied. Devising probability sampling in space and place is more elaborate (and more intriguing) than probability sampling in the thematic or temporal domains, because spatiality is two- or three-dimensional instead of

[6]The frequent desirability of straight transects is one reason that physical geographers such as biogeographers are fond of driving along wilderness or mountain service roads to do their sampling and measurements. Erecting power lines is costly in labor and materials, so utility companies go to great lengths to minimize their total length by building straight paths through forests and over rolling topography.

[7]Discretizing fields into self-contained sampling frames is difficult when processes flow into and out of the frame boundaries. The study of "island biogeography" is so important, for instance, because islands bound processes that would not be bounded on larger landmasses.

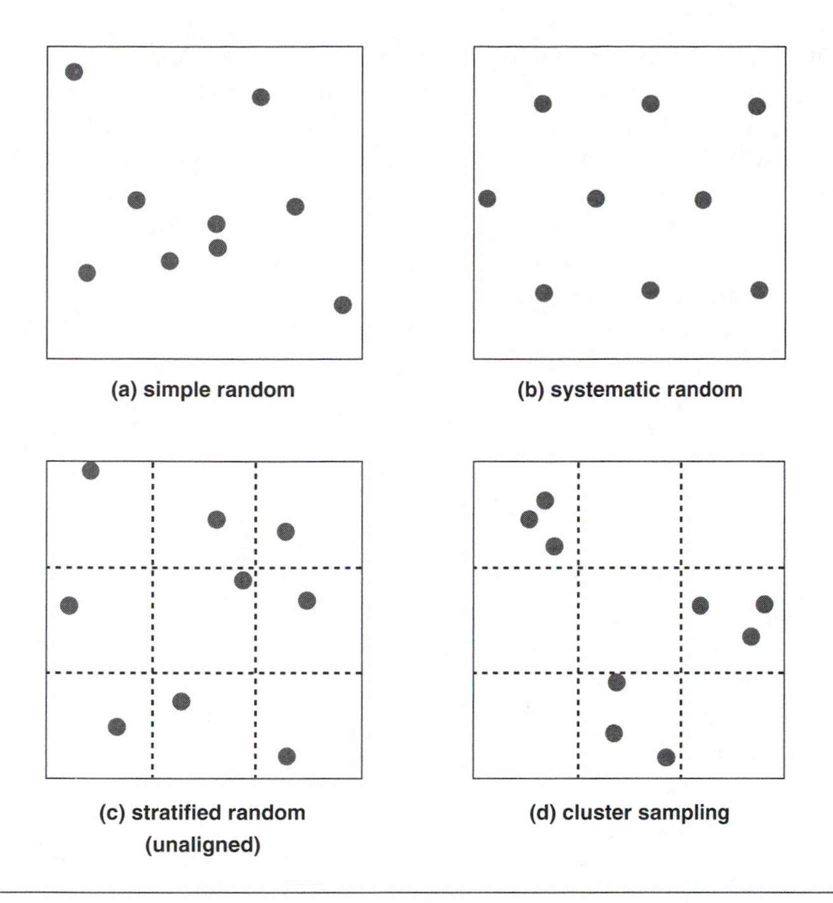

Figure 8.2 Spatial sampling of points from fields.

one-dimensional. Figure 8.2 shows four examples involving point features sampled within a bounded sampling frame. The points could be sampled according to a simple random design. Implement this by imposing a coordinate grid over the total sampling frame, randomly choosing an "X" and a "Y" value for each point you need. You could sample the points according to a systematic random design, where you choose one point at random and then sample the rest at a set distance away to be equally spaced. Sometimes geographic researchers break continuous space into discrete polygonal features shaped like squares—called **quadrats**. These quadrats are analogous to the nonspatial strata we discussed above. Point locations are sampled from within the quadrats. Figure 8.2 shows stratified random sampling, whereby one point is chosen within each quadrat. The location of the points within different quadrats is random, so this is "unaligned" sampling. Finally, cluster sampling can be carried out by sampling quadrats first, then sampling points within each chosen quadrat. This is the same as the cluster sampling discussed above that started with discrete features to begin with.

These approaches to probability sampling all ignore the actual distribution of features or properties of interest; they are examples of **independent spatial sampling**. Think about it for a moment. If you wanted to sample the topography of

South America, its profile of elevation features like mountains and canyons, would it be most efficient to sample and measure elevation from a set of equally spaced points? Would the Amazon Basin require as many points as the Andes Mountains to produce a representative sample of elevations? No, it certainly would not. Often, geographers do not sample from fields as if they were homogeneous. They concentrate their sampling where they believe their property of interest exhibits more variation in reality. That is, geographers sometimes sample on the basis of a model of patterns or **trends** in the spatial distribution of their property of interest; they practice **nonindependent spatial sampling.** In that case, sampling is focused on locations of greater change in the trend. The location and/or orientation of transects, for example, is often done nonrandomly (or at least not fully randomly). It is done in a reasoned way, based on knowledge of the feature distribution being sampled. Thus, transects are placed at right-angle orientations across streams or rivers, up mountainsides rather than along them, and so on.

After they complete sampling and measurement, geographers will typically want to make inferences back to the continuous field. This is the subject of statistical inference from samples to populations that we discuss in detail in Chapter 9. In the context of the *spatial* statistical inference we are concerned with here, this process is called **spatial interpolation.** Spatial interpolation is a simple concept to understand; however, the mathematical techniques used to accomplish it can be rather sophisticated. To appreciate the concept of spatial interpolation, suppose you must provide an estimate of the temperature in Denver, Colorado, if the temperature at Fort Collins (~60 miles north of Denver) is 100°F and the temperature at Colorado Springs (~60 miles south of Denver) is 90°F. If you guess 95°F, you have succeeded at simple spatial interpolation.

We touch on the topic of spatial interpolation again in Chapter 12. Interpolation is a form of educated guessing, so naturally we should be curious as to how accurately it can actually regenerate the continuous field from which samples are taken. The accuracy of spatially interpolating the surface—the spatial distribution of property values—of a field from point measurements depends on[8]

a. the accuracy of measurement

b. the density of sample points (the number of points per area)

c. the spatial distribution of points (largely due to the sampling design)

d. the particular interpolation procedure used (see Chapter 12)

e. the actual spatial distribution of the surface being measured and interpolated.

In order to sample the full detail of a field, it is necessary to sample enough points (to sample at a sufficient density) so that the distance between points is less than the size of any relevant features on the surface you want to capture. According

[8]MacEachren, A. M., & Dsavidson, J. V. (1987). Sampling and isometric mapping of continuous geographic surfaces. *The American Cartographer, 14,* 299–320.

to the "sampling theorem," this distance is one half or less of the length of the smallest feature, whether a "hill" or a "valley" in the values of your property of interest.[9]

Geographers sampling from continuous fields must, therefore, answer questions about how many points should be sampled, the locations from where they should be sampled, and how they should be interpolated. Perhaps more important, however, is the question of whether interpolation should be carried out in the first place. Is the field model really an appropriate conceptualization of your phenomenon? This is the fundamental question of the "ontology" of geographic reality, to which we return in Chapter 12.

Sample Size

Besides issues that concern how samples are obtained, creators and consumers of research must always address the question of *how large* samples of cases should be. The answer to this question is a compromise between two competing motivations—the benefits versus the costs of larger samples. On one hand, larger samples are more likely to be representative of the sampling frame. Larger samples allow a researcher to test more variables and more interaction effects among variables (see Chapter 7), and they allow the researcher to conduct analyses within more subcategories of types of cases. More cases increase the precision of estimation and the power of hypothesis testing (as do particular sampling frames and designs, see footnote 2). Clearly, larger samples bring many benefits to the research enterprise. Given unlimited resources, researchers would almost always want larger samples. But researchers don't have unlimited resources; that's one of the main reasons they sample in the first place. So against the motivation to obtain larger samples is the cost of obtaining them. Larger samples generally cost more money, more time, and more effort. At some point, the additional precision and power gained by increasing sample size is minimal (the marginal benefit becomes tiny), but the increase in cost is not likely to be minimal. Furthermore, it's impossible in some situations to obtain larger samples; for example, a geographer studying community responses to tsunamis will probably have difficulty finding many such communities.

Taking the competing motivations into account, we can give some approximate guidelines on desirable minimum sample size. It is important to recognize that desirable sample size varies greatly with one's research goal, and that different scientific disciplines, as well as subdisciplines of geography, have quite different traditions with respect to sample size. As we discussed in Chapter 7 on research design, quality research is sometimes done with a single case. Early exploratory phases of research often focus on just one or a few cases. In fact, we strongly recommend the wisdom of trying out one's study procedures (instructions, measurement tools, question wordings, stimulus materials, analysis procedures, and so on) in a trial pilot study employing a small sample of cases similar to those you will use in the regular study,

[9]See, for example, Tobler, W. (2000). The development of analytical cartography. *Cartography and Geographic Information Science, 27*, 189–194.

no matter how large a sample you eventually intend to obtain. With some research questions, only one case is logically required to achieve a respectable level of confidence in one's conclusions. For example, if you wanted to show that it is possible for an artificial reef to sustain populations of a particular fish species, you would need only one such reef to establish this. In contrast, large samples of one to two thousand or more are required to estimate the parameters of some population with great precision. This number must be especially large if the makeup of the population is very heterogeneous with respect to your variable of interest.[10] Political pollsters, for instance, must obtain large samples of a few thousand in order to predict national election outcomes accurately (although as a percentage of all voters, this is surprisingly small). Between these two sample-size extremes, researchers who are testing hypotheses about causal relationships—theories—traditionally try to get at least 20–30 cases.[11] This recommendation derives from a tradition of making sure to have enough power, which increases with increasing sample size, to achieve statistically significant results given a weak to moderate relationship in the data (a correlation of approximately .3). If more than two variables are to be analyzed, or more complex comparisons are to be made (for example, nonlinear as well as linear relationships are to be analyzed), this number should be increased.

Formal **power analysis** techniques have been developed, and they can be modified a little to perform **precision analysis** for statistical estimation. A primary purpose of these techniques is to determine how large a sample is required to obtain statistically significant results or a confidence interval of desired width, given assumptions about the size of relationships in the population and the amount of noise in the data. Hopefully, these assumptions are based on careful reasoning, perhaps on prior research. The size of the relationship expressed as a proportion of noise in the data is called an **effect size**. In some situations, one can decide on an effect size that would be meaningful in a particular research context and aim for that effect size. These techniques are also used to evaluate nonsignificant results; as we discuss in Chapter 9, statistical nonsignificance does not prove that there is no relationship in the population. It may be that the effect is too small to be picked up robustly with a given number of cases. These considerations are part of the quantitative technique of reviewing research called "meta-analysis," which we discuss in footnote 1 of Chapter 13.

---

[10]Contrary to most people's initial intuition (including our own), the sample size required to estimate a population parameter precisely does not depend on the size of the population.

[11]This number is based on the simple situation of calculating relationships by correlating two variables measured on each case. Alternatively, in experimental studies, relationships are often measured by comparing the means of one variable across groups or conditions of another variable (these concepts were covered in Chapter 7). In the simplest such experiment, the means of two groups of cases are compared (for example, the readability of two map designs is compared). In this experimental design, 20–30 cases per group are considered a desirable minimum. If the same set of cases can be placed in both of the groups at different times (that is, a within-case design—see Chapter 7), this number can revert back to a total of 20–30 cases or even less.

Review Questions

- What are populations and samples?
- Why do geographic researchers sample? What are some things that geographers sample, and why do they sample them?

Sampling Frames and Sampling Designs

- What are sampling frames and sampling designs, and how do they relate to populations and samples?
- What is the distinction between probability and nonprobability sampling designs, and what are the implications of this distinction for research?
- What are the following types of sampling: simple random, systematic random, stratified random, cluster, multistage area?

Implications of Sampling Frames and Designs

- What is sample representativeness, and how does it influence generalizability?
- How common are nonprobability-sampling designs in geographic research, and what are the implications of using such sampling designs?
- What are nonparticipation and volunteer biases, and what are some approaches to minimizing their negative effects on research?

Spatial Sampling from Continuous Fields

- What are some implications of sampling from sets of discrete objects versus sampling from continuous fields?
- What is the general approach usually taken to sampling from continuous fields, and what are some difficulties that arise from this approach?
- What is spatial interpolation, and what are some factors that influence its accuracy?

Sample Size

- What are the competing motivations for larger and smaller samples?
- What are the techniques of power analysis and precision analysis, and how can they be used to help decide on sample size?

Key Terms

cluster sampling: specific type of probability sampling design in which cases in the sampling frame are grouped into spatial areas or clusters, and then cases are selected randomly from each cluster to be in the sample, usually proportionately in number to the cluster size

convenience sampling: specific type of nonprobability sampling design in which researchers take every case from the sampling frame they can conveniently get hold of to be in the sample, until their sample is large enough

effect size: the size of a statistical relationship expressed as a proportion of noise in the data

generalizability: the validity with which you can draw conclusions about sampling frames from samples, or about populations from sampling frames; depends on representativeness

independent spatial sampling: spatial sampling design that ignores the actual distribution of features or properties being sampled

multistage area sampling: specific type of probability sampling design in which cases in the sampling frame are grouped into spatial areas or clusters (like cluster sampling), smaller areas are defined and randomly selected within the larger areas, possibly more than once at increasingly smaller scales of area, and cases to be in the sample are finally selected randomly from each of the smallest areas

nonindependent spatial sampling: spatial sampling design that is based on the actual distribution of features or properties being sampled

nonparticipation bias: the degree to which a sample is not representative of a sampling frame because potential cases who refuse to participate in the study are different than those who do

nonprobability sampling: any sampling design in which the probability of a particular case being selected from the sampling frame is unknown ahead of time

population: the entire set of entities of interest, including cases, measures, settings, and so on; sometimes called a "target" population

power analysis: formal technique to estimate the statistical power in a particular hypothesis test; primarily used to estimate the sample size required to achieve statistical significance, given a population effect and error variance of particular sizes

precision analysis: formal technique, similar to power analysis, to estimate the width of the confidence interval in a particular statistical estimation; primarily used to estimate the sample size required to achieve an interval of a given width

probability sampling: any sampling design in which the probability of a particular case being selected from the sampling frame is known ahead of time

quadrats: rectangular polygonal features created to sample from continuous field distributions

representativeness: the degree to which the make-up of the sample resembles that of the sampling frame, or the make-up of the sampling frame resembles that of the population; determines generalizability

sample: incomplete subset of the entire set of entities of interest, including cases, measures, settings, and so on; necessarily smaller than the population from which it is drawn

sampling: any way of selecting a sample of entities from a population

sampling design: the procedure used to identify cases from the sampling frame to go into the sample

sampling frame: the subset of the population from which cases are actually drawn to become part of the sample; usually smaller than the target population

selecting with replacement: choosing a subset of outcomes or entities from a larger set in such a way that the same outcome or entity can be chosen more than once; the entity or outcome is essentially "replaced" after it is chosen so it can be chosen again

selecting without replacement: choosing a subset of outcomes or entities from a larger set in such a way that the same outcome or entity cannot be chosen more than once; the entity or outcome is not "replaced" after it is chosen

simple random sampling: specific type of probability sampling design in which each case in the sampling frame, and each subset of cases, has an equal chance of getting selected into the sample

snowball sampling: specific type of nonprobability sampling design in which researchers use cases they have already put into the sample to find out about further cases that could be selected

spatial interpolation: mathematical technique in which unknown data values at locations within a field are inferred from values that are actually measured; essentially spatial statistical inference from a sample of measurements to a population field

stratified random sampling: specific type of probability sampling design in which cases in the sampling frame are grouped into thematic classes or "strata," and then cases are selected randomly from each class to be in the sample, usually proportionately in number to the class size

systematic random sampling: specific type of probability sampling design in which a first member is randomly selected from the sampling frame to be in the sample, and then every "nth" member after that is selected; as compared to simple random sampling, each case in the sampling frame has an equal chance of getting selected, but each subset of cases does not

transects: linear features created to sample from continuous field distributions

trend: pattern in the spatial distribution of properties of interest that strongly influences the adequacy of different spatial sampling frames and designs

volunteer bias: the degree to which a sample is not representative of a sampling frame because potential cases who volunteer to participate in the study are different than those who do not; also known as self-selection bias

Bibliography

Fowler, F. J. (2002). *Survey research methods* (3rd ed.). Thousand Oaks, CA: Sage.

Haining, R. (1990). *Spatial data analysis in the social and environmental sciences.* Cambridge: Cambridge University Press.

Murphy, K. R., & Myors, B. (2004). *Statistical power analysis: A simple and general model for traditional and modern hypothesis tests* (2nd ed.). Mahwah, NJ: Lawrence Erlbaum Associates.

Webster, R., & Oliver, M. A. (1990). *Statistical methods in soil and land resource survey.* Oxford: Oxford University Press.

Weisberg, H. F., Krosnick, J. A., & Bowen, B. D. (1996). *An introduction to survey research, polling, and data analysis* (3rd ed.). Thousand Oaks, CA: Sage.

CHAPTER 9

Statistical Data Analysis

Learning Objectives:

- How do the graphical and mathematical techniques of data analysis help to achieve the four scientific goals of description, prediction, explanation, and control?
- What are descriptive statistics, and what are the main properties of data sets we might want to describe?
- What is a statistical relationship, and what are some of the different types of relationships?
- What are inferential statistics, and what is the basic logic of their two main approaches, estimation and hypothesis testing?
- What are some of the properties of geospatial data that make their analysis particularly rewarding and particularly difficult?

Data analysis is the set of display and mathematical techniques, and attendant logical and conceptual considerations, that allow us to extract meaning from systematically collected measurements of our phenomena of interest and communicate it to others. Data analysis thus helps us achieve the four scientific goals described in Chapter 1: description, prediction, explanation, and control. Data analysis helps to efficiently identify and describe patterns in large amounts of data, patterns that improve our predictions about the unknown, explain the causes of relationships, and allow us to exert influence over phenomena we wish to control.

Although we treat them in separate chapters, in practice the display and mathematical techniques of data analysis are not entirely separate but make up a single set of procedures for extracting and communicating meaning. In this chapter we discuss the techniques of **statistical data analysis**; we cover the display techniques, such as mapping and graphing, in Chapter 10. We must emphasize at the outset, however, that the present chapter is *not* intended to be a detailed tutorial on how to statistically analyze data. Any number of textbooks provide detailed treatments of data analysis in geography and other disciplines (we list a few in the Bibliography), and in any case, all geography students need to take courses specifically focused on data analysis, both the statistics-based and display-based versions. Our purpose in this chapter is to provide an introduction and overview that describe the conceptual forest of data analysis rather than its technical trees.

Mathematical data analysis in geography is usually treated as being statistical (probabilistic or stochastic) in nature. As we pointed out in Chapter 2, geographers in most subfields assume their data sets do not reflect constructs and their interrelationships in a simple deterministic manner but in a complex and imperfect way that is influenced by factors difficult to control completely, in all the senses of control discussed in Chapter 7. There are three primary reasons most geographers treat data in a statistical rather than deterministic fashion. First is that data are usually considered to be an incomplete sample of a larger population of data of interest; different samples from the same population thus vary from each other (the sampling error we mentioned in Chapter 8). Second is that measurement in all scientific fields is imperfect, necessarily involving error; thus, the values we have in our data set vary at least a little from the true values they would have if they were perfect reflections of our constructs of interest. The third reason geographers treat analysis statistically is that their phenomena of interest are typically expressions of complex sets of many interacting variables, all of which are not likely to be identified or measured in any single piece of research. A statistical approach allows geographers to interpret the meaning of one or a few variables in this background of complex multivariate reality; it provides an approach to "statistical" control, as we discussed in Chapter 7. In fact, 20th-century developments in our understanding of "chaotic systems" suggest that some unpredictability is inevitable in complex systems and will never be avoided, even in theory.

Statistical Description

As we discussed in Chapter 8, if we were feasibly able to measure the entire target population of cases in which we were interested, we would do so. We would then interpret the resulting data set, looking for patterns of high and low values in the variables we measured, typical and atypical values, values measured on the same cases that seem to vary together in some systematic way, and so on. That is, we would want to describe patterns in our data. How would we do that? We could simply print out an unorganized list of raw data measurements or view them on a computer screen. Surely that would be a difficult way to see patterns, however. We could organize the raw data into a matrix with columns that represent variables and rows that represent cases. But it would still be difficult to see patterns in this matrix, especially if the data set was larger than a few rows and columns.

Instead, we would go further by displaying the data in graphs, tables, maps, and other displays. And we would try to summarize various potentially relevant properties of our data by calculating summary indices designed to reflect these properties in an efficient manner. This is **descriptive statistics** (statistical description). We would describe patterns in our set of population measurements by displaying the data in various ways and by calculating summary indices of the population data called

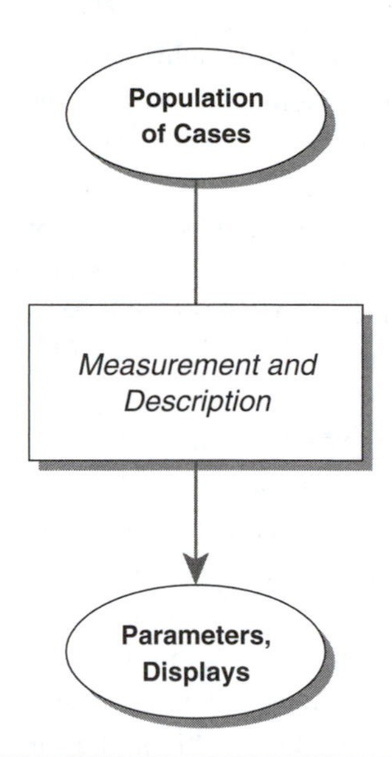

Figure 9.1 Data collection and statistical analysis with populations of cases. If possible, researchers would always take this direct approach to learn about parameters.

parameters (see Figure 9.1). We would interpret the patterns in these displays and calculated parameters by combining what we see in our data with what we know about our cases, our constructs, our measures, the places and times we are studying, and so on. Then we would tell other people what we found and what we interpreted it to mean. Then we would be finished.

What are the properties we might find relevant or interesting to summarize in our population data? The first one that comes to most people's mind is a measure of the average or most representative value in the data, a descriptive property known as the **central tendency**. There are many different ways to identify or calculate a central tendency. They reflect the "central value" in somewhat different ways conceptually, and they are more or less appropriate depending on the measurement level that characterizes your data (see Chapter 2). The three most common are the mode, median, and mean. The **mode** is the most frequently occurring value in your data set. It is a good measure of central tendency for nominal variables, such as soil type or religious affiliation. The **median** is the middle value in your data set; half the scores in the set are higher than the median and half are lower. It is a good measure of central tendency for ordinal variables, such as rivers ranked by their discharge volume or cities ranked by their position in an urban hierarchy. The **mean** is the "average" value in your data set. The standard type is the **arithmetic mean** (there are several others useful in geography), calculated by adding up all the scores and dividing the sum by the number of scores. It is a good measure of central tendency for interval and ratio (metric) variables, such as counts of tree rings or the lengths of straight segments in street networks in different regions. However, metric variables are sometimes strongly skewed—unevenly distributed towards high or low values—because of extreme values in one direction, high or low (we discuss skew, a property of the form of a distribution, below). When this is so, the median is a better descriptive measure of central tendency.[1] Monetary variables, such as real estate prices in a region, provide a good example. The mean of four houses valued at $82,000, $56,000, $133,000, and $77,000 is $87,000. If a single million-dollar property were added to this set, the five houses would have a mean of over $269,000, which is not very representative of the set of houses. The median of $82,000 is better.

After the central tendency, we would probably want to know how scores in our data differ from the central tendency, a property known as **variability** or dispersion. After all, none of the measures of central tendency tell us anything about how close the individual scores are to the central value or to each other. Variability is interesting not only in its own right, but because the variability of a variable helps in interpreting its central tendency; for example, a measure of central tendency is more representative when the variability is low. Three common measures of variability are the range, variance, and standard deviation. The **range** is a simple index reflecting the distance between the highest and lowest values in the data set. It is a good measure of variability for ordinal or metric variables, but it is limited insofar as it is based on only two scores in the distribution. The **variance** is based on an average of **deviations from the mean**, which are calculated for each individual score by subtracting the arithmetic mean from it. Thus, high scores will have large positive deviations, low scores will have large negative deviations, and scores near the mean will have small deviations. To calculate the variance, the deviations are squared, summed, and divided by the number of scores; that is, the variance is the "mean squared deviation from the mean." Because the deviations are squared (if they were not, the deviations from the mean would always sum to zero), the variance is "blown up" relative to the original distribution of scores, so its square root is often used instead for descriptive purposes.[2] That is the **standard deviation**.

A third property we might want to describe about our data is their overall **form**, essentially the shape of the distribution. Form is easier to understand if you think of data in terms of a graph rather than a list or table. There are a variety of form properties we might want to describe about our data. **Modality** refers not to the value of the most commonly occurring score, which is central tendency, but to the number of "local" modes a distribution has; a local mode is not necessarily the most common in the entire distribution, but it is more common than values just below and above it. As we mentioned above, **skewness** is a property of form that describes "unevenly" distributed scores. A distribution with "positive" skew has mostly low and medium scores with a few extremely high scores that are not balanced by an equal number of extremely low scores; "negative" skew is the opposite pattern

---

[1]We emphasize "descriptive" here. Especially when calculating various inferential indices, the mean is often favored over the median, even for skewed variables, because of various statistical reasons beyond the scope of this book. Again, check out an advanced statistics course.

[2]Again, we emphasize descriptive. As in footnote 1, the variance is typically preferred over the standard deviation for inferential indices, for statistical reasons beyond the scope of this book.

(see Figure 9.2). Skewness contrasts with **symmetry**, a property of nonskewed distributions whose two sides around the central tendency are mirror-image reflections of each other. One common symmetric distribution form is the unimodal **bell-shaped distribution**. A particular bell-shaped distribution that has a specific proportion of scores within any given range from the central tendency is the important **normal distribution.**[3]

Central tendency, variability, and form are properties of entire data sets. Another approach to describing data is the calculation of **derived scores** that describe properties of individual scores by expressing their value relative to the rest of the data set. A simple one is to express scores in terms of their rank within the data set; for example, the highest score is "1," the next highest is "2," and so on. A more sophisticated version expresses scores in terms of their **percentile** rank. This is the percentage of the data set that is less than the score in question. The highest score is at the 99th percentile (a score cannot be greater than 100% of the scores). The median is thus at the 50th percentile. Sometimes it makes sense to convey raw scores in terms of ratios or proportions, which can be expressed in terms of a decimal fraction from 0 to 1 or a percentage from 0% to 100%. Finally, derived scores are sometimes calculated from metric-level data so that the values of individual scores are expressed by dividing their deviation from the mean by the standard deviation of their data set. This ***z* score** expresses the score in terms of standard deviation units above or below the mean of the data set.

A final property of data we often want to describe concerns pairs of variables or larger sets, rather than single variables. It is the property of **relationship**, which is when cases show systematic (consistent) patterns of high or low values across pairs of variables, each variable measured on a common set of cases. The simplest form of relationship is a **linear relationship**, which comes in one of two types (see Figure 9.3). A **positive** (or direct) **relationship** between two variables means that cases with high values on one variable—high relative to that variable's central tendency, that is—tend to have high values on the other; cases with low values on one variable tend to have low values on the other. A **negative** (or indirect) **relationship** between two variables means that cases with high values on one variable tend to have low values on the other, and vice versa. **Relationship strength** is the strength of these patterns, whichever direction they take. Weak relationships tend only weakly to show the systematic pattern, whereas strong relationships tend strongly to show it. *No relationship* describes when there is no systematic tendency for high or low values on one variable to go with high or low values on the other variable: a case with a high value on one variable is just as likely to have a high value on the other variable as it is to have a medium or low value.

[3]The normal distribution is so important because many variables naturally approximate such a distribution; the heights of adult female German Americans and of Jack Pine trees are two examples. But even more important, it has been proven that both random measurement errors and the sampling error for statistics as estimates of parameters (see the section on Statistical Inference in this chapter) are normally distributed. In many situations, however, other theoretical distributions besides the normal are useful to geographers.

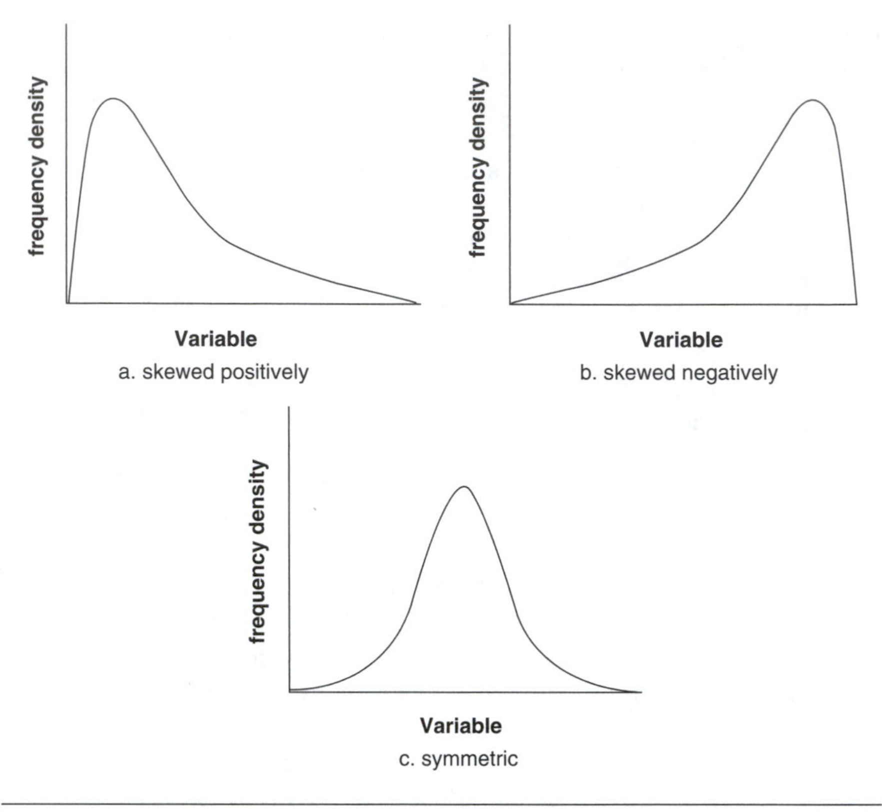

Figure 9.2 Distributions that are (a) skewed positively, (b) skewed negatively, and (c) symmetric. This symmetric distribution is a normal distribution, which is bell-shaped.

Relationship is usually quantified by some type of **correlation coefficient** (strictly speaking, the square of the correlation is usually the index of relationship strength). When dealing with linear relationships, the correlation coefficient is calculated by a formula designed to produce a value of 1.0 when there is a perfect positive relation, –1.0 when there is a perfect negative relation, 0.0 when there is no relationship at all, and some intermediate value when there is a relationship of intermediate strength, which is pretty much always true with actual data. But there are some other statistical indices of relationship. In areas of geographic research where true experiments are conducted (see Chapter 7), relationships are often statistically revealed by comparing the central tendency (usually the mean) of one variable across discrete levels or experimental conditions of another variable. For example, there is a relationship between "Cloud Seeding" and "Rainfall" if the mean rainfall produced by a cloud seeding procedure is higher or lower than that produced when no seeding is carried out (in fact, to the best of our knowledge, the evidence for cloud seeding is rather equivocal).

When researchers are especially interested in the *form* of the relationship in addition to its strength, which is usually the situation, they often apply a statistical technique called **regression analysis** (we discuss the reason it is called "regression"

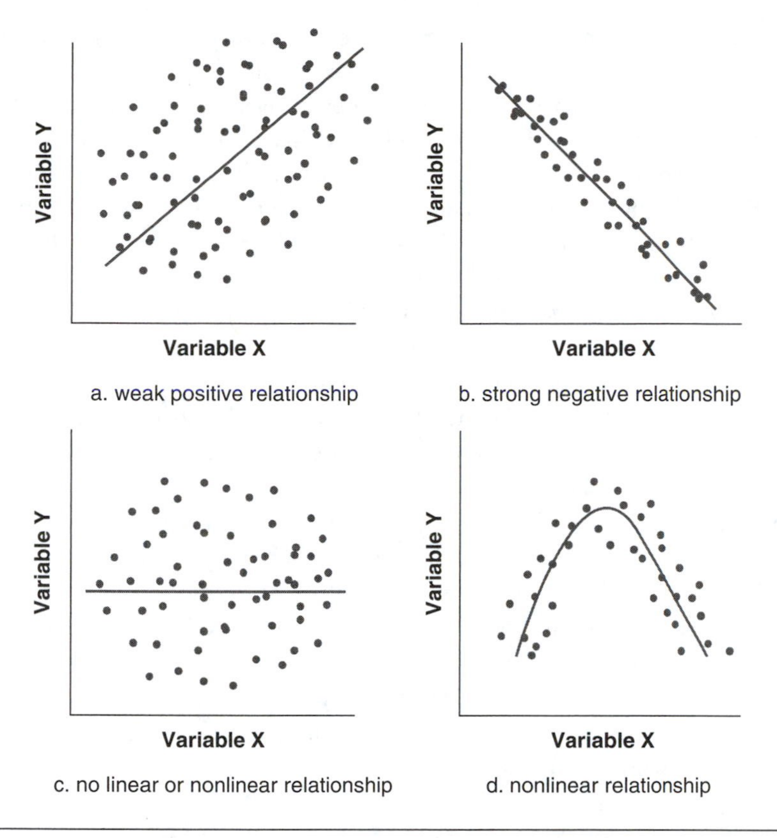

Figure 9.3 Scatterplots and regression lines depicting strong and weak relationships, either positive or negative, between two variables. The absence of a linear relationship may or may not indicate there is no relationship of any form.

in Chapter 11). In regression analysis, a statistical model of relationship form is investigated. This model is a simplified representation of the form of the relationship between two variables, or among larger sets of variables.[4] It expresses relationship as an equation that predicts the values of one or more variables (the **criterion variables**, usually labeled *Y*) as a function of the values of one or more other variables (the **predictor variables**, usually labeled *X*). When possible values for the predictor variable are "plugged in to" the regression equation, the resulting predicted values for the criterion variable can be graphed, resulting in a picture of the modeled relationship form. The simplest case is the linear relationship we spoke of above, such as graphs *a* and *b* in Figure 9.3. These relationships are expressed by the equation for a straight line:

$$\hat{Y}_i = a + bX_i$$

[4]Relationships between *pairs* of variables and the statistical techniques for their analysis are called **univariate.** Relationships and techniques for larger sets of variables are called **multivariate.** There are many multivariate techniques used in geographic research, including principal components analysis, multidimensional scaling, and cluster analysis.

This is a simple linear equation resulting from ordinary least squares (OLS) regression. Values of the Y variable are predicted to fall along a straight line created when values of the X variable are multiplied by a weighting constant b and added to another constant a. The i subscripts indicate that the individual scores for each variable are to be entered and output one at a time. As is familiar from high-school math, the weighting constant is the slope of the line, and the additive constant is the place where the line intercepts the Y axis. The "hat" symbol "^" is placed over Y to indicate that the outputs of the equation are predicted values for the criterion, not its actual values determined empirically. In Figure 9.3, the predicted Y values fall along the straight or curved lines, whereas the actual Y values are the data points that fall above or below the lines. The vertical distance of the points from the predicted line is the **error of prediction** of the model, which is assumed to be random from case to case. If the errors of prediction are not random, the model has been inappropriately formulated.

In some areas of research, geographers are not particularly interested in the specific values for the constants of the linear equation, perhaps because the quality or quantity of sampling or measurements is insufficient to justify faith in the specific values. Instead, they may just be interested in whether the relationship does or does not tend to go in only one direction, either up or down, even if it is not exactly a straight line. Such a relationship is called **monotonic.** In contrast, in many other situations, geographers are interested not just in linear relationships but in **nonlinear relationships.** Regression analysis and other statistical techniques for studying relationships can investigate a potentially unlimited variety of such nonlinear relationships that do not follow straight lines (a "quadratic relationship" is shown by graph *d* in Figure 9.3). Alternatively, nonlinear relationships can sometimes be "linearized" by subjecting the variables to a **transformation** that applies some mathematical operation to each of the raw scores; logarithmic, trigonometric, and square-root operations are examples. But these important topics go beyond our scope here; get to those statistics courses!

Statistical Inference

Now that we have calculated parameters to describe our population of data, displayed it with the techniques of Chapter 10, looked at it over and over in different ways, and finally communicated it to others using the techniques of Chapter 13, we are finished—except for one problem. We don't *have* a population of data. We're scientists who want more general truths than our specific data set by itself reveals directly, so we have only a sample. This leads us to the thorny (you may apply your favorite invective here) topic of **inferential statistics** (statistical inference): drawing informed guesses about likely patterns of data in a population on the basis of evidence from samples drawn from that population. It's thorny because it's conceptually quite difficult, as compared to descriptive statistics; we know this not only intuitively but from years of teaching statistics. Even worse, it's thorny because it makes our interpretation of data and the decisions we make about our research ideas far more uncertain than they would otherwise be. Ultimately, in fact, this need to sample makes interpretation and decision-making in science fundamentally and irrevocably uncertain. Like we said—a problem.

So we are not finished just yet. Given the data set we obtained by sampling from our population, we will again want to describe patterns, by calculating summary indices of the sample data called **statistics**[5] (see Figure 9.4). These are conceptually identical to the population parameters we covered above, but in many cases, the statistics are calculated with slightly different formulas than their corresponding parameters[6] (the sample indices of central tendency are important exceptions; they are calculated in exactly the same way as the parameters, although different symbols are used for them). The typical difference in the formulas for statistics arises from the fact that whereas parameters are considered to reflect properties of populations that are fixed, statistics reflect properties of samples that fluctuate from sample to sample. We learned in Chapter 8 that this fluctuation is called **sampling error**. Perhaps "sampling variability" would be a better term insofar as sampling error is not a *mistake* that one can even potentially avoid if one is sampling. The necessity of sampling error and the resulting fact that we can never know for sure how representative our sample is of our population is the root of the problem we lamented above.

Given our sample statistics, we next use them to infer our parameters. Here we should probably balance our negativity about inferential statistics with a little appreciation for what we can actually do with them. This is where the real beauty and power of the statistical theory developed over the last couple centuries really shines. You see, we don't simply make guesses about parameters from sample statistics. We use statistical theory *to assign probabilities to our guesses,* probabilities that we are right or wrong, or at least probabilities that we are close to being right. That is the heart of inferential statistics, and it is a very big deal. It allows to us to optimize our reasoning under necessarily uncertain circumstances. We appreciate that some of you may not readily grasp the incredible implications of this. Think of

---

[5]This is a narrow technical meaning of the term "statistics." The term is frequently used broadly as shorthand for the entire set of logical and mathematical techniques of statistical data analysis that is the topic of this chapter. Some people (not us) use "statistics" colloquially to refer to the raw data set, as in "the weather forecaster failed to predict the weather accurately even though she had all the temperature and pressure statistics."

[6]The way to calculate the sample standard deviation, for example, is based on the population formula, but with one change. Instead of dividing the sum of squared deviations from the mean by the sample size, as is done when calculating the population standard deviation, you divide by the sample size *minus 1*. This is called the **degrees of freedom**. If it were not done this way, the sample standard deviation would be biased low as an estimate of the population standard deviation; on average, its expected value would be a little less than the actual population standard deviation.

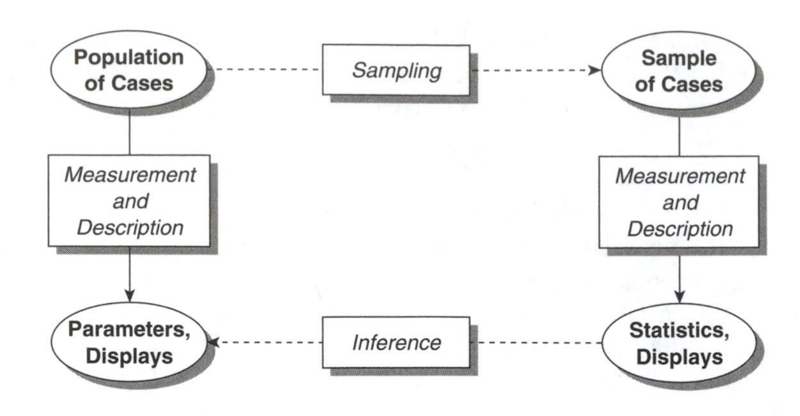

Figure 9.4 Data collection and statistical analysis with samples of cases. Almost always, researchers, especially those doing basic science, must take the indirect approach of sampling and performing inferential analyses to learn about parameters.

it like a gamble. Suppose your friend Martha tells you about a company that is publicly announcing a big business decision in two days. She strongly urges you to invest $10,000 in the company right away, because their stock is "very likely to go way up in value soon." Would you find it useful to know that her "very likely" actually means a 42% chance, and that in addition, there is a 27% chance that the stock will *drop* in value? We thought you would.

So both the power and the difficulty of inferential statistics comes from deriving probabilities about how likely it is that sample patterns, which vary from sample to sample, reflect population patterns, which are constant. The probabilities come from a distribution of the different sample patterns that describe all the possible samples of a given size that could be sampled from a population (strictly speaking, from a sampling frame—see Chapter 8). This is called the **sampling distribution** for a given statistic. For example, if I took a sample of 20 cases and calculated its mean, I would have one guess about the mean of the entire population of cases. If I took another sample of 20 that had at least one different member, I would have another sample mean that would be at least a little different than the first. I could keep doing this, recording each new sample mean as I calculated it, until I had managed to look at every different sample of size 20 that could be drawn from the population. I would then have a sampling distribution of sample means. This distribution would show the entire range of values that the sample mean could take, given the population distribution and the sample size in question. It would also show that certain values of the sample mean would occur more often than other values; for example, a sample with a mixture of high, medium, and low scores is more likely than a sample with all high scores or all low scores. Thus, the sampling distribution would show the probability that a single sample of size 20 would have a mean within some given range of values. We use these probabilities to determine the probabilities for the inferences we make from samples to populations.

If we actually constructed our sampling distributions "manually," as in our example, we would know for sure exactly what the probabilities would be for certain values of our sample statistics. But of course, that's not only impossible to do in most cases but stupid.[7] Impossible because there are far too many different samples that would have to be taken in any population of a decent size. Given a population of only 100 cases, for example, one could draw over 500 *quintillion* (5 followed by 20 zeroes) different samples of size 20. And this is just a tiny population of a hundred. That leads us to the reason we say that manually deriving the sampling distribution would be stupid. In order to generate and measure all possible samples, you would need to be able to access and measure all members of the population. If you can reasonably do that, however, by all means do it and skip this inference stuff. With a population of a hundred, you should certainly do that. But as we discussed in Chapter 8, one almost never has such small populations, and often the populations in basic science are extremely large and indefinite in size.

Luckily, we don't have to use a manual approach to generating sampling distributions. Instead, we use statistical theory. For example, the "central limit theorem" says the sampling distribution of the sample mean will be a normal distribution with a mean (**expected value**) equal to the actual population mean and a standard deviation (**standard error**) equal to the population standard deviation divided by the square root of the sample size. There are a variety of other theoretical ideas about sampling distributions for other statistics.[8] However, the theories that allow generation of sampling distributions involve assumptions for their valid application. Such assumptions include:

1. ***Distributional assumptions.*** These are assumptions about properties of the population from which the sampling distribution is generated. They include normality and "homogeneity of variance"—the assumption that the variances of the populations from which separate experimental samples are taken are equal. Actually, these assumptions apply only to so-called **parametric statistical tests.** There are tests called **nonparametric** that do not require these assumptions because they assume data are only nominal or ordinal, not metric. We return to this distinction in Chapter 11.

2. ***Independence of scores.*** This is the assumption that individual data values from separate cases or measurement events are independent of each other. Independence

---

[7]Some of our colleagues actually do generate sampling distributions in this manual fashion (with a computer anyway), and they are not stupid. They do this because they are statistical researchers who specialize in developing and testing ideas about statistical tests.

[8]In practice, most inferential statistics are not done directly with the raw sample statistics but with derived indices called **test statistics.** These are typically indices that combine the statistics in question with estimates of their variability in the sampling distribution, in order to increase the interpretability of sample patterns as reflections of population patterns. For example, inferences about means are often evaluated with "t scores," calculated by dividing the mean by its standard error.

is the property that separate scores in a data set cannot be predicted from each other—they are uncorrelated. As we discuss below, data in geography often show nonindependence as a function of spatial relations on the earth's surface (as a function of temporal relations too).

3. *Correct specification of models.* These are assumptions about the appropriateness of the statistical models fitted to data. For example, do the data being fitted with a linear model really follow a linear pattern? Are all the relevant predictor constructs included in the model?

We return to the statistical assumptions in Chapter 11, where we will see that some of the assumptions are more important than others for the valid use and interpretation of inferential statistics.

Estimation and Hypothesis Testing

There are two major approaches to inferring parameters from statistics. The first is **estimation**. This is the inferential approach of choice when you do not have a particular value that you want to evaluate for the parameter but only want to make your best guess of the parameter value. Estimation has two parts, the **point estimate** and the **confidence interval** around the point estimate. The point estimate is the guess about the specific parameter value. For example, our point estimate of a population mean is usually the sample mean. The confidence interval is a range of values that is distributed, usually symmetrically, around the point estimate; it is a guess, with a specified probability of confidence, that the true value of the population parameter falls somewhere within the range of values. This confidence probability is most often .95 (95%). When pollsters tell you that candidate *X* is favored by 44% of likely voters, "plus or minus 3%," the 44% is the point estimate, and the "plus or minus 3%" is a confidence interval from 41% to 47% that you should be 95% confident contains the actual percentage of likely voters in the population who favor candidate *X*. Of course, given a particular confidence probability, we prefer narrow intervals to wide intervals; that is, we prefer greater **precision of estimation**. We do that by collecting more data or by reducing random noise in the data.

The second major approach to inference is **hypothesis testing**. This is the inferential approach of choice when you *do* have a particular value you want to evaluate for the parameter. That value is expressed in the **null hypothesis**, symbolized H_0. H_0 is a hypothesis about the exact (point) value of a parameter, or set of parameters. In statistical hypothesis testing, we use our sample statistics to make an inference about the probable truth of our null hypothesis. If we decide that our sample data indicate that the probability of the null being true is too low, we accept that the **alternative hypothesis** must instead be true. The alternative, symbolized H_A, is the hypothesis that the parameter in question does *not* equal the exact value hypothesized by the null; the alternative thus always hypothesizes a range rather than an exact value.[9]

So hypothesis testing is used to evaluate the probability that the null hypothesis is true. The alternative hypothesis is not directly tested, even though it's typically what we believe is actually true in the population. That may sound a little confusing, but there is a good logical reason we directly test the hypothesis we do not believe is true. It's because hypothesis testing is a statistical version of a classic form of logical reasoning that allows us to disprove a tested hypothesis but not prove it. This form of logical reasoning is called **modus tollens**. In classic form, modus tollens is the conditional logic ("syllogism") of consequences. In the abstract:

–If A is true, then B is true

–B is not true

–Therefore, A is not true

A is the "antecedent" proposition, and B is the "consequent" proposition. For example, A might be the statement that "it is raining today" and B could be "I take my umbrella to work today." So this logical argument states the two premises that "If it is raining today, then I take my umbrella to work today" and "It is not true that I have taken my umbrella to work today." The valid conclusion is that "Therefore, it is not raining today." Now consider what happens if the second premise is true rather than false:

–If A is true, then B is true

–B is true

–Therefore, ??

[9]Typically, the alternative specifies the entire range of values that does not include the null value. Thus, evidence against the null is provided whenever the sample statistic is far lower *or* higher than the null value. This typical nondirectional test is called "two-tailed." If you have a prior reason to be certain the true parameter value differs from the null in one direction but not the other, some texts will recommend a directional "one-tailed" test that proposes an alternative covering the range either lower or higher than the null value, but not both. This has some effect, usually very little, on the probability with which you can reject the null hypothesis. It can work to your benefit as long as you predict the proper direction for the difference from the null, but it also prevents finding differences in the other direction if you predict incorrectly. Some researchers consider one-tailed tests to be little more than a way to help salvage weak results.

What do you conclude, now that you know that the consequent has been found to be true? If you conclude, "Therefore, A *is* true," you are making a very common and widespread logical error, the **fallacy of affirming the consequent**. In fact, the valid conclusion is:

–Therefore, no inference can be drawn

To see this, consider that although I do take my umbrella to work every day that it is raining, I also take my umbrella to work in anticipation of rain that never comes. Or I might take my umbrella to make a demonstration in class about the logic of hypothesis testing. None of this violates the truth of the two premises. In other words, this common form of conditional reasoning is not "bidirectional" with respect to the truth of the consequent. However, it is often used incorrectly in everyday reasoning as if it were bidirectional in this way.

Hypothesis testing employs modus tollens conditional logic, with the statistical twist that H_0 plays the part of A, and R_0, the likely range of the sampling distribution if the null is actually true, plays the part of B. Also, because we now must deal with statistical uncertainty, as we discussed above, we must draw conclusions that are at best "probably" rather than "definitely" true.

–If H_0 is true, then R_0 is likely	–If H_0 is true, then R_0 is likely
–R_0 is not true	–R_0 is true
–Therefore, H_0 is probably not true	–Therefore, no inference can be drawn

Thus, because of its reliance on modus tollens logic, hypothesis testing is useful for disconfirming ("disproving" is too strong a word for probabilistic reasoning) null hypotheses but not for confirming them. This fact is usually not a severe problem, however. That's because the null hypothesis is very often a guess that there is no relationship (the relationship is "null") in the population. For example, one might hypothesize that a correlation or the difference between two means in a population is zero. These hypotheses are equivalent to saying there is no relationship in the population. The alternative hypothesizes that there is a relationship.

In Table 9.1, we overview the steps of statistical hypothesis testing. In all such tests, an empirically obtained test statistic, calculated from sample data, is compared to a range of values we consider to be likely if the null were really true. That range of values is the portion of the null sampling distribution around the most likely value—the null hypothesized value. This portion can vary from test to test, but by convention it is most often the portion that ranges over 95% of the distribution around the null value; it's computed just like the 95% confidence interval in estimation, but with the null value in the center rather than the sample estimate. Every hypothesis test ends with one of two decisions. If the empirical test statistic is quite far from the null value, outside the null range, we consider the null hypothesis to be unlikely. We reject the null hypothesis and accept the alternative. This is

Table 9.1 Steps of Statistical Hypothesis Testing

1. State your null (H_0) and alternative hypotheses (H_A).
   a. null is equality of a parameter or set of parameters to a point value; for example, H_0: *Population mean = 120.*
   b. H_0 is never about the value of a statistic—you can just look at your data to find that out; H_0 is also never about "getting significance"—you will know that for sure at the end of your test.
   c. alternative is all other values; for example, H_A: *Population mean = 120.*

2. Determine your appropriate test statistic and its sampling distribution if H_0 is true (the "null sampling distribution").
   a. appropriate test statistic depends on which parameters are in your hypotheses; for example, "*t* scores" are often used to test hypotheses about one or a pair of population means.
   b. the value hypothesized by H_0 becomes the expected value of the null sampling distribution.
   c. a middle portion of the null sampling distribution, often the middle 95% like the confidence probability in estimation, is considered the "likely" range (R_0).

3. Calculate the test statistic from your sample data.

4. Compare the empirically obtained test statistic to the null sampling distribution.
   a. if the test statistic is so different from the null expected value that it falls outside the likely range, then conclude that the null is probably false ("reject the null") and the alternative is probably true ("accept the alternative")—statistical significance at a given rejection *p* level.
   b. if the test statistic is close enough to the null expected value that it falls within the likely range, then conclude that either the null or alternative might be true or false ("fail to reject the null" or "retain both hypotheses")—statistical nonsignificance at the given *p* level.

called statistical **significance** at the rejection ***p* level** of "1 minus the likely probability"; given the common likely probability of 95%, the *p* level is usually 5%. On the other hand, if the empirical test statistic is not so far from the null value and is within the null range, we conclude that we cannot reject either the null or the alternative hypotheses. We "fail to reject the null" or "retain both hypotheses." *We do not accept the null*—that would be the fallacy of affirming the consequent. This is called statistical **nonsignificance** at the applied *p* level.

In Figure 9.5, we show that when you perform a hypothesis test, you always end up making either a correct inference or a mistake. Of course, since that depends on the true value of the parameter in the population, you can never know with certainty whether you have made a mistake. But you can put a number on the chances that you have made a correct decision or a mistake. As Figure 9.5 shows, when the null hypothesis is actually true, the probability of mistakenly rejecting it, called a **Type I error**, is just the rejection *p* level, also symbolized by the Greek letter **alpha** (α). The probability of correctly retaining the null (and the alternative) in this situation is thus $1 - \alpha$. When the null hypothesis is actually false, the probability of

Two Possible Truths

Two Possible Decisions	H_0 **is true**	H_0 **is false**
Reject H_0, Accept H_A	Error, Type I Prob = α "Significance Level"	Correct Decision Prob = $1 - \beta$ "Power"
Retain Both H_0 and H_A	Correct Decision Prob = $1 - \alpha$	Error, Type II Prob = β

Figure 9.5 When conducting hypothesis tests, two decisions are possible and two actual truths are possible. Thus, four outcomes are possible in hypothesis testing, two of which are correct decisions and two of which are errors.

mistakenly retaining it, called a **Type II error**, is symbolized by the Greek letter **beta** (**β**). The probability of correctly rejecting the null (accepting the alternative) in this situation is thus $1 - \beta$; this probability is charmingly referred to as **power.** Unlike α, which is set by convention, β and power are determined by the size of α, by the size of one's sample, by the amount of random noise in the data, and by the actual difference of the parameter in question from the value hypothesized by the null. In other words, it is mostly *not* just set by convention or choice, and is rather difficult to estimate accurately. The techniques of "power analysis" mentioned in Chapter 8 can be used to estimate power.

Perhaps a concrete example will help clarify all of this. Imagine that you want to know whether a particular coin is "fair" for flipping. Your null hypothesis would be that the probability of heads coming up equals the probability of tails: .5 or 50%. Suppose you sample the coin's fairness by flipping it six times, and it lands "heads" every time. The probability of that happening if the coin really is fair is roughly .015 or 1.5%. From the perspective of hypothesis testing, that is rather unusual if the null hypothesis is actually true. At an α-level of .05, you would reject the null and conclude the coin was unfair. According to Figure 9.5, if the coin were actually fair, you would be making a Type I error by rejecting it; this is sort of like unfairly convicting an innocent defendant in a court trial. If the coin were actually unfair (one side was weighted), you would be making a correct decision on this test, like appropriately convicting a guilty defendant. Suppose instead that you flip the coin six times and it lands "heads" four times and "tails" two times. The probability of that happening if the coin really is fair is roughly .234 or 23.4%. That is not so unusual if the null hypothesis is actually true. At an α-level of .05, you would retain both hypotheses and conclude you did not show the coin was unfair. If in fact the coin were actually fair, you would be making a correct decision by retaining the null, like appropriately acquitting an innocent defendant. If the coin were actually unfair, you would be making a Type II error on this test, akin to finding a guilty person to be not guilty.

A final observation about hypothesis testing is in order. Although the null hypothesis most often states that there is no relationship in the population, such a hypothesis is rarely if ever going to be true for any variables in any population in any domain of reality. After all, a population correlation of .01 means the null hypothesis that the correlation is .00 is wrong; remember that positive correlations range from .01 to 1.00. In practice, however, researchers always have limited power in their data, so that a tiny population relationship is usually found to be non-significant. In most areas of geographic research, therefore, sample relationships are not considered real in the population unless they are suitably large. For example, with a sample size of 30 independent data points (considered small in some research areas, adequate in others), you would need a sample correlation of between .30 and .40 to conclude that there really was a relationship in the population. In essence, null hypothesis testing usually does not just identify false null hypotheses—it identifies null hypotheses that are *quite* false. This is a good thing.

Data in Space and Place: Introduction to Geospatial Analysis[10]

Data in geography often have a property that notably differentiates them from many data in other sciences—they are spatially distributed, and that spatiality is theoretically relevant. Geographic features have location, extent or size, shape, pattern, connectivity, and more. As we discuss further in Chapter 12, geographic data represent natural and human earth-surface features and processes, their properties, and their

[10]The large and diverse literature on geospatial data analysis tends to use the term "spatial statistics" when the features being studied are conceived of as discrete entities—points, lines, or polygons. Thus, it is the common term for most applications in human geography and other social sciences. In contrast, "geostatistics" is used when the phenomena being studied are conceived of as fields—continuous two-dimensional surfaces. It is the commonly used term in physical geography and other earth sciences. As we discussed in Chapter 8, and return to in 12, however, the discrete-continuous distinction certainly does not map perfectly on to the distinction between human and physical geosciences, and analytic techniques from each tradition have application in the other.

spatial distributions. Even when other sciences have spatial data, which they do more often than they sometimes acknowledge, that spatiality is typically not a focus of interest like it often is in geography. For example, "central place theory" explains where cities of different sizes are located, or should be located, as a function of the influence of distance on the interactions of economic agents (such as shoppers or retailers) with the economic institutions of the cities. To analyze the spatiality in data, a variety of descriptive spatial indices can be calculated that are analogous to the nonspatial descriptive statistics we overviewed above. Spatial central tendency, variability, form, and so on can be examined. For example, spatial means or medians can be calculated, as can indices of feature clustering that are somewhat analogous to variability measures. Other spatial properties can be analyzed that do not have obvious nonspatial analogues. For example, geographers sometimes want to analyze whether spatial patterns take particular shapes, such as hexagons or circles.

Spatiality is often important to geographic data analysis even when it is not the focus of interest, because its existence in data strongly influences the accuracy of inferential statistical analyses of nonspatial variables. In particular, geographic data usually exhibit a fairly robust pattern of dependence as a function of their spatial arrangement on the earth's surface (often their arrangement in time too). That is, features or processes at one place are more (or less) similar to those at other places than would be expected by chance. Such patterns of spatial dependence are known as **spatial autocorrelation**. Most often, closer places are more similar, an expression of **distance decay** or the so-called First Law of Geography: *Everything is related to everything else, but near things are more related than distant things.* This common pattern of spatial dependence is known as "positive" spatial autocorrelation. The less common reverse pattern, where closer things are less similar, is "negative" autocorrelation. And any number of hybrid or more complex patterns of autocorrelation are at least conceptually possible, if rarely investigated. Whatever its specific pattern, such spatial dependence constitutes a violation of the important statistical assumption of independence we introduced above. Independence of scores is an important requirement for the accuracy of statistical significance testing, as it is normally done. Spatiality usually violates independence in a big way that radically distorts statistical inference. It must be accounted for when spatially interpolating fields from discretely sampled data, discussed in Chapter 8. Much of the effort of geospatial analysis goes toward identifying patterns of spatial autocorrelation and accounting for them in statistical tests.

Consider, for example, a network of rain gauges set up to measure rainfall. Rainfall measurements are strongly and positively spatially autocorrelated. If it is raining in my back yard, I can say with a high degree of confidence that it is raining in my neighbor's back yard, but my level of confidence that it is raining across town is lower, and my level of confidence that it is raining 300 miles away is even lower still. This distance decay of statistical relatedness is often characterized by a **variogram** (or semi-variogram). One characteristic of spatial autocorrelation identified by a variogram is the "range" or distance over which relatedness operates. If this range for rainfall were 10 miles, it would mean that knowing it is raining at a particular place would provide me with some information about the likelihood of rain within a radius of 10 miles from that place. Beyond 10 miles, my estimate of whether or not it was raining would return to near chance—random guessing.

The various ways that spatiality plays a role in geographic data analysis often inspires the cliché geographic witticism that "spatial data are special." And special they are—sometimes a special difficulty. Identifying and dealing with the specific pattern of autocorrelation in data can be quite challenging. As another example, distance frequently plays a leading conceptual role in geospatial analysis, but there are many ways to conceptualize and quantify distance,[11] even if it is restricted just to physical separation rather than temporal, monetary, or other forms of separation, a restriction that is often unacceptable. Other difficulties include the fact that several descriptive spatial indices are insensitive to the overall pattern of the data, such as the **Gini coefficient**, which is used to calculate economic or demographic segregation. Some indices depend in theoretically uninteresting ways on the size of the surrounding area used to delimit, typically somewhat arbitrarily, the scope of a problem. Some delimitation is required; after all, we can't use the entire earth surface or the universe as the spatial backdrop for every test we carry out.

One of the special difficulties of spatial analysis arises from questions about which areal units should be used to analyze geographic data. It is common in geography to have data organized into areal units (zones or regions) that are at least partially if not totally arbitrary with respect to the researcher's phenomenon of interest. As we saw in Chapter 6, this almost always characterizes census data. Researchers usually don't believe that the causal factors underlying their phenomenon of interest operate at the census tract level, but that's one of the common ways the data are packaged. Most other secondary sources of data are like this too, and we have already noted that geographers use a lot of secondary data. For most questions of interest in basic research, units like nature preserves or U.S. states are pretty arbitrary, but that is what's available. And in a disturbingly large number of situations, researchers will take data off of a continuous representation like a map, but break the map into areal pieces so that they can apply discrete spatial analysis techniques (this is analogous to the discretization discussed in Chapter 8 that geographers carry out in order to sample from fields). This practice is even more dubious given that researchers are typically ignorant about how the data used to make the map were obtained and treated in the first place. How's an honest researcher to choose an appropriate way to break up the space when there are an infinite number of possibilities?

The rub with all this is that changing the number, size, shape, and/or location of areal units can change the results of analyses, often dramatically. The phenomenon known as **gerrymandering** provides a great example of this. Gerrymandering refers

---

[11]There are numerous (potentially infinite) "metric geometries" that calculate distance differently. Some of these have been applied in geography, including the city-bock and spherical metrics (after all, the earth is very nearly a sphere). If that weren't enough, consider that physical separation is often best considered in terms of such things as counts of the number of stops a subway train makes, rather than metric distances.

to the design of electoral-district boundaries to concentrate certain types of voters into certain districts so as to give particular candidates an advantage in the election.[12] Both racial and political-party gerrymandering are alive and well in the United States. For example, the state of Pennsylvania in the year 2000 had a majority of voters registered as Democrats; however, resulting in part from gerrymandering of the congressional districts, the majority of congressmen elected from the state of Pennsylvania were Republican (it has certainly worked out for the Democrats in other situations). Figure 9.6 demonstrates the profound effect that redesigning district boundaries can have on election outcomes. The effect that theoretically arbitrary areal geometries can have on the results of geographic analyses is known as the **Modifiable Areal Unit Problem**, or MAUP for short.

We mentioned size as an aspect of areal units that contributes to the MAUP. The question of the proper size for the areal units to be used in geographic analyses really goes beyond just the MAUP. It is an aspect of the fundamental issue of scale in geography. In Chapter 2, we pointed out that scale concerns time and theme as well as space. We also distinguished between phenomenon and analysis scale (and also discussed cartographic scale), noting that geographic phenomena are very often scale-dependent. That is, theories or models often apply at one scale, or a range of scales, and not at others. In order to observe and study a phenomenon accurately, researchers must match their scale of analysis to the actual scale of the phenomenon. That is, researchers must identify the scale of a phenomenon so they can collect and organize their data in units of that size.[13]

Given spatial units of a particular size, one can readily aggregate or combine them into larger units; it is not possible without additional information or theory to disaggregate them into smaller units. A great deal of geographic data is aggregated from data gathered at a finer spatial resolution. U.S. census data again provide a good example. As we saw in Chapter 6, census data are summarized at several levels, from the whole country to blocks, yet these levels of analysis are derived by aggregation from individual responses to the census form. The level at which the data are aggregated can seriously influence statistical patterns identified in the data and the ultimate conclusions drawn about their meaning. As a general rule, the correlation between two geographically distributed variables increases with their level

[12]Gerrymandering was named by the artist Gilbert Stuart (his portrait of George Washington is on the U.S. dollar bill) in an 1812 political cartoon depicting the complexly shaped Massachusetts voting district designed by Governor Elbridge Gerry to concentrate his Federalist opponents in one district. Stuart depicted the sinuous arching district as a salamander—or "gerry-mander." It looks more like a winged dragon, if you ask us.

[13]Using data at one scale to make inferences about phenomena at other scales is known as the **cross-level fallacy**. A specific instance of this is making inferences in aggregated form from data that were measured on individual people; this was identified as the "ecological fallacy" in a classic paper: Robinson, W. S. (1950). Ecological correlation and the behavior of individuals. *American Sociological Review, 15,* 351–357. The reverse error might be called the "atomistic fallacy."

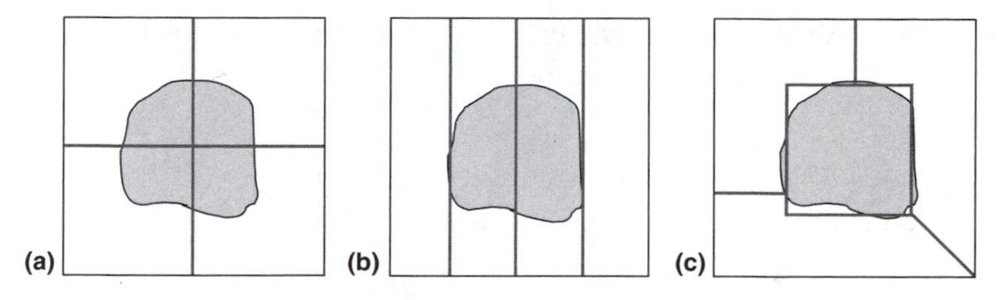

Figure 9.6 Gerrymandering as an example of the MAUP. The gray region represents an area in which the residents are predominantly of the same group, such as members of a racial or ethnic group, or a particular political party (the surrounding white region would be made up of residents that are not members of the group). The straight lines represent alternative voting district boundaries that could be designed. In (a), the group is broken up so that it is a distinct minority in each of the four districts. In (b), the group is broken up so that it is completely missing from two districts but constitutes nearly half of each of the other two districts. In (c), the group is not broken up at all, so it is completely missing from three districts but constitutes a strong majority of the fourth district. Of course, none of this matters for an election unless members of the group are in fact more likely to vote for the same candidates than would be expected by chance (and the candidates must be different than those the residents in the white region would vote for).

of aggregation. For example, assume you run a simple regression between median household income and education level for several different levels of spatial aggregation: for example, states, counties, and census tracts. The highest correlation will likely occur at the state level and the lowest at the tract level. But the potential severity of this aggregation effect can be even more disturbing than that, especially when other aspects of the MAUP apply in such situations as well. Given the right arrangement and aggregation level of data units, the correlation between two variables distributed across the earth's surface can be calculated to take on almost any value from +1.0 to −1.0!

Problems of the scale and arrangement of data units are deep and fundamental in many domains of geographic research, and it appears that many of them will not be unambiguously solved any time soon. Ideally, one would like a theory that specifies the scale and arrangement of spatial (as well as temporal and thematic) units at which structures and processes actually exist. Unfortunately, such theories are few and far between. Lacking this, as geographers typically do, it is often recommended that empirical "trial-and-error" approaches be used to try to identify the appropriate units at which a phenomenon should be analyzed. Computer tools exist that allow analysts to try many alternative regionalizations of the data space and hope for similar results across the regionalizations, or at least identify systematic changes across regionalizations that can be characterized.

However, the best way to deal with these problems remains contested in the field today.[14]

Review Questions

- How does data analysis contribute to the four scientific goals of description, prediction, explanation, and control?
- Why is data analysis in geography usually conceptualized in statistical (probabilistic) terms?

Statistical Description

- What are the following properties that describe distributions of data, and what are some specific indices for expressing each one: central tendency, variability, form, derived scores, relationship?
- What do we mean by the strength and form of statistical relationships?
- What are monotonic, linear, and nonlinear relationship forms?

Statistical Inference

- What is the purpose of statistical inference, and why are statistical inferences necessarily and ultimately uncertain?
- Why do we say that sampling error is not an *error* in the sense of a mistake?
- What are sampling distributions, and how do they relate to population and sample distributions? In scientific research, sampling distributions are generated by statistical theory rather than actual repeated sampling; why is this so?

Estimation and Hypothesis Testing

- What are the similarities and differences between the two inferential procedures of estimation and hypothesis testing?
- What are point estimates and confidence intervals in estimation?
- Why is hypothesis testing ultimately useless for confirming null hypotheses?

---

[14]Tobler offered the view that MAUP effects result from inappropriate methods of analyses, and that spatial analysts should find "frame-independent" analytic tools that produce the same results regardless of the partitioning of the data space. See Tobler, W. R. (1990). Frame independent spatial analysis. In M. Goodchild & S. Gopal (Eds.), *Accuracy of spatial databases* (pp. 115–122). London: Taylor & Francis. This solution is not very satisfying, at least with respect to scale aspects of the MAUP, because many of us consider the frequent scale-dependent nature of geographic phenomena to be a substantive reflection of their true nature rather than just an artifact of human analysis.

- What are the steps of hypothesis testing, including the two possible decisions that it can lead to?
- What are the two types of correct decisions and two types of errors possible when hypothesis testing?

Introduction to Geospatial Analysis

- What are some spatial properties of phenomena that might be of interest when analyzing data?
- What is spatial autocorrelation, what forms can it take, and why is it so important to geographic data analysis?
- What is the MAUP, and what is a specific example of MAUP? How is gerrymandering an example of MAUP?

Key Terms

alpha (α): the probability of mistakenly rejecting the null hypothesis in hypothesis testing when it is actually true; the same as the significance "*p* level"

alternative hypothesis: the hypothesis that expresses all possible values for the parameter not specified by the null hypothesis; symbolized H_A, it is always a range guess about the value of a parameter or set of parameters

arithmetic mean: the most common type of mean, calculated as the sum of the values of all scores divided by the number of scores

bell-shaped distribution: distribution of data that has the form of being symmetric and unimodal

beta (β): the probability of mistakenly retaining the null hypothesis in hypothesis testing when it is actually false

central tendency: the descriptive property of the average or most representative value in a data set

confidence interval: one's guess of the range of values the parameter probably has in estimation, with some level of confidence probability

correlation coefficient: index of relationship strength; when dealing with linear relationships, it is calculated to equal 1.0 when there is a perfect positive relation, –1.0 when there is a perfect negative relation, and 0.0 when there is no linear relationship at all

criterion variables: the variables chosen to be predicted by the values of the predictor variables in a regression model, usually designated *Y*

cross-level fallacy: drawing inferences about phenomena at one scale from measurements made at smaller (the ecological fallacy) or larger (the atomistic fallacy) scales

data analysis: the set of display and mathematical techniques, and attendant logical and conceptual considerations, used to extract meaning from data and communicate it to others

degrees of freedom: in data analysis, the number of independent pieces of information represented by a given sample data set; usually just a little fewer than the sample size

derived score: way to describe properties of individual scores in a data set by expressing their value relative to the rest of the data set

descriptive statistics: the branch of statistical data analysis that uses mathematical and display techniques to describe patterns in data sets

deviations from the mean: the basis for the variance and standard deviation, calculated for each score by subtracting the arithmetic mean from the score

distance decay: positive spatial autocorrelation, wherein similarity between phenomena decreases as distance between them increases; famously referred to as the First Law of Geography

distributional assumptions: assumptions about properties of the population from which the sampling distribution is generated, including normality and homogeneity of variance; theoretically must be valid for the valid conduct of parametric tests

error of prediction: the vertical distance between data points and the predicted line in a regression model of relationship; assumed to be random from case to case

estimation: one of two major approaches to inferential statistics (hypothesis testing being the other), appropriate when you do not have a particular value that you want to evaluate for the parameter but want only to make your best guess of its value

expected value: the arithmetic mean of the sampling distribution of a given statistic

fallacy of affirming the consequent: common logical mistake in modus tollens reasoning in which a true consequent, the second premise of the argument, is taken as evidence for a true conclusion

form: the descriptive property of the shape of the distribution of a data set, easier to grasp when the data are graphed

geospatial analysis: data analysis that explicitly takes account of the spatiality in geographic data; variously called “geostatistics” or “spatial statistics”

gerrymandering: the design of electoral-district boundaries to concentrate certain types of voters into certain districts so as to give particular candidates an advantage in an election; expression of the MAUP in an electoral context

Gini coefficient: statistical index of the concentration of some property within areal units (regions), as compared to other units; typically used to quantify residential segregation based on wealth or ethnicity

H_0: common symbol for the null hypothesis in hypothesis testing

H_A: common symbol for the alternative hypothesis in hypothesis testing

hypothesis testing: one of two major approaches to inferential statistics (estimation being the other), appropriate when you have a particular value that you want to evaluate for the parameter

independence of scores: the property that individual scores in a data set cannot be predicted from each other—they are uncorrelated; it is an important assumption in inferential statistics that is nonetheless frequently violated in geography

inferential statistics: the branch of statistical data analysis that attempts to infer patterns in populations of data from the evidence of samples of data; in addition to generating accurate guesses, inferential procedures generate probabilities that the guesses are correct or close

linear relationship: the simplest form of relationship, in which the values of two variables tend to follow a straight line when graphed against each other

mean: descriptive index of central tendency calculated as the average of all values in a data set

median: descriptive index of central tendency calculated as the middle value in a data set

modality: descriptive index of form calculated as the number of "local" modes in a data set, which are values that occur more commonly than values just below or above them

mode: descriptive index of central tendency calculated as the most frequently occurring value in a data set

Modifiable Areal Unit Problem: the effect that theoretically arbitrary areal geometries can have on the results of geographic analyses; "MAUP" is its common acronym

modus tollens: classic form of the conditional logic of consequences that provides the basis for statistical hypothesis testing

monotonic: approximate relationship form that goes in only one direction, either up or down, but does not necessarily follow an exact straight line

multivariate: relationships and the techniques for analyzing them that involve more than a pair of variables

negative relationship: linear relationship in which high values on one variable tend to go with low values on the other variable, and low values tend to go with high values; also called "indirect" relationship

nonlinear relationship: any relationship form that follows a pattern other than a straight line

nonparametric statistical tests: class of inferential statistical tests appropriate for nonmetric data and for metric data that violate the distributional assumptions; they are less powerful and less flexible than parametric tests

nonsignificance: the outcome of hypothesis testing that does not allow you to reject the null hypothesis but forces you to retain both hypotheses at a given rejection probability *p*

normal distribution: particular and important bell-shaped distribution of data that has a specific proportion of its scores within any given range from the central tendency

null hypothesis: the hypothesis that is tested when doing hypothesis testing; symbolized H_0, it is always a point guess about the value of a parameter or set of parameters

parameters: summary statistical indices calculated to describe properties of population data

parametric statistical tests: class of inferential statistical tests for metric data that satisfy the distributional assumptions

percentile: derived score calculated as the percentage of the data set that is less than the score in question; for example, the median is at the 50th percentile

***p* level:** the rejection probability in a hypothesis test, which equals "1 minus the null likely probability (confidence probability)"; set by convention, it is most often 5%

point estimate: in estimation, one's guess of the precise value of the parameter

positive relationship: linear relationship in which high values on one variable tend to go with high values on the other variable, and low values tend to go with low values; also called "direct" relationship

power: the probability of correctly rejecting the null hypothesis in hypothesis testing when it is actually false; equal to $1 - \beta$

precision of estimation: the width of the confidence interval in estimation, given a particular confidence probability

predictor variables: the variables chosen to predict the values of the criterion variables in a regression model, usually designated *X*

range: descriptive index of variability calculated as the difference between the highest and lowest values in a data set

regression analysis (model): statistical model of the form of the relationship between two or more variables

relationship: descriptive property of how values of pairs (or larger sets) of variables vary across cases in systematic ways

relationship strength: the degree to which systematic relationship patterns hold across all cases

sampling distribution: distribution of a sample statistic based on all possible samples of a given size taken from a given population; unless one is teaching or doing research specifically *on* statistical analysis, the sampling distribution is created from theory rather than actual repeated sampling

sampling error: the way different samples from the same population vary from each other

significance: the outcome of hypothesis testing that allows you to reject the null hypothesis and accept the alternative hypothesis at a given rejection probability p

skewness: descriptive index of form that is the deviation from symmetry caused by an uneven distribution of extreme score values to the high or low side; "positive" skew has extreme scores to the high side, "negative" skew has extreme scores to the low side

spatial autocorrelation: nonindependence among measurements of phenomena as a function of their location relative to other phenomena; commonly observed in geographic data, it may be positive, negative, or some combination of the two

standard deviation: descriptive index of variability calculated as the square root of the variance

standard error: the standard deviation of the sampling distribution of a given statistic

statistical data analysis: approach to mathematical data analysis that treats data as reflecting phenomena of interest in a probabilistic way, rather than a deterministic way

statistics: summary statistical indices calculated to describe properties of sample data and infer properties of population data

symmetry: descriptive index of form for distributions of data sets with two sides that are mirror images around their central tendency; a symmetric distribution has no skew

test statistic: derived indices used to conduct inferential statistics, calculated by combining a particular statistic with estimates of the variability in the data

transformation: mathematical operation applied to each raw score in a data set to produce a new data set that has particular desired properties; for example, a nonlinear relationship of a certain form can be linearized by taking the logarithm of each score

Type I error: mistakenly rejecting the null hypothesis in a hypothesis test when it is actually true

Type II error: mistakenly retaining the null hypothesis in a hypothesis test when it is actually false

univariate: relationships and the techniques for analyzing them that involve a single pair of variables

variability: descriptive property of how values in a data set differ from the central tendency or each other; also called "dispersion"

variance: descriptive index of variability calculated as the sum of squared deviations from the mean, divided by the number of deviation scores; the square of the standard deviation

variogram: graphical display used to identify patterns of spatial autocorrelation; it shows average dissimilarities between measurements as a function of distance between them

***z* score:** derived score for metric-level data, calculated by dividing a score's deviation from the mean by the standard deviation of its data set

Bibliography

Clark, W. A. V., & Hosking, P. L. (1986). *Statistical methods for geographers.* New York: Wiley.

Fotheringham, A. S., Brunsdon, C., & Charlton, M. (2000). *Quantitative geography.* Thousand Oaks, CA: Sage.

Games, P. A., & Klare, G. R. (1967). *Elementary statistics: Data analysis for the behavioral sciences.* New York: McGraw-Hill.

Haining, R. (1990). *Spatial data analysis in the social and environmental sciences.* Cambridge, U.K.: Cambridge University Press.

Isaaks, E. H., & Srivastava, R. M. (1989). *An introduction to applied geostatistics.* New York: Oxford University Press.

Openshaw, S. (1983). *The Modifiable Areal Unit Problem.* Norfolk, U.K.: Geo Books.

Winer, B. J., Brown, D. R., & Michels, K. M. (1991). *Statistical principles in experimental design* (3rd ed.). New York: McGraw-Hill.

CHAPTER 10

Data Display

Tables, Graphs, Maps, Visualizations

Learning Objectives:

- What are tables best used for, and what are some design alternatives for their construction?
- What are graphs best used for, and what are some design alternatives for their construction?
- What are some principles for good graphing?
- What are some of the special powers of maps, and what are some of their special pitfalls?
- How are new computer technologies being applied to displaying data in innovative and powerful ways?

In Chapter 13, we point out that scientific communication occurs not only verbally but also graphically. In Chapter 9, we similarly pointed out that data analysis includes not just mathematical techniques but also graphical techniques. We refer to these graphical techniques for communication and analysis as **data display**. Data displays *depict* data patterns rather than literally *describe* them. This gives data displays exceptional power to help us understand and communicate data. The spatial medium of graphical representation takes advantage of the human visual and spatial cognitive systems, including perception and memory, and the unique semiotic qualities of spatiality—the particular way it symbolically represents information. When depicting information about phenomena that *are* inherently spatial, as geographers so often do, data displays can do this at whatever scale and viewing perspective is convenient—you can see it all in a glance. Even nonspatial and

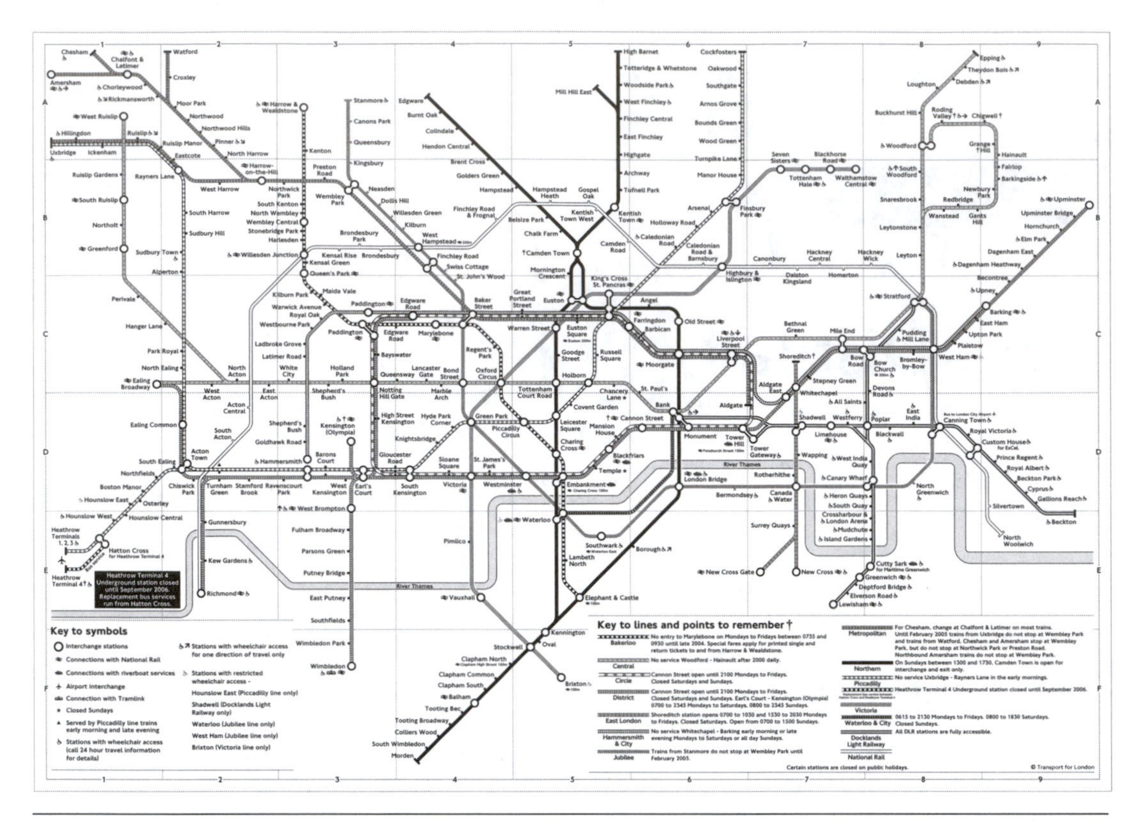

Figure 10.1 The London tube (subway) map. By focusing on line connections and stop sequences, this famous image exemplifies the power of graphical displays to highlight and clarify the relevant while omitting or downplaying the irrelevant, all in the interest of serving human communication. (Reprinted by permission of the Transport for London.)

nonperceptible information can be spatially and visually displayed. What's more, data displays allow us to highlight and clarify the relevant properties of phenomena and omit or downplay the irrelevant. The London tube (subway) map and countless others like it around the world provide a wonderful example of this point (Figure 10.1).

Scientific researchers use data displays for several purposes: to examine data initially, to interpret their meaning, and to communicate the data and their meaning to others. As part of their initial examination, researchers use displays to begin the process of intuitively grasping patterns in their data. How are the different variables statistically distributed, how are different values of the variables spatially distributed, and how do pairs or larger sets of variables relate to each other? Data display is quite useful for initially evaluating how "well-behaved" your data are. Are statistical assumptions about distributions, such as normality or homogeneity of variance, viable? Do the values of variables more or less fit the values you expect them to have? Are there unusual or extreme values—**outliers**—in your data? Do variables take on values that should be impossible, which is a signal that data have been miscoded or misrecorded somehow?

Initial examination merges with data interpretation as data display and examination continues. Dive right in—think of this process as "submersing" yourself in your data. Look at your data in an as many different ways as you can think of. Constructing a variety of tables can help, but for most people, the true power of data display derives from turning values into pictorial (graphical) displays. Try several different types of displays, including and excluding different variables and ranges of values, in a flexible manner that highlights different aspects of your data. Modern computing software and hardware makes this relatively easy to do. During the 1970s, this submersive, graphical, and ad hoc approach to data interpretation was dubbed **exploratory data analysis**. There is now a great deal of interest in this approach to data interpretation in many scientific disciplines, especially those like geography that generate very large amounts of data that are spatially and temporally distributed. In any event, whether you enthusiastically advocate exploratory data analysis or not,[1] the wisdom of exploratory submersion in your data as *part* of the total process of data interpretation and analysis has always been solid and commendable scientific practice. At the end of the chapter, we discuss recent computer display techniques that facilitate exploratory analysis and the interpretation of complex spatio-temporal data.

Of course, data displays are also invaluable for the communication of data to other people. They are a critical part of both written and oral communication in science, including published articles and books, theses and other student papers, grant applications, talks given at conferences and job interviews, and so on. As compared to displays made for initial examination and interpretation, those made for communication, especially if they are to appear in published scientific literature, function in a more "archival" fashion. They will be viewed more frequently, by more different people, and as relatively permanent records. Thus, displays made for final communication should be made with greater care, following more closely the guidelines for the design of effective displays we cover below.

The question arises as to whether one can ever use *too many* data displays. The answer is yes. Data displays are powerful in part because they are relatively concrete,

---

[1]A potentially important criticism of the exploratory analysis approach is that it is largely an ad hoc data-driven approach to science rather than an a priori theory-driven exercise. The exploratory approach definitely risks capitalizing on chance findings, even when findings are established to be statistically significant. Clearly, such an approach should offend anyone who advocates "hypothetico-deductive" reasoning in science, in which prior theories predict empirical results that can subsequently confirm or disconfirm the theories. It is true that patterns found via exploratory analysis maintain a definite conditional status until their further replication and confirmatory analysis. However, as we pointed out in Chapter 1, it is historically inaccurate to claim that scientists have ever stuck purely to one logical approach to theoretical and empirical reasoning and activity, and we do not believe any single approach is advisable or even feasible. Just remember that exploratory data analysis is but *part* of an optimal approach to data analysis and scientific progress.

so if you want to communicate something more abstract, words are often a better choice. Displays also consume quite a bit of space on a page, so you should not use them if they are only communicating a small amount of information. We recommend that you use displays when you have more than a handful of data values to report, whether the values are raw scores or aggregate measures like averages. That is, a general rule, which should sometimes be broken, is to use tables or graphs or maps when you need to communicate more than three or four values. If you are reporting fewer than this, most of the time it works best simply to list them verbally or numerically in the body of your text. "There were 35 inches of rain this season, which is far above the normal rainfall of 19 inches." Does that call for a table or picture? Could you make a good display for this information that would take up less space than the sentence does? Even if you have more than a handful of values to report, it may be best to communicate highly familiar information or information that can be summarized easily in verbal form. "Average annual precipitation follows a smooth gradient decreasing from east to west across the plains region of the U.S. and Canada, whereas temperature follows a smooth gradient decreasing from south to north." Would you prefer a picture of that?

Finally, as we note in Chapter 13, scientists do not use graphic displays, or **figures**, only to depict data. A variety of figures, including photographs and drawings, are used to communicate information other than data patterns. They are also used in scientific communication to depict the equipment and material employed in research, procedures followed, settings, example cases, the structure of models of all kinds, and more. As we suggest in Chapter 13, graphical displays are especially nice to use during oral presentations, and not just because they communicate efficiently and concretely; they also entertain and stimulate audience members by allowing them to activate other parts of their minds besides the verbal system. Indeed, we use figures in most chapters of this book to communicate things besides data patterns.

Guidelines for Designing Displays

We have seen that data displays help in the initial examination and ultimate interpretation of data, not just in their final communication to others. But even the examination and interpretation of data are really examples of communication—they are essentially processes of *self*-communication. Therefore, guiding principles for the design and use of data displays really boil down to one idea: effective communication.[2] Communication is about transferring ideas to, or stimulating ideas in, sentient or information-processing entities. In our case, those entities are individual human beings or groups, like families, communities, or policy-making bodies.

[2]In the past, the archival function of data displays was quite important; they served to store information of various kinds. This function has largely been subsumed by digital files and databases. Graphical displays are now almost exclusively about facilitating human communication.

To communicate effectively is to communicate a great deal of truthful and relevant information that a person or group should know, or wants to know, as quickly and easily as possible. That is, displays should be designed to depict valid and relevant information, in a manner that is clear, accurate, and unambiguous, and in a manner that is as efficient as possible. Displays should give access to the complex; they should not complicate the simple.[3]

The design of displays unquestionably affects what and how well they communicate. A particularly consequential example of how graphic design can make the simple complex and thereby influence communication is provided by the voting ballot for the 2000 U.S. presidential election in Palm Beach County, in the state of Florida. Palm Beach County employed a double-column or "butterfly" ballot (see Figure 10.2) instead of the much more common single-column ballot. George W. Bush, the top candidate on the left side, won the election in that county by a small margin; he was eventually declared the victor in an extremely close Florida vote count and consequently, in an extremely close national election. Some commentators observed that a surprisingly large number of votes were cast in Palm Beach County for Pat Buchanan, the top candidate on the right side—surprising considering the demographics of the county and the outcome of a previous election in which Mr. Buchanan ran. A look at the ballot suggests—at least it did to us the first time we saw it—that it would probably cause confusion for some voters. In order to vote for the *second* candidate on the left side, Al Gore, one would have to punch the *third* hole; punching the second hole would count as a vote for Mr. Buchanan. Even given the arrows that apparently indicated to most voters how to negotiate the confusing ballot layout, a significant minority of voters apparently did vote mistakenly for Mr. Buchanan when they intended to vote for Mr. Gore. In fact, experimental research conducted shortly after the election provided evidence that such a ballot would indeed cause confusion for some voters.[4] As a consequence, graphical miscommunication likely had a large effect on the outcome of this extremely close election (various other kinds of measurement errors certainly influenced vote counts elsewhere, although perhaps not with such a dramatic influence on the final outcome).

What about the role of aesthetics in data displays? Stressing communication like we do has sometimes been criticized as leading to a disregard for aesthetics. This may be somewhat true, but perhaps it reflects a misunderstanding of the power of beauty to increase the effectiveness of communication. Attractiveness induces people to look at a display more, which enhances communication. And displays that are cognitively effective in their communication tend for that reason also to be aesthetically pleasing. A quality piece of music or a quality piece of pie share something in common with a quality graph or map—they have a purity of focus on a meaning, whether sensory, emotional, or cognitive, and they avoid irrelevant and distracting ingredients that don't harmonize and contribute to that meaning. So we believe that beauty is valuable to data displays, but we don't think a display can be beautiful *as a*

---

[3]As noted by Wainer (1997).

[4]Sinclair, R. C., Mark, M. M., Moore, S. E., Lavis, C. A., & Soldat, A. S. (2000). An electoral butterfly effect. *Nature, 408*, 665–666.

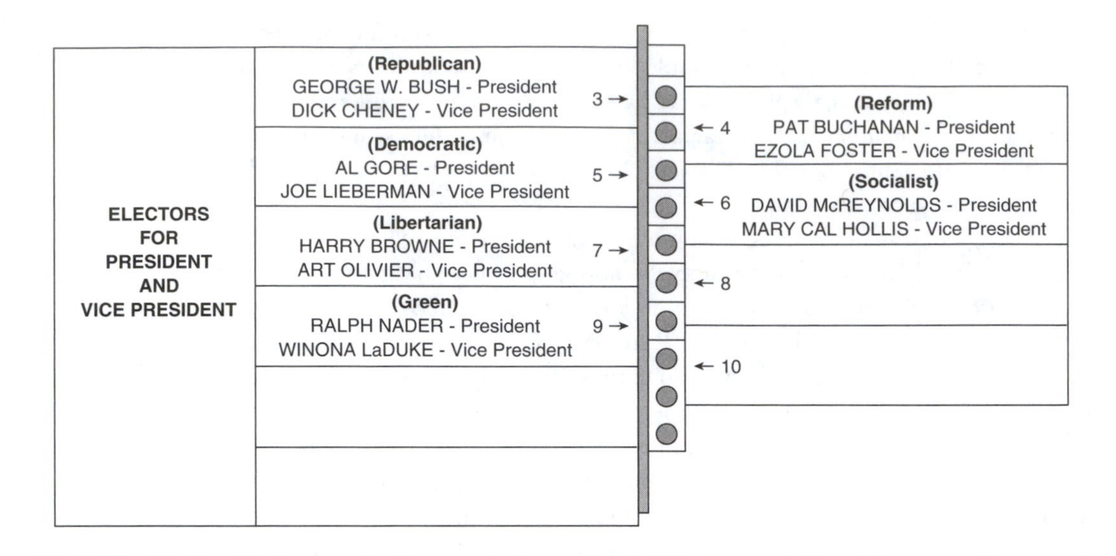

Figure 10.2 Recreation of the relevant portion of the 2000 presidential double-column "butterfly" ballot from Palm Beach County, Florida, U.S. A variety of evidence (as well as graphical logic) suggests a significant minority of voters who intended to vote for the *second* candidate on the left list punched the *second* hole from the top, which was actually a vote for the top candidate on the right list.

data display if it doesn't facilitate communication of the data's meaning. Below, we consider some display adornments that are, in the end, "unattractive" in this way.

Tables

Tables are organized lists, arrays, or matrices of data. They show data values directly with numbers (or class labels for nominal variables), and unlike the other display types we discuss in this chapter, tables make only minimal use of spatiality. For instance, data values are often sequentially ordered in table space according to their numerical magnitudes, and variables usually correspond to spatial dimensions. That is, distributions of single variables map directly onto one-dimensional lists (vectors), and joint distributions of two variables together map directly onto two-dimensional arrays or matrices. Given the two-dimensional "flatness" of tables, joint distributions of three or more variables have to be shown indirectly with hierarchically articulated two-dimensional displays, such as two neighboring tables that represent the two levels of a third variable. With respect to spatiality in tables and graphs, the **stem-and-leaf display** is an interesting case in point (see Figure 10.3). Unlike a standard table, it uses spatial magnitude (line length) to communicate perceptually the frequencies of values in particular ranges of the data. This is the kind of communication device a graph uses, and for that reason, we think of the stem-and-leaf display as "hybrid" between tables and graphs.

As we advised above, when you have only a handful of values, you should probably just verbally state the values without a data display. If you decide you do want

Ozone Concentration in Hundreds of PPB
Downunder, Queensland
October 11–December 5, 2003

Stems	Leaves
0	477
1	002559
2	0114577889
3	000112344566689
4	1224569
5	026
6	01578
7	245799
8	3
9	–

Figure 10.3 A stem-and-leaf display, essentially a hybrid display between tables and graphs. The fictitious data values are daily ozone concentrations in hundreds of parts per billion for the equally fictitious town of Downunder in Australia.

to display data, tables are often a good display choice when you want to show data values precisely and in detail, although even this becomes hard to do effectively with a table when you have a very large number of data values to communicate. In fact, although it is useful to record data values precisely and in detail when the table is functioning archivally, it is usually not so useful when tables function to communicate. One more often wants to communicate general patterns of data. As such, values in tables should be adequately rounded in order to support effective communication of general patterns. At the least, the spurious precision defined in Chapter 2 should be avoided in tables. Believe it or not, one almost never cares to see the fifth digit after the decimal, perhaps not even the first or second.

Tables depict data in one of two ways. The first is to present the overall distribution of the data by showing all of the values in the data set and how many scores have each value. That is, such **distribution tables** show the **frequency** of data values; alternatively, they can show **relative frequency** (as proportions or percentages), **cumulative frequency**, or **cumulative relative frequency**. Distribution tables can also present data in terms of derived scores, such as percentiles or *z* scores (see Chapter 9). A special type of distribution table called **contingency tables** can show relationships between nominal variables or metric variables that can be grouped into discrete classes; they show the frequencies of cases across levels of one variable cross-tabulated against frequencies of a second variable. The second way tables depict data is to present summary descriptive indices calculated on the data. Such **descriptive index tables** can present measures of central tendency, variability, relationship, and so on.

As we suggested in Chapter 9, the limiting example of a table is an unorganized printout of raw data values. Most people would not consider this much of a table until the data values were organized or summarized in some informative way. A typical way to do this is simply to order the values from lowest to highest, or vice versa. When displaying nominal variables, there is no natural numerical order to the data, even if the nominal classes have been given numerical labels. You can use arbitrary ordering; with some nominal variables, it might make sense to use an alphabetical ordering. But often, you can communicate more effectively by ordering the nominal classes according to the magnitudes of frequency within the classes or some other logic; you might organize country names according to country proximity, for example.

Finally, metric-level data values in tables as well as graphs are often grouped into classes or intervals in order to facilitate efficient communication of the pattern of the data. For example, one often sees incomes expressed in terms of something like "less than $20,000," "at least $20,000 but less than $40,000," "at least $40,000 but less than $60,000," "at least $60,000 but less than $80,000," "at least $80,000 but less than $100,000," and "over $100,000." There are a great variety of ways to create these **class intervals.** To begin, there is the question of how many classes there should be. It is not possible to answer this formulaically, but there should probably be at least three or four classes but no more than 10 or so; more can sometimes be appropriate, however. Either all intervals are bounded with minimum and maximum values, or upper or lower intervals may be open-ended, as in our example above. All intervals may be of the same width or different widths. Likewise, intervals may contain the same number of cases or different numbers. The widths and frequency counts of intervals depend on the design rule you use to construct the intervals.[5] For example, if you decide to let the data pattern itself dictate where interval breaks go (so-called natural breaks), you are almost certain to have intervals of varying widths *and* counts. Often, when constructing classes for variables that are bounded on only one end, such as a ratio-level variable with a zero value on one end like in our income example, interval widths should perhaps increase in a nonlinear fashion as one moves toward the unbounded end. After all, the effective difference, whether economic, social, or psychological, between making $20,000 and $40,000 is greater than between $80,000 and $100,000.

Graphs

Graphs are pictorial representations of data. They show data values by metaphorically mapping them onto the spatial and nonspatial properties of images. For example, a higher value in the data is usually shown as being higher on the graph image, or a particular vegetation region is shown by a particular line pattern. Graphs typically include numerical and verbal symbols as well. But the use of spatial properties such as location, size, distance, and direction to represent data is at the heart of what a graph is and how it functions. Well-made graphs are particularly effective for communicating general rather than precise patterns of data values, although some researchers place precise data values redundantly on their graphs, next to the symbol whose properties also encode the data values. Using graphs to communicate data patterns is especially useful with large data sets.

[5]In a cartographic context, the way class intervals are created is recognized as an especially important issue because it affects the visual impression created by, especially, choropleth maps. In the case of choropleth maps, however, perceptual considerations definitely keep the effective, and therefore recommended, number of classes down to around seven.

Table 10.1 lists some principles of good graphing, organized into three dictums. The first is to clearly and sufficiently label the graph and its parts. This includes labeling the units of variables on the axes with **tic marks**. It is possible to use too few or too many tic marks and labels; you want to be adequately informative while avoiding unnecessary clutter. As Table 10.1 points out, a general rule is to aim for something like 4–10 tics. The second dictum is to avoid uninformative and content-free graphic marks. Graphs should draw attention to their data rather than themselves. The third dictum is to fill the graph space with data marks. This should be done whenever there is no constraint on what values should be placed at the lower and upper ends of the axes. It stretches the data pattern out as far as possible, facilitating the display of variation over the graph space. Of course, if you want to emphasize a lack of variation, you should ignore our advice and choose axis values far above and/or below the values in your data.

In other words, you can manipulate the range of your axes in order to exaggerate or minimize data variation. Is that propaganda? Yes, trying to direct people to focus on what you want them to focus on and away from what you don't want them to focus on is at least "impression management." But, arguably, there is no way to avoid directing people like this, accidentally if not intentionally. It doesn't seem insidious to us as long as your ultimate objective is communicating truth, as you understand it. And as long as you have followed the other principles of good graphing, particularly those under the first dictum to label clearly and sufficiently, you are not deceiving consumers of your graph. At this point, people often recall the quote attributed to the statesmen Benjamin Disraeli: "There are three kinds of lies—lies, damned lies, and statistics." It is certainly true that statistics (data) can be displayed in ways that communicate deceptively. However, although amusing, we find this quote misleading. Data don't lie to people; *people* lie to people—with intentionally poor data collection, analysis, and especially display.

There are a variety of more or less standard graph styles available to the researcher, and new styles continue to be invented. Here are some guidelines for choosing among them; as *guidelines*, they should probably be ignored on occasion. The first consideration is whether you are graphing the distribution of one or more variables, or the relationship between two or more variables. **Distribution graphs** depict the distributions of variables, most often employing a two-dimensional space whose mapping onto data values is defined by two axes that meet at a right angle[6] (but see our discussion below of dimensionality in graphs). The values or ranges of values of the variable are typically displayed on the horizontal **X-axis** or **abscissa.** The frequency (or relative frequency, and so on, as discussed above with tables) of each value or range of values is displayed on the vertical **Y-axis** or **ordinate. Relationship graphs** depict the form and strength of relationships between pairs of variables, again most often with a two-dimensional space. The values or ranges

[6]Not all graphs should use the standard **Cartesian coordinate axes** (X and Y) meeting at right angles. When you have cyclic data, such as directions in space or measurements in repeating time periods, you may want to use **polar coordinate axes.** The fact that normal Cartesian axes meet at a right angle implies that X and Y are uncorrelated. If they are in fact somewhat correlated, you can suggest that by using oblique Cartesian axes.

Table 10.1 Principles of Good Graphing

Dictum 1: Label Clearly and Sufficiently.

- Always include a title or caption (sometimes put in a caption).
- Always label the variables that are expressed on each axis, and their units (as appropriate).
- Mark each variable's units on its axis (stay away from excessive precision); label the class names of nominal variables and the numerical values, or ranges of values, of ordinal and metric values.
- Indicate positions of the values of variables on the axes with tic marks; as a general rule, aim for something like 4–10 tics.

Dictum 2: Avoid Uninformative and Content-Free Graphic Marks.

- Marks should provide useful or necessary information, not just "decoration."
- This includes dots, lines, patterns, labels, and so on.
- Reduce the size, brightness, and attention-grabbing hues of less important marks that do not show data patterns of central importance.
- Avoid visual clutter, vacuous complexity, moiré patterns, 3D symbols for 2D data.

Dictum 3: Fill the Graph Space with Data Marks.

- Stretch out the data pattern as much as reasonable within the graph space.
- Do this by choosing minimum and maximum units on the axes so there are values somewhere within the graph near the lowest and highest values on the axes.
- Use nonlinear axes when appropriate (for example, logarithmic scales).
- Truncate axes to help accomplish this, as appropriate; if a truncated axis represents a ratio-level variable (a metric variable with a true "0"), use **interruption tics** (something like "∔") on the axis to indicate that it has been truncated.
- When two or more graphs are to be directly compared, use identical graph formats to facilitate their direct comparison (for example, small multiples).

of values of one variable are displayed on the X-axis; those of the other variable are displayed on the Y-axis. More complex relationships, such as the relationship between two variables separately for each level of a third variable, can be displayed by using three spatial dimensions, using nonspatial symbols like color hue, showing separate two-dimensional graphs for each level of a third variable, and so on.

Two additional considerations when choosing a graph style are whether variables are discrete or continuous, and the level at which they were measured. We present these together because their meanings, including the way they influence graph choice, are partially related; remember from Chapter 2 that nominal and ordinal variables must be discrete, but interval and ratio (metric) variables can be either discrete or continuous. Figure 10.4 shows schematic examples of various common graph styles. Distribution graphs of nominal or ordinal variables should generally be made with a discrete graph style, such as the **bar graph** (bar chart). Like tables of nominal variables, graphs of nominal variables should also order values according to some useful communication logic such as class magnitudes. Distribution graphs of discrete metric-level variables can use a discrete graph style such as the **histogram**, which is a bar chart whose bars' widths represent the range of a quantitative class interval for a metric variable. For example, spectral frequency data generated by remote sensing are traditionally graphed with a histogram densely filled with bars. A style of distribution graph that necessarily shows relative frequencies (proportions) is the **circle diagram**, known by many people as the "pie chart." They are a poor choice for variables above the nominal level because they force a confusing circular logic onto the linear sequential logic of ordinal and metric variables; cyclic variables such as measurements over 360° of direction or 12 months of a year might work fine, however. Distribution graphs of continuous variables should generally use continuous graphing styles, the most common of which is the **line graph** (polygon or "connect-the-dot" graph). That deserves emphasizing: *Line graphs are strictly correct only for graphing continuous variables.* That's because a line itself is continuous, so that as a visual metaphor it implies continuously filled intervals between any two data values; it's probably a good idea for those filled intervals to be conceptually possible. Finally, we should point out that distribution graphs sometimes show a statistical model of a distribution that has been fitted to the data (Chapter 9), perhaps by overlaying a line or bar graph over the data itself; these are **curve-fitting graphs.**

In addition to styles for distribution graphs, there are a few graph styles specifically appropriate for relationships. Showing relationships between two classed variables, especially nominal variables, is almost always done with the contingency table we mentioned above instead of a graph. If the *Y*-variable is metric, either bar or line graphs can be used to show relationships, depending on whether the *X*-variable is classed or continuous. Probably the most common way to graph relationships with any type of data is the **scatterplot.** A dot is placed at the intersection of imaginary lines emanating from each data point's values on the X- and Y-axes. This produces a "scatter" of dots; as shown in Figure 9.3 of the previous chapter, the width and shape of this scatter indicates relationship strength and form. As Figure 9.3 also shows, relationship models are sometimes depicted on scatterplots by overlaying straight or curved lines on the dots.

There is one specialized relationship graph appropriate specifically for showing relationships among three variables that are mutually interdependent so that their proportions must sum to 100%. This is the **ternary diagram**, also known as the "trilinear" or "triangle" plot. It is the standard way to show the relative proportions of sand, silt, and clay that determine soil texture (Chapter 4). Ternary diagrams manage to depict three spatial dimensions of information with a two-dimensional figure, and because of this, they can be tricky to interpret at first. As Figure 10.5 shows, there are three Y-axes on a ternary diagram, one for each of the three variables whose proportions are shown. The three axes are not the three sides of the triangle, however; they are the three intersecting bisector lines within the equilateral triangle. To read off the proportion of sand in a particular batch of soil, for example, focus on the bisector line that starts from the center of the side labeled "Sand." That line starts at soil with 0% sand and ends at the opposite vertex, which is soil with 100% sand; it makes sense that a vertex represents 100% of one of the components because

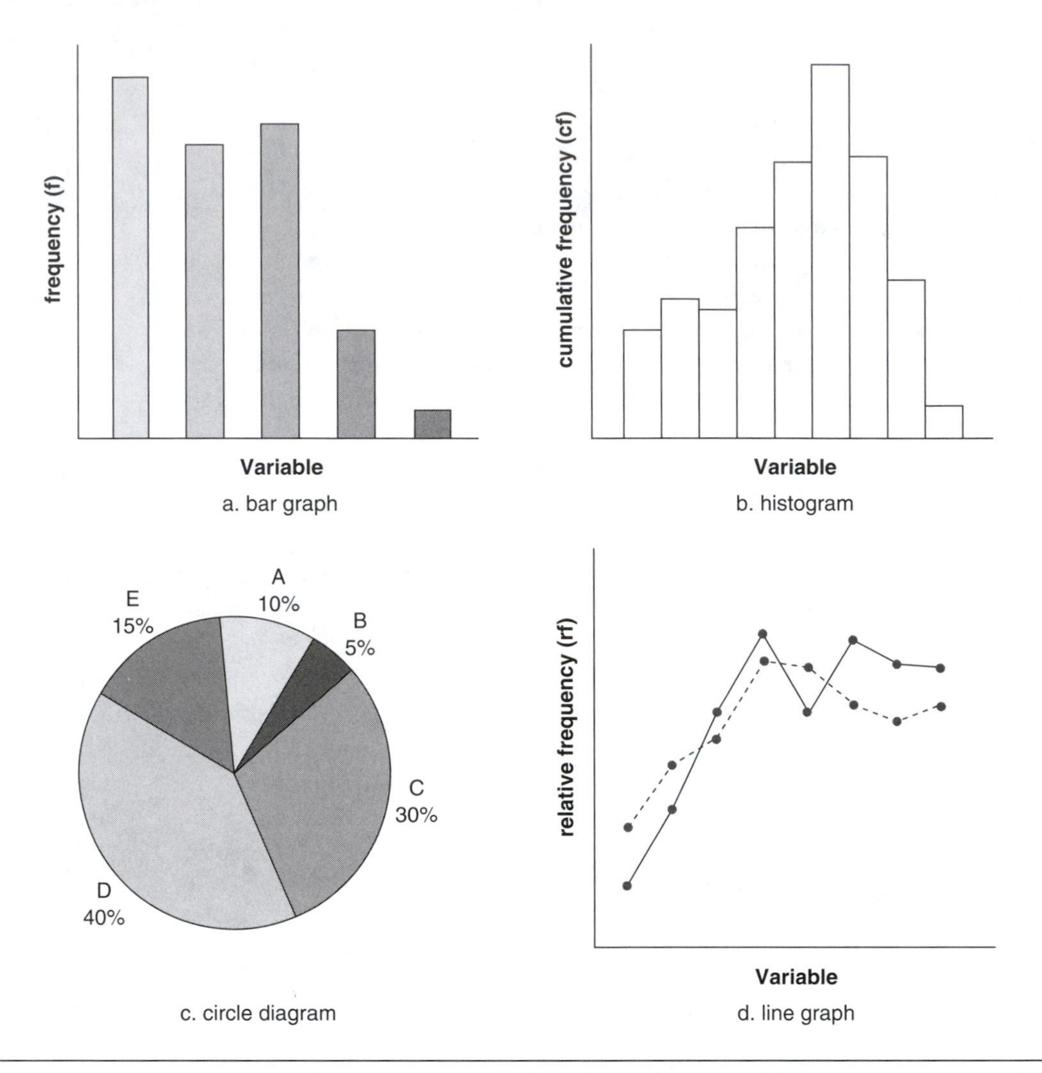

Figure 10.4 Schematic examples of graph styles (real graphs need captions, labels, and so on, as described in the text).

vertices are always at the 0% baseline of the other two variables. Any line parallel to the sand baseline shows all soils with intermediate proportions of sand, according to where that line intersects the vertical sand axis. Likewise, the proportions of silt and clay in a particular soil can be read off the vertical axes that bisect from the center of those sides. When these vertical bisector lines are left out of the diagram, as is sometimes done, interpreting ternary graphs can become confusing for the novice, who typically tries to interpret the sloped sides of the triangle as axes.

Above, we pointed out that the typical graph employs two dimensions of space, as coded by the two axes X and Y. In fact, even traditional flat graphs have three dimensions of space available to encode data, and any number of additional dimensions that can be thematically encoded. Spatially three-dimensional graphs are sometimes made, and sometimes they are a good idea, but not too often. Perceptually,

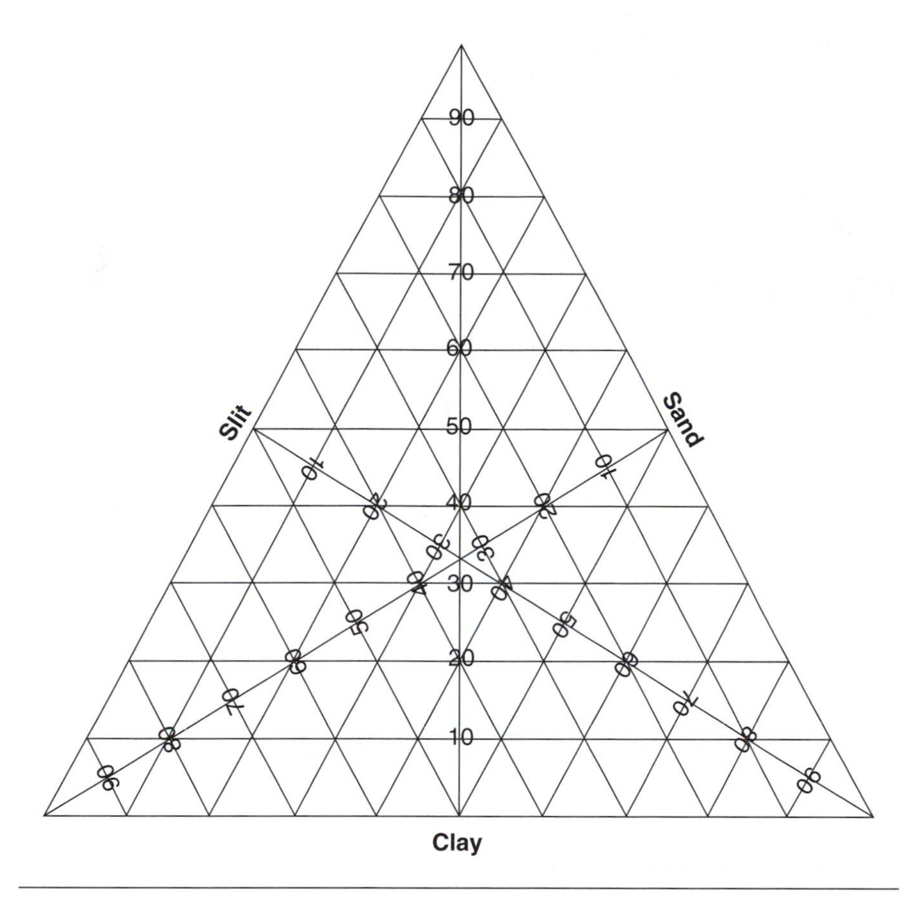

Figure 10.5 The classic example of a ternary diagram showing the relative components of sand, silt, and clay that determine soil texture (graphic by M. V. Gray).

locations displayed on the X- and Y-axes are seen quite differently than locations on the protruding/receding "Z-axis" of the third dimension. For example, there is foreshortening of the third dimension. Graphically, it is challenging to display information on the "virtual" third spatial dimension of a two-dimensional image, whether paper or CRT screen, in the same way it can be displayed on the "real" first and second dimensions. Using three spatial dimensions to depict information probably works much better on a dynamic display that can be manipulated by the viewer, because which dimension plays the role of Z can be alternated. On a static image, it is usually a better idea to depict a third variable by the use of hierarchically grouped bars, multiple lines, or various thematic codes such as size, colors, patterns, or texture[7] (yes, we know size is a spatial property but not one that requires locational coding on a graph axis). A particularly effective graphing style for geographic

[7]We do not like **stacked bar graphs** as a way to show more variables of data, however. They obscure direct comparisons across classes except for the class at the bottom of the bars, because only that class starts at the same baseline location on each bar.

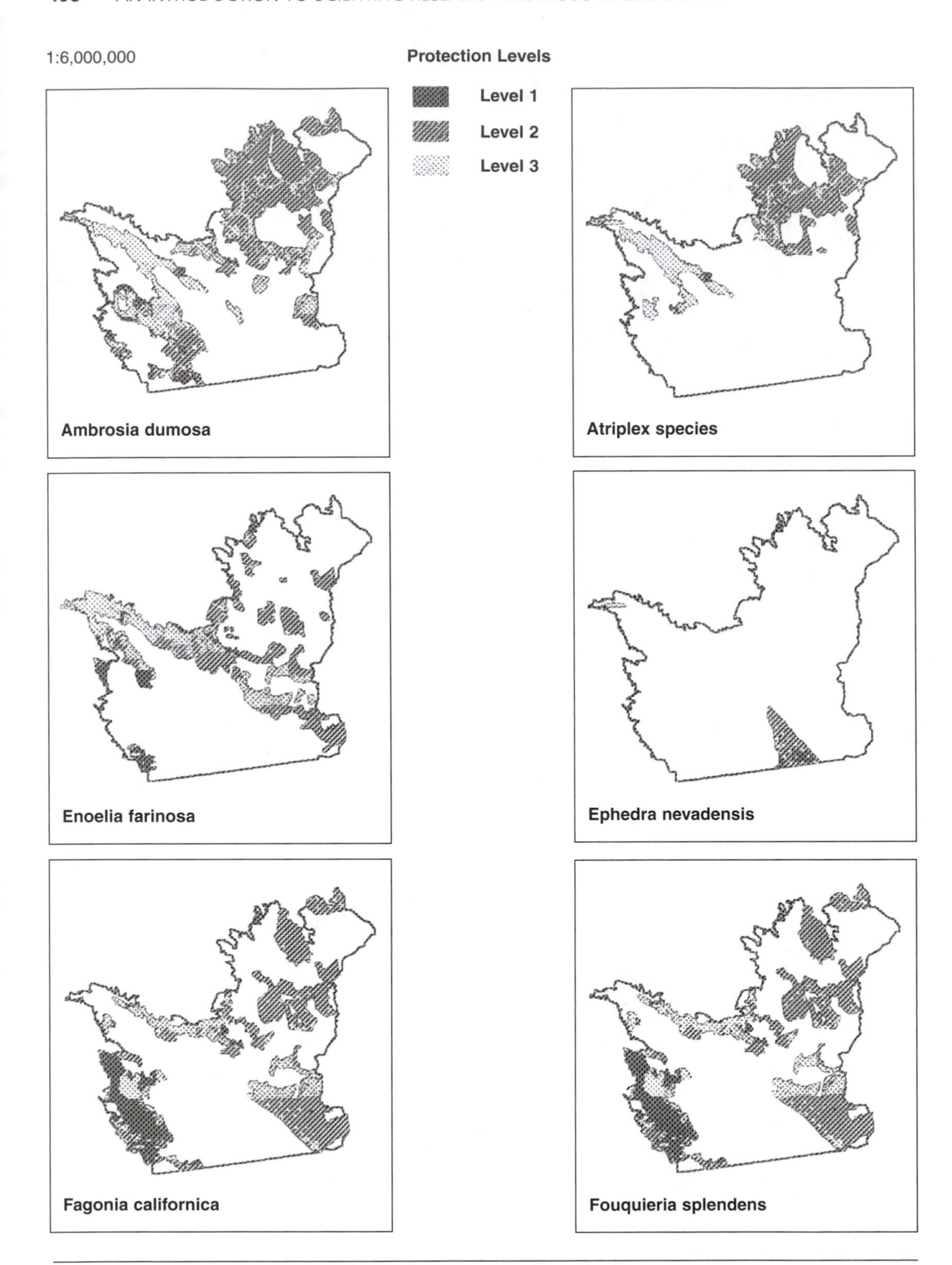

Figure 10.6 Example of a small multiples graph/map; depicts the management status for species of desert-scrub shrubs in the Colorado (Sonoran) Desert of southeastern California. (Graphic by M. V. Gray. Reprinted with permission.)

data in many situations is the **small multiples** graph (Figure 10.6), which is really a type of thematic map when its repeating graph space is an earth-surface base map like in this figure.

We offer a final admonishment about graph dimensionality. Many people, especially new researchers, have a predilection for graphing data with three spatial dimensions when they are only depicting two dimensions of information. For example, just about any graph style can be converted into **simulated 3D**; 3D bar graphs using rectangular prisms are a common example. We agree with the advice of most graphical experts, such as those whose books are included in the Bibliography, that the use of the third dimension in this way for purely "decorative" purposes, when it carries no information, is bad graphing practice.

Maps

Maps are graphic displays that depict earth-referenced features and data (geographic information), using a graphic space that represents a portion of the earth's surface.[8] The earth surface depicted in this graphic space is designed to resemble the actual earth surface more or less closely. But either way, as a map rather than a graph, the actual spatial layout of features on the earth's surface is always shown directly as a spatial layout on the map. Maps are obviously the quintessential geographic display, and geographers, specifically cartographers, are the experts of the map as a data display tool. Therefore, just as we apprised you of the need to take specific data analysis courses in Chapter 9, we urge you to take at least one course that focuses on cartographic display and interpretation. There's much more worth learning about designing and using maps than we cover here.

There is an important distinction between reference maps and thematic maps. **Reference maps** focus on depicting a variety of actual earth-surface features as accurately and precisely as possible. In an effort to achieve general-purpose functionality, reference maps tend to show the most significant features that a person could potentially perceive directly while moving about the planet. Feature locations are encoded with a **coordinate system**; the latitude-longitude **graticule** is just one of many. "Significant" features are typically things that are large, relatively stable over space and time, relevant to the basic shape of the earth's surface from a human perspective, and so on. Standard examples include ground surface topography, water bodies of sufficient size, forested areas, cities, transportation networks, and so on. In truth, though, whatever individual or organization chooses what to include and what to exclude on a map is intentionally or inadvertently working from a

---

[8]We refer here to a traditional conception of a cartographic map. Terms like "map" and "mapping" are extremely broad and are used in many disciplines to refer to a spatial or structural representation of information of any type. Not only are there maps of other planets, of the universe, of molecules and electrical circuits, there are "maps" in the mind and in the brain, in families, in cultures, in administrative and economic institutions, in the genetic code, and so on. Indeed, central intellectual and informational concepts such as symbol, representation, semantics, and metaphor all involve a notion of "mapping" from one thing to another.

conception of what map viewers do or do not need to see, or should or should not be allowed to see. As with graphs, you can call it propaganda or call it informational impression management.

Thematic maps, in contrast, are special-purpose displays. They focus on showing the spatial distribution of one or a few thematic variables, often variables like "AIDS incidence" or "average insolation" that are not easily perceptible from a single place or at a single moment, if at all. Unlike reference maps, thematic maps depict very little actual earth-surface detail, and locational accuracy and precision tend not to be their strengths. Thematic maps are really hybrid "map-graphs" that use a simple geographic base map for the graphic space so that thematic variables can be "geo-referenced" (Chapter 12). See Figure 10.8 for an example of a thematic map; we discuss the map further below.

Having said that maps always depict aspects of spatial layout directly is not to claim that this spatial layout is necessarily undistorted on the map—far from it. Maps always distort the earth's surface and the features and variables arrayed thereon, including spatial layout and virtually all other properties and attributes. Perhaps foremost, maps always require **selectivity**. There is no way to show all the details of a portion of the earth's surface on a map, no matter how big your supercomputer is, so decisions must always be made as to what to include and what to exclude. Even to the degree it is possible to include great detail on a map, it is typically unwise to do so. That's because selectivity and carefully applied distortion is a considerable source of the effectiveness of maps as communication displays. Remember the London tube map in Figure 10.1?

Even if it did not serve the purpose of communication, all spatial properties of earth features, including size, shape, distance, direction, and connectivity, would never be shown accurately on a map. That's because they *cannot* be. It has been proven, and it may strike you as intuitively evident if you think about it, that you cannot flatten the curved earth surface onto a piece of paper or a computer screen without distorting one or more of the spatial properties listed above. The standard demonstration of this is to imagine peeling an orange in one piece and then flattening it on a table—you will have to tear the peel up pretty well to get it all flat. The topic of cartographic **projection** concerns the many different ways of "developing" (flattening) the earth surface, or any portion thereof, and the patterns of spatial distortion that result. Figure 10.7 shows just two examples of projections, the Mercator and the sinusoidal, and their patterns of distortions. There are in fact more than a hundred projections that cartographers have studied over the centuries, and an infinite number are possible. Don't be misled by the fact that our figure shows *global* projections. It is true that necessary spatial distortion is greater as more of the earth surface is depicted (smaller cartographic scale), but all maps distort spatiality, even maps of your neighborhood. It's just that the necessary distortion is so small on very large-scale maps that you don't see it. Different projections should be chosen for different map uses, so that important information is relatively undistorted or distorted in acceptable ways, whereas less important information can be more distorted. The Mercator map was created for use by navigators on the high seas, because it shows constant compass directions as straight lines on the map. To achieve this considerable benefit, it must greatly exaggerate the size of landmasses toward the poles relative to those toward the equator; did you know that Brazil is actually a little over four

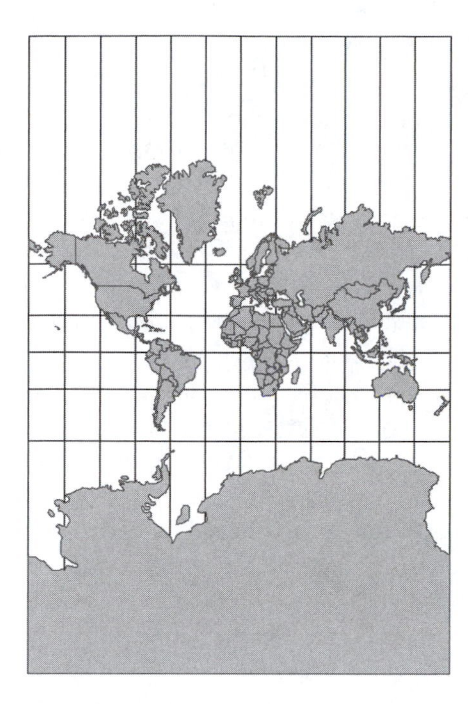

a. Global Mercator projection.

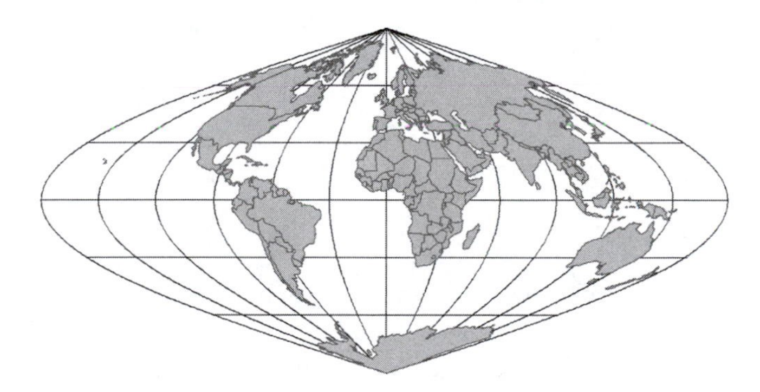

b. Global sinusoidal projection.

Figure 10.7 Two examples of global map projections, (a) Mercator and (b) sinusoidal. The Mercator greatly distorts the size of landmasses relative to each other, and the sinusoidal greatly distorts their shapes (both projections distort other spatial properties too). (Graphic by Sarah Battersby. Reprinted with permission.)

times *larger* than Greenland? Because of this exaggeration, the Mercator is a poor choice as a general world map for atlases or K–12 education, a job to which it has unfortunately been put at times in many parts of the world.

Maps distort spatial and nonspatial properties for other reasons than projection. As we mentioned above, maps cannot avoid simplifying reality, and we wouldn't have it any other way. **Generalization** refers to the amount of detail included on features as they are mapped. It is essentially an issue of simplification, but also includes aspects of selection and enhancement of features of particular interest. For example, maps show rivers with fewer meanders and coastlines with fewer crenulations than they actually have. Thematic maps show regions, such as climate or cultural regions, as much more homogeneous and crisply bordered than they actually are. Furthermore, there tends strongly to be more simplification of detail as one represents larger portions of the earth surface, that is, as one makes smaller-scale maps. Features are also exaggerated for graphic legibility and to direct attention to more relevant information. Take a look at a road map of a country or continent. Measure the width of the highway and use the scale factor to translate it to the actual width it would have in the real world at that graphical width. You'll come up with an answer in the miles or tens of miles. But if it were shown at accurate scale, it would be hard to see.

The symbols on maps can be relatively easy or difficult to interpret, and when they are interpreted, they are often *mis*interpreted to some degree. Maps represent features and properties of the earth using spatial and nonspatial graphic properties, including location, size, shape, patterns, textures, color hue, color value, and more. Above, we pointed out how maps resemble the reality they represent more or less closely. That is, symbolic representation on maps is relatively **iconic** (mimetic) or **abstract** (conventional). The fact that maps show spatial layout on the earth with spatial layout on the image is an iconic aspect of how they represent. More distant features are usually shown as further apart on the map, features that are connected are usually shown as connected, and features to the southwest of other features are usually shown to the lower left (assuming a north-up map orientation). Other symbols are more or less iconic. Larger **graduated circles** represent larger cities, green represents vegetation and blue represents water, and a pick and shovel represents a mine. Some symbols, in contrast, are quite abstract. A bunch of contour lines does not look like a hillside, a checkered pattern does not look like high fertility, and a small square does not resemble the scene of a tornado touchdown. Even vegetation does not always look green, and water certainly does not look blue all the time. The most abstract of symbols on maps are the words on it. How does the phrase "tourist attraction" look like Carhenge, a simulated Stonehenge in western Nebraska made from buried automobiles?

Just as there are principles of good graphing, there are principles of good mapping. In fact, such principles are more complex for maps and harder to present as a simple list. That is why we urge you to take a course or two that covers cartographic design and read books like those we list in the Bibliography. The rationale behind good map design is the same as it is behind the design of other displays. Maps should be designed to facilitate effective and efficient communication to one's self and other people. A variety of general pieces of advice about map design follow from this, including that you should choose relevant and high-quality data, show them clearly and truthfully, highlight the important and downplay the less important, focus on the data and not on mere decoration, and so on. One of the central issues in the design of maps, especially thematic maps, is to make good choices for map symbols to represent data variables and their values. This is the same concern we have with graphs when we choose a bar graph over a line graph, or vice versa, because of the nature of the underlying variables being graphed. But the issues are more involved when it comes to maps. For example, graphs usually do not incorporate color except for decorative purposes, whereas maps nearly always involve various aspects of color (**hue, value, saturation**) as communication symbols, or **visual variables**. And of course, graph space is abstract and shows metaphorical "data space," whereas map space always corresponds to earth-surface space, with greatly varying amounts of detail, precision, and realism called for. Because maps show the earth, they require more sophisticated spatial annotation, such as compass roses, distance scale factors, and coordinate system grids. These features are called **map elements** because they describe basic characteristics of the mapped portion of earth surface.

Take the example of **choropleth maps**. This is a widely used style of thematic mapping in which discrete regions, such as countries or census tracts, are filled with colors, shades of gray, or textures that represent the values of some quantitative variable (see Figure 10.8). Since a finite number of classes are mapped, the data must be grouped into classes or intervals; if you try to avoid the grouping by mapping a continuous visual variable onto a continuous data variable, your visual system will do the grouping anyway, resulting in a *de facto* set of about 5–10 classes that your visual system decides how to create. All the usual questions result about how many and which classes to use. What visual variables should be used? Hue (red, yellow, blue, and so on) is usually a poor choice here, with some notable exceptions, because people generally interpret it to correspond to qualitatively different classes rather than quantitatively spaced amounts; in other words, hue is good for nominal variables, as in Figure 10.8. Area patterns like dots or hatchings are usually a poor choice for the same reason. Color or gray tone values typically work well. However, choropleth maps are not too good if you are mapping a thematic variable that is assessed on something other than spatial area. It's fine for precipitation because that variable is assessed per areal unit like square kilometers. It's not good for population-based variables, such as birth or death rates, because large areas with low populations (such as Siberia) are given far too much visual emphasis compared to small areas with high populations (such as the city of Moscow). Many authors recommend an alternative in these cases, such as the **population cartogram** shown in Figure 10.8.

New Trends in Scientific Visualization

Humans have made forms of reference maps for several thousand years. In comparison, thematic mapping is a rather modern invention, having been around only since the 18th century or so. But since the 1970s and 1980s, there has been a bloom of innovation in the display of geographic information—indeed, in the display of all types of information. These innovations have of course been spurred on primarily by computer hardware and software. Computer technologies, and other technologies dependent on the computer, have made possible a variety of computationally intensive analytic techniques, and a variety of new display formats and "information interfaces." They have also led to a stunning increase in the amount and complexity of data and other information we have to comprehend and communicate in the first place. This has provided us with great challenges even as it has provided us with great opportunities.

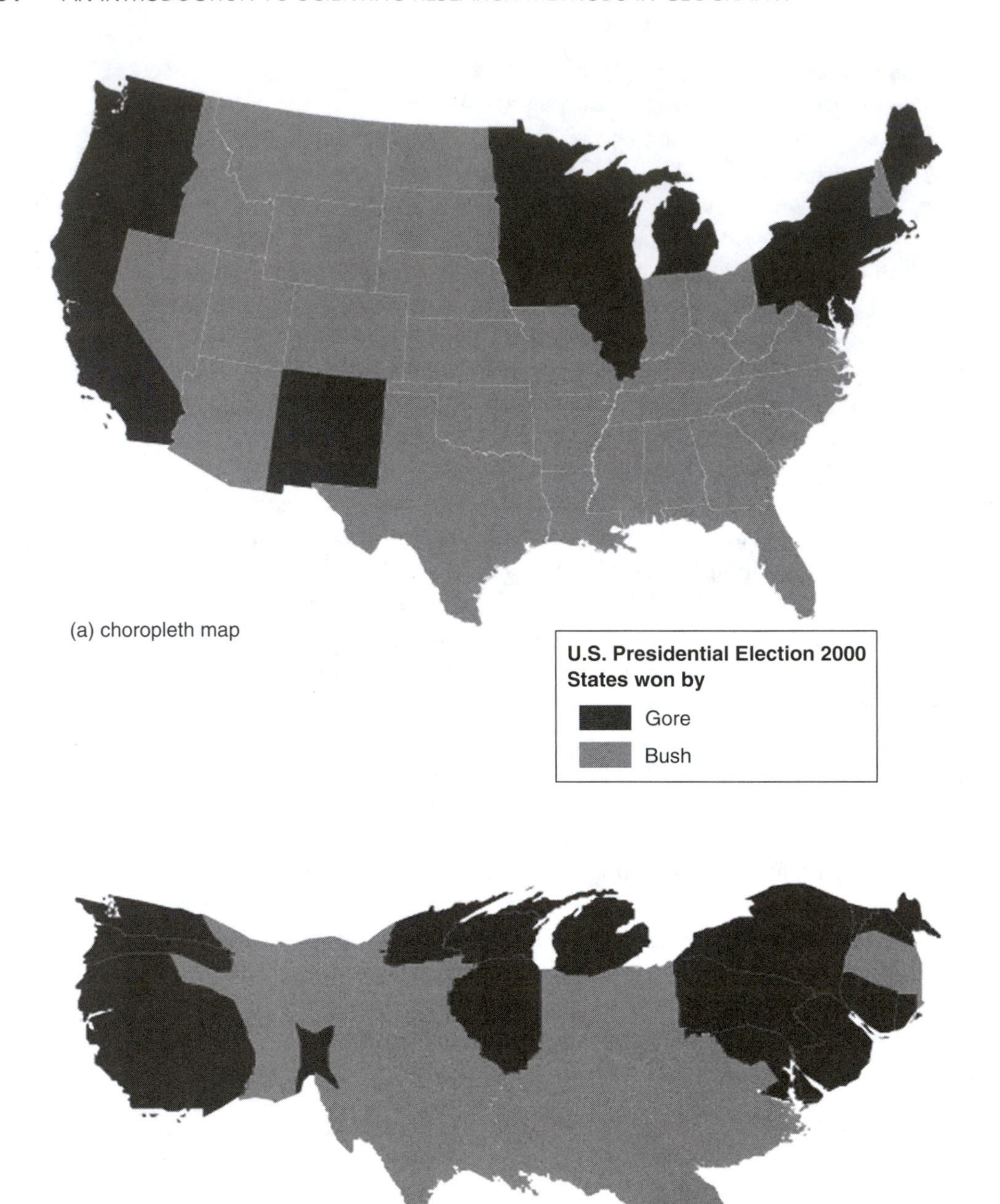

Figure 10.8 Thematic maps of the conterminous United States showing the breakdown of the electorate by state in the 2000 U.S. presidential election. The top map (a) is a traditional choropleth map showing geographic areas accurately but distorting each state's population. The bottom map (b) is a population cartogram map showing each state's population accurately but distorting geographic areas. Cartographic principles would suggest using less saturated colors, but intense red and blue have become standard symbols for the Republican and Democratic parties in the United States since this election. (Graphic by Sara Fabrikant. Reprinted with permission.)

We noted above that geographers and other scientists are quite interested in the techniques of exploratory data analysis. **Information visualization** is a key to exploratory data analysis. Such techniques are especially valuable with large data sets that are spatially and temporally distributed. In Chapter 12, we discuss the power of visualizing information against its geographic background with the help of computer geographic information systems; this is often called **geo-visualization.** Given the power that spatiality lends to the search for patterns in data, as when graphs display numerical information pictorially, it has been proposed that the interpretation of very large data sets that are not necessarily spatial (such as a corpus of news stories) could benefit from display in spatial and even geographic form, such as in the form of simulated landscapes; this technique is called **spatialization.** Other technologies of the last couple decades allow researchers to generate multiple views of data, at multiple scales and from multiple perspectives, relatively quickly and easily. The speed with which the visualizations can be created is important, because the nature of human perception and working memory requires that information must be accessed within particular time parameters if it is to be mentally integrated, compared, and so on, by an analyst.

Technologies allow some truly novel ways of looking at data. Manipulable or interactive graphics can be made, graphics that incorporate pop-up menus, "slider bars," "brushes," "magic lenses," and more; see texts listed in the Bibliography. **Animations** allow dynamic displays so data that include change over time can be directly visualized. In fact, even nontemporal information can be expressed with temporal visual variables, allowing us to increase the information dimensionality of our displays by adding rate, rhythm, regularity, and other temporal properties to our toolbox of visual variables. **Augmented reality** combines digital displays of data superimposed and appropriately keyed and updated to the actual surrounds in which researchers find themselves; it might be viewed via electronic "data goggles." This could give researchers access to data and analysis in new ways and in new situations, for instance, while they are in the field. Going further, **virtual reality (VR)** is a way of simulating places in natural or built environments that offers the prospect of allowing people access to places without the cost of actually traveling there. As a research tool for data collection in human behavioral geography, VRs also allow one to exert empirical control (see Chapter 7) over variations in places that is difficult or impossible to achieve otherwise. There are a variety of such interactive, real-time, 3D graphical displays that display information from a first-person perspective and change appearance appropriately in response to movements by users (so-called **active control**). Various virtual systems include desktop displays, projected displays, "caves," and fully immersive systems. And while we're at it, why stop at *visual* display? There is great interest in the development of nonvisual and multisensory display technologies, and not just for people with visual disabilities. Researchers who specialize in studying and developing geographic information techniques have developed or at least proposed **sonifications** (sound displays), **tactilizations** (touch displays), and more ("smell maps" anyone?).

Of course, there is no guarantee that all of these technological possibilities really work well as display techniques. In fact, we are confident that some of the proposed and developed techniques do not work well. Remember, effective and efficient communication to humans is the point. Novelty and the "gee whiz!" factor might help you get funding, but by themselves, they are not adequate reasons to actually use something to interpret and communicate data. Educated, talented, and thoughtful judgment can certainly help decide whether a display method is a good idea. Researchers who specialize in studying and developing display techniques, including geographers, are also applying a scientific approach to help understand and evaluate the display innovations of today and tomorrow.

Review Questions

- What is data display, and what are some purposes of data display in research?
- What should we consider when deciding whether to use displays, as opposed to other forms of communication?
- What does the design principle of "effective communication" mean, and what are some considerations that stem from this principle?

Tables

- When does it make most sense to use tables to display data, rather than other forms of displays?
- What are distribution and contingency tables? What are frequency, relative frequency, cumulative frequency, and cumulative relative frequency?
- What are some choices for how to organize metric-level data into class intervals?

Graphs

- When does it make most sense to use graphs to display data, rather than other forms of displays?
- What are some principles of good graphing and specific design guidelines that derive from them?
- What are distribution and relationship graphs, and what are specific styles of each?
- What are the basic issues of dimensionality in graphing (including spatial, temporal, and thematic dimensionality)?

Maps

- When does it make most sense to use maps to display data, rather than other forms of displays?
- What is the distinction between reference and thematic maps, and why is it so important to considerations of map design?
- Selectivity, projection, generalization, and varying symbol abstractness are always involved in mapping. Why are they always involved, and why are they potentially misleading to map viewers?

New Trends in Scientific Visualization

- What is information visualization (including geo-visualization), and how does it go beyond the traditional graphical display methods of graphing and mapping?
- What are animation, augmented reality, virtual reality, sonification, and tactilization?

Key Terms

abscissa: the horizontal axis on a Cartesian coordinate graph; also known as the X-axis

abstract symbols: map representations not resembling the earth phenomenon they stand for; also known as conventional symbols

active control: property of information displays that change appearance appropriately in response to movements by users

animation: information visualizations in which change over time or other properties of data are displayed with moving or changing displays

augmented reality: interactive, real-time, graphical displays that display information superimposed over a person's actual surroundings as seen in the visual field

bar graph: graph style for discrete variables showing the frequencies or some other variable for each value on the X-axis by means of the heights of rectangular bars; also known as a bar chart

Cartesian coordinate axes: standard graph dimensions consisting of X- and Y-axes that meet at a right angle

choropleth map: widely used style of thematic mapping in which discrete regions are filled with colors, gray shades, or textures that represent the values of a quantitative variable

circle diagram: distribution graph for nominal variables showing relative frequencies; also known as a pie chart

class intervals: categories constructed to group metric-level data values into discrete ranges

contingency table: table showing relationships between discrete variables by cross-tabulating frequencies of the individual variables against each other

coordinate system: mathematical system for coding locations on the earth surface, particularly on reference maps

cumulative frequency: the frequency of each value in a data set accumulated as one moves from the lowest to the highest value, so that the frequency of the final value is the total number of scores in the data set

cumulative relative frequency: the relative frequency of each value in a data set accumulated as one moves from the lowest to the highest value, so that the relative frequency of the final value is 1.0 or 100%

curve-fitting graph: graph showing a statistical model that has been fitted to a data set

data display: graphical techniques for data communication and analysis, including tables, graphs, maps, and other visualizations

descriptive index table: tables presenting summary descriptive indices, such as central tendency, variability, or relationship

distribution graph: graph presenting the distribution of a data set by showing all of its values and their frequencies, relative frequencies, or cumulative frequencies

distribution table: tables presenting the distribution of a data set by showing all of its values and their frequencies, relative frequencies, or cumulative frequencies

exploratory data analysis: submersive, graphical, and ad hoc approach to data interpretation that emerged during the 1970s

figures: pictorial graphics used in scientific communication, including data displays, drawings, and photographs

frequency: how many scores have each possible value in a data set

generalization: the amount of detail included on features as they are mapped; primarily a function of simplification, it also includes aspects of selection and feature enhancement

geo-visualization: geographic information visualizations that include depictions of the earth-surface context of the information

graduated circles: common thematic map symbol in which a circle is placed at a location on the map so that its area is equivalent to the value of a quantitative variable

graphs: pictorial representations of data that metaphorically map data values onto the spatial and nonspatial properties of images

graticule: the latitude-longitude coordinate system on the earth or a representation of the earth

histogram: bar chart whose bars' widths represent the range of a quantitative class interval for a metric variable

hue: quality of color corresponding to its wavelength or mixture of wavelengths, including red, yellow, blue, and so on

iconic symbols: map representations resembling the earth phenomenon they stand for; also known as mimetic symbols

information visualization: any graphical or pictorial technique for displaying data, especially referring to computer techniques that take advantage of the computer's capabilities to create displays difficult or impossible to simulate on a traditional flat and static image

interruption tics: small marks used to indicate that a ratio-level Y-axis has been truncated

line graph: graph style for continuous variables showing the frequencies or some other variable for each value on the X-axis by means of the heights of points connected by straight line segments; also known as a polygon graph

map: graphic display depicting earth-referenced features and data with a graphic space that represents a portion of the earth's surface

map elements: symbols on maps, such as compass roses and coordinate grids, that describe basic characteristics of the portion of earth surface mapped

ordinate: the vertical axis on a Cartesian coordinate graph; also known as the Y-axis

outliers: unusual or extreme values in a data set

polar coordinate axes: circular graph dimensions consisting of a radius axis of varying length and angle; appropriate for cyclic data, such as directions in space or measurements in repeating time periods

population cartogram: style of thematic mapping in which discrete regions are expanded or shrunk so their areas represent the values of a quantitative variable

projection: geometric or mathematical solution to developing (flattening) the earth surface on a map

reference map: type of map that focuses on depicting a variety of significant earth-surface features as accurately and precisely as possible, with the intent of achieving general-purpose functionality

relationship graph: graph depicting the form and strength of relationships between variables

relative frequency: the frequency of each value in a data set, as a proportion or percentage of all the scores

saturation: quality of color corresponding to the brightness or intensity of its hue, varying from a pure gray tone to a pure chromatic hue

scatterplot: common way to graph relationships with any type of data; for each case, a dot is placed in the graph space at the intersection of each pair of values of the variables placed on the X- and Y-axes

selectivity: the fact that maps must necessarily depict less than all possible detail in the world

simulated 3D: graphing data with three spatial dimensions that depict only two dimensions of information; most authors advise against the practice, even though most computer graphics packages facilitate it

small multiples: type of graph or thematic map that repeats a framework graphic space several times in order to show the same variables according to instances of some repeating factor, such as data at several times or in several places

sonification: information visualization that incorporates sound as a symbolic medium

spatialization: information visualization using a simulated spatial entity, such as a geographic landscape, to metaphorically represent nonspatial information

stacked bar graph: bar graph simultaneously showing the distributions of two discrete variables by breaking each bar of the X-axis variable into the classes of a second variable

stem-and-leaf display: hybrid table-graph displaying a list of data values with the second digit of data precision (the leaves) arrayed in a line whose length spatially communicates frequencies, like a bar graph

tables: organized lists, arrays, or matrices of data

tactilization: information visualization incorporating haptic stimuli, such as texture or pressure, as a symbolic medium

ternary diagram: specialized three-variable relationship graph showing three dimensions arranged as the three intersecting bisector lines of an equilateral triangle, appropriate when the three variables are mutually interdependent proportions that must sum to 100%; also known as a trilinear or triangle plot

thematic map: special-purpose type of map that focuses on depicting the spatial distribution of one or a few thematic variables, usually with limited spatial accuracy and precision; essentially a hybrid graph-map

tic marks: very short line segments like hyphens indicating where particular data values fall along graph axes

value: quality of color corresponding to its lightness, varying from pure white through various shades of gray to pure black; characterizes both chromatic (with hue) and achromatic (black-to-white tones) tones

virtual reality (VR): interactive, real-time, 3D graphical displays displaying information from a first-person perspective and changing appearance appropriately in response to movements by users

visual variables: spatial and nonspatial properties of the appearance of map symbols that symbolically represent values of mapped variables

X-axis: the horizontal axis on a Cartesian coordinate graph; also known as the abscissa

Y-axis: the vertical axis on a Cartesian coordinate graph; also known as the ordinate

Bibliography

Cleveland, W. S. (1994). *The elements of graphing data* (revised ed.). Murray Hill, NJ: AT&T.

Dent, B. D. (1999). *Cartography: Thematic map design* (5th ed.). Dubuque, IA: William C. Brown.

Huff, D. (1993). *How to lie with statistics.* Illustrated by I. Geis. New York: Norton.

MacEachren, A. M. (1995). *How maps work: Representation, visualization, and design.* New York: Guilford.

Monmonier, M. (1991). *How to lie with maps.* Chicago: University of Chicago Press.

Muehrcke P. C., Muehrcke J. O., & Kimerling, A. J. (2001). *Map use: Reading, analysis, and interpretation* (rev. 4th ed.). Madison, WI: JP Publications.

Tufte, E. R. (1983). *The visual display of quantitative information.* Cheshire, CT: Graphics Press.

Wainer, H. (1997). *Visual revelations: Graphical tales of fate and deception from Napoleon Bonaparte to Ross Perot.* Mahwah, NJ: Lawrence Erlbaum Associates.

CHAPTER 11

Reliability and Validity

Learning Objectives:

- What is the meaning of reliability in scientific measurement, and what are the three main approaches to assessing reliability?
- What is the meaning of validity in scientific research, and what are the four main types of research validity?
- What is the relationship of reliability to validity?
- What are some ways to increase each of the four types of validity in research?
- What are researcher-case artifacts, and what are some ways to minimize their effects?

In this chapter, we discuss two fundamental concepts—reliability and validity—that are relevant to the collection and interpretation of all types of data in all areas of scientific research, including geography.

Reliability

Reliability is the "repeatability" of scores or measured values of variables. Perfect reliability occurs when you measure something the same way twice and get exactly the same value each time, assuming the thing is actually the same at the time of the two measurements. For example, reliably measuring a person's weight means that the scale reads exactly the same value on two occasions when the person actually weighs exactly the same both times. In the language of Chapter 2, high reliability occurs when the manifest variable (the variable as measured) remains nearly unchanged whenever the latent variable (the underlying construct) remains unchanged. When two or more measurements of something are quite different, even though the thing has not changed much, the measurement is of low reliability.

According to measurement theory, any observed score or data value can be partitioned into two components: (1) a **systematic component** that reflects whatever underlying construct we are consistently picking up with our measurement operations and (2) a **random component** that reflects all influences on our measurements that vary nonsystematically from measurement event to measurement event. Low reliability is caused by the random component of measurement. An example would be the way census counts of populations in particular census tracts can vary a bit depending on which day and time of day a census worker attempts to count people. It's important to recognize that only random or nonsystematic errors of measurement cause lowered reliability. Errors in the systematic component are consistent across measurement events; they cause biased or inaccurate measurement, not unreliable measurement. If I step on an old-fashioned bathroom scale to weigh myself, the fact that I read off a slightly different number on the basis of the angle at which I happen to view the dial each time would cause lower reliability. In contrast, the fact that the springs in my scale are old and stretched out would cause the dial to indicate that I weigh some 15–20 pounds lighter than I really weigh, and it would do this consistently each time I step on the scale. As long as the reading is too light by the same amount each time, it is a reliable measurement that is inaccurate. In other words, this scale would be biased low—a poor measurement tool but good for the ego.

Low reliability has a variety of detrimental effects in research. Unreliable data contain more **noise**, which is random error. Noise in data decreases the precision of estimation and the power of hypothesis tests (Chapter 9). Noise weakens ("attenuates") measured relationships between variables; a correlation between two variables will be smaller when measurement reliability is poor, even when the relationship of the underlying constructs is the same. In the extreme, completely unreliable measurement precludes finding any relationships at all between variables—random variables are uncorrelated with one another. Ultimately, low reliability limits the validity of your research conclusions, although high reliability does not by any means ensure valid research conclusions. In other words, saying that data are reliable in the technical sense does not mean that you can necessarily *rely* on their truthfulness. High reliability is necessary but not sufficient for high validity.

Given this, it is important to think about measurement reliability in research. In many types of research, it's a good thing to explicitly assess reliability. If it is too low, you can take steps to increase it. If it is satisfactorily high, you can report that, which will help convince reviewers of your research that you are measuring something in a high-quality and dependable manner. Reliability is quantified as some type of correlation or proportion of agreement between two (or more) sets of measurements of the same thing. This is a little tricky. According to measurement theory, repeated measurements of perfect reliability agree exactly only when the measurements are done in exactly the same way on the same entity at exactly the same time, which is impossible (sort of the measurement equivalent of the cryptic wisdom that you cannot step into the same stream twice). For example, I can't put two rain gauges in exactly the same place at exactly the same time. In practice, researchers assess reliability by applying a compromise approach to measuring something more than once. They take one of three specific approaches to assessing reliability via repeated measurement, none of them ideal:

1. ***Remeasurement reliability.*** In this approach, the researcher simply measures the same variable using the same measurement procedures on the same cases, but not at exactly the same time. In the context of explicit reports with humans, it is usually called "test-retest reliability." It is not an ideal measure of reliability in every situation insofar as the repeated measurements are not done at exactly the same time. This is problematic insofar as the underlying construct may change over time so that you end up measuring something a little different on the two occasions; for example, measurements of air temperature change over the course of a day. In some situations, you may need to take steps to deal with the fact that measurement can change the case being measured. If so, use alternate ways of measuring that are considered equivalent on repeat measurements. In the extreme, measurement can even destroy the case; for example, as we mention in Chapter 14, biogeographers have sometimes killed their cases in order to study them, which would obviously preclude this approach to assessing reliability.

2. ***Internal consistency reliability.*** This approach can be applied when a single construct is operationalized via multiple measurements to begin with. For example, place attitudes are typically measured with a survey that asks people several questions concerning their beliefs toward a place, not just one. The effectiveness of a geographic information display would be based on several questions or several tasks involving the display, not just one. The multiple measurements are meant to be redundant, partially overlapping, measures of the same construct. If so, scores on the separate measurements should largely agree with each other. For instance, responses to half the measurement items could be correlated with responses to the other half, an approach known as "split-half reliability." The most sophisticated form of the internal consistency approach correlates the score for each item with that of all the other items combined; these are averaged to produce a single **item-total correlation**[1] ("coefficient alpha," which is unrelated to the α probability introduced in Chapter 9).

3. ***Inter-rater reliability.*** The final approach to assessing reliability is applicable whenever one's data are based on judgments by researchers or their assistants. Such data usually come from the coding of open-ended explicit reports or nonreactive measures such as physical traces, behavioral observations, and some archives. With such data, reliability is assessed by looking for the amount of agreement between

---

[1]Incidentally, item-total correlations are very useful for developing the test or survey in the first place. Items that have very low correlations with the other items, below .3, are poor items and probably should be omitted. Negative item-total correlations mean that the item has to be reverse-scored before it is combined with the other items.

two (or more) observers or coders who work on the same subset of records (that is, the same plots of land, the same videos of behaviors, and so on). This approach to reliability is sometimes called "inter-observer" or "inter-coder" reliability. We considered this approach in some detail in Chapter 5, where we discussed the coding of open-ended records.

Perfect reliability would be great, but it is unattainable. That's not true just in human geography, or in other social and behavioral sciences, but in any basic or applied field of science, including physical geography. But *high* reliability is certainly attainable. What constitutes "high" does vary from scientific field to scientific field; it is generally higher in the physical sciences than in the life sciences, where it is often higher than in the social and behavioral sciences. Even within specific disciplines, reliability standards vary quite a bit across topical domains, which is almost entirely because of the fact that potential reliability varies so much as a function of what and how you are measuring. The depth of the ocean within a particular area just offshore can be measured more reliably than the average concentration of a pollutant in that area. The number of people living in a household can be measured more reliably than their attitudes about using mass transit. It is nice to measure with reliability of well over 90% agreement across measurement occasions, and even in the less reliable domains of human geography, at least 80% agreement is preferable. When reliability is lower than this, steps should usually be taken to improve it. Again, we stress that in some domains, high reliability is elusive and very difficult to attain. If reliability is too low, however, there is no point to measuring and doing science in a particular domain—random numbers do not make useful data.

How can we increase measurement reliability? A basic dictum of measurement theory is that combining multiple imperfect measurements of the same thing produces a single measurement that is more reliable, as long as the measurements are not *completely* imperfect. So you can increase reliability by increasing the number of items, observations, tasks, raters, and so on. Reducing the amount of noise in each measurement is obviously a good way to increase reliability. Thus, you can increase reliability by increasing the quality of your measurement procedure—use better sensors on satellites, make sure the thermometer is placed away from intermittent influences such as occasional direct sun, clarify survey questions to get respondents to understand them more uniformly, clarify coding procedures for open-ended data, and so on. Another approach to increasing reliability is to increase and improve the training of all people who help generate data, including assistants who count, interview, code, and so on.

Validity

Validity is the "truth-value" of research results and interpretation. Given that research is largely an attempt to increase the truthfulness of our understanding of the world, validity is a core concern for researchers. A widely used typology of validity in research includes four classes: internal, external, construct, and statistical conclusion.

Internal Validity

Internal validity concerns the truth of conclusions about causal relationships. As we discussed in Chapter 7, causal conclusions are thrown into doubt when there are possible alternative causal variables other than the variable we conclude is causing patterns in our data. That is, "threats to internal validity" occur when researchers interpret a statistical relationship between two variables in their data as indicating a causal relationship, when in fact some other variable is the real cause. We also pointed out in Chapter 7 that increasing internal validity is the primary purpose of empirical control in research. Our discussion there showed that the likelihood of high internal validity in a particular study varies greatly with its research design; internal validity is generally maximized in experimental studies as opposed to non-experimental studies. Furthermore, particular threats to internal validity, rather than others, are more or less likely to operate in a study depending on its research design.

There are many specific potential threats to internal validity in research studies. In Chapter 7, for instance, we discussed some specific threats that occur in developmental research designs, including history, cohort effects, and differential mortality. There is one especially interesting and subtle threat to internal validity that we want to consider in detail here, however. **Statistical regression to the mean** is the phenomenon that extreme scores, those far from the central tendency of the data set, tend to be less extreme when remeasured (often called "regression" for short, the phenomenon is the basis for the name of the general statistical technique of regression analysis mentioned in Chapter 9). For example, a year in which a city has an exceptionally high number of new housing starts will probably be followed by a year with somewhat fewer housing starts, other things being equal. In the same way, a year in which the city experiences an exceptionally low number of new housing starts will probably be followed by a year with more.

The critical point is that these changes in housing starts from one year to the next will tend to occur even without an economically or socially substantive cause, such as a change in interest rates or the city's population. The principle that extreme measurements tend to be followed by less extreme measurements—they *regress* back toward the mean—applies over and above any substantive factors that may operate to cause less extreme values. The phenomenon was identified and named by Francis Galton, the 19th-century English scientist and cousin of Darwin. He observed that sons of very tall fathers tend to be shorter (sons of short fathers tend to be taller as well). This phenomenon happens even when mothers themselves are not exceptionally short. Another standard example is the batting average of baseball players, which is the percentage of "at-bats" in which the player gets on base, over the course of the season. After a month of the season, a small handful of major league players may have averages over .400, which is a very high average (showing how tough it is to hit a major-league pitch). But as the season continues, these averages come down one by one, so that by the end of the season, no player has an average over .400; at least it has gone that way since Ted Williams finished the 1941 season with an average of .406. Why does this seasonal pattern nearly always happen? Is it because pitchers come to figure out hitters they have seen earlier in the season? Is it because the days get hotter, which allows pitches to move to the plate faster? Do batters tire as the season wears on? Although some of these factors may operate, the major explanation is regression to the mean. The good news is that exceptionally poor batting averages will probably increase as the season goes on, at least for those players who don't get benched.

Why does regression to the mean happen? The answer is that the random component of a set of measurements expresses itself in a likely manner—moderately—after having expressed itself previously in an unlikely and extreme manner. One variable can show regression over repeated measurements whenever its reliability is less than perfect. Even more generally, a variable can show regression with respect to another variable, so that cases with extreme scores on one variable have less extreme scores on the other variable. This can happen whenever the two variables are less than perfectly related; as we learned above in this chapter, unreliability is a special case of a variable being imperfectly correlated with itself. When relationships are imperfect, one variable can only partially be used to predict the other. The two variables share some but not all of their variation. The part they do not share is random variation—noise. It's this noise that leads to regression, first by randomly acting unusually and contributing to a very high or low score, then by randomly acting more usually, neither very high nor very low.

But why is moderately valued noise more common than high or low noise? If the noise is random, aren't all of its possible values equally likely, as implied in our discussion of random sampling in Chapter 8? The answer is no, because the noise is not one thing but a *collection* of things; in the parlance of probability theory, a noise value is an "event" rather than an "elementary outcome." Take a thoroughly mixed jar of 50 red balls and 50 white balls, all balls equal in size. Given a random selection process, each individual ball has an equal chance of being selected—1% on the first pick. (Because our jar has an equal number of red and white balls, the events "red" or "white" on the first pick are also equally possible, namely, 50%.) But every possible combination of 10 balls after 10 picks, classed in terms of their color, certainly does not have an equal chance of occurring. Applying some basic probability algebra for the joint probabilities of discrete outcomes sampled without replacement (you don't put the ball back in the jar after you pick it), it turns out that picking 10 balls that are all red is very low, less than 0.1 % (one-tenth of a percent). Picking all white balls has the same tiny probability. But picking five red and five white balls is rather likely, with a probability just over 25%. Similarly, picking four reds and six whites, or vice versa, is also quite likely, at just over 21%. So very high or low noise is like a very red or very white sample of balls—unlikely. Moderately valued noise, in contrast, is like a balanced sample of nearly equal numbers of red and white balls—quite likely. (This is the same reasoning that explains why, as we pointed out in Chapter 9, a sample with a mixture of high, medium, and low scores is more likely than a sample with all high scores or all low scores.) Regression happens, therefore, because extreme scores on one variable or after one measurement are partially due to unlikely noise, which will probably be less extreme on the other variable or the second measurement by chance alone.

Given that regression is the expression of a random phenomenon, its chance of occurring in a given data set increases as the number of measurements increases.

It is likely with one or a few measurements but effectively certain with a large aggregate of measurements. Regression to the mean is a threat to internal validity because its attendant rise or drop in scores is often misinterpreted as being caused by something meaningful and substantive, not the whims of chance. In our housing example above, no politician worth his or her salt would miss an opportunity to take credit for a jump in a slumping market or blame his or her incumbent opponent for a decline in a booming market. Similarly, an exceptionally hot or wet year is virtually always followed by a cooler or dryer year, even if no long-term climate change is occurring as a result of human activity or any other systematic cause.[2] We hope you now know better than to accept such claims at face value.

External Validity

We pointed out in Chapter 9 that researchers want to generalize their results and conclusions beyond the cases, measures, settings, times, and so on, that they actually use in their studies to other cases, measures, settings, times, and so on. **External validity** concerns the truth of these generalizations. In other words, external validity is the validity of inferences drawn from samples to populations (Chapter 8). External validity is clearly influenced by all aspects of how you sample, because they influence the resulting relationship of your sample to your sampling frame and the relationship of your sampling frame to your target population. And as we made clear in Chapter 8, it is not just the sampling of cases that is relevant but places, times, measures, and all other aspects of research. Thus, external validity is increased by large and representative samples of cases, places, times, measures, and so on.

However, external validity is also influenced by how the particular research settings and materials you use are specifically like the settings and materials involved in the phenomenon you study under its "natural" conditions. That is, in many research domains within geography, questions about external validity are thought to depend on how realistic the research setting was—to what degree were the conditions under which empirical observations were made similar to the normal conditions under which the phenomenon of interest exists or expresses itself. This issue of research realism or "verisimilitude" is often called **ecological validity.** It is called *ecological* because it is based on the idea that certain phenomena express themselves in a certain way because of the totality of the context in which they exist, including aspects of context that are spatial, temporal, or thematic. A geographer who studies the spread of plant communities in a laboratory might misunderstand aspects of this phenomenon because the laboratory fails to mimic the solar or atmospheric conditions that normally hold in the real world. Or a geographer who studies how analysts interpret remotely sensed images might misunderstand this phenomenon

[2]Just to be extra clear on this point, the reality of the phenomenon of regression to the mean by no means precludes the possibility that real substantive and meaningful changes over time in the phenomenon of interest are taking place. The fact that climate naturally changes over time, for instance, does not prove that humans are having no effect on climate themselves—it just proves that the phenomenon of climate change per se does not prove anthropogenic influence.

because the tasks given to the analysts seem "artificial" to them and thus do not engage their normal reasoning processes. Thus, ecological validity is primarily a concern for domains in which phenomena exist within a context that has important implications for the expression of those phenomena. In other words, it is a concern for most geographers; those who work in relatively abstract areas of physical geography close to physics or chemistry are probably safest in ignoring context, or at least in having a clear understanding of which aspects of context are relevant and need to be controlled.

A concern for ecological validity, as a way to ensure the external validity of conclusions about phenomena to the "real" world, is one reason to collect data in the natural context of phenomena, using methods that do not change the phenomena as a result of measurement (for example, field observations). It must be recognized, however, that a research setting high in ecological realism is not necessarily highly generalizable to many different contexts. The geographer who mimics the ecological conditions in one particular watershed may be able to learn about phenomena in that watershed but may be unable to generalize confidently to other watersheds. In fact, high ecological realism may interfere with generalizability by creating research conditions that are too narrow and specific. Thus, high ecological validity does not ensure high external validity, and high external validity need not require high ecological validity. That is, high ecological validity is neither necessary nor sufficient for high external validity. Furthermore, it should be recognized that generalizing to normal or natural conditions is not always the primary goal of a scientist. Testing theories about causal relationships across a range of conditions may require carrying out research in highly unusual or even hypothetical contexts. Or one may carry out research to show what is *possible* rather than what is likely or normal.[3]

Ultimately, questions about external and ecological validity lead us to one of the ubiquitous dilemmas of science. Scientists want to make statements of truth that are *general*, as much as possible. That is an important implication of the characteristic belief in simplicity shared by scientists (see Chapter 1). But all research studies that are actually carried out are finite, and most of them are in fact quite modest in scope. Scientific researchers must always use particular places, cities, rivers, measurement instruments, and other research entities in their studies, even though they want to conclude things about larger sets of these entities. That is, scientific research requires sampling, as we discussed in Chapter 8. So scientists always want to say something that goes beyond the direct evidence they actually have. Such inferences cannot be avoided even though they can always be mistaken.

Construct Validity

Construct validity concerns the truth of how variables as operationalized for measuring represent the theoretical constructs they are supposed to represent; that

[3]A thought-provoking and humorous critique of the idea that research should always be designed to generalize widely or is necessarily faulty if it fails to imitate "natural" reality closely is provided by Mook, D. G. (1983). In defense of external invalidity. *American Psychologist, 38*, 379–387.

is, it concerns how well manifest variables reflect latent variables. In Chapter 2, we learned that constructs or latent variables are the hypothetical entities that we attempt to measure in research. They are pieces of the idealized world that make up the subject matter of our theories. Measured or manifest variables are these entities as they are expressed by actual measurement procedures.

The idea of construct validity reflects the recognition that the scores in our data sets are *attempts* to assess the values that cases actually possess on our constructs of interest, attempts that are nearly always a little mistaken, maybe very mistaken. Above, we pointed out that any observed data value can be partitioned into a systematic component and a random component, and that low reliability is caused by large error in the random component. In contrast, construct validity is about the systematic component in data, specifically the *meaning* of the systematic component. Low construct validity is a mismatch between what you are trying to measure according to your conceptual framework and what your measurement procedure actually measures consistently. That is, only reliable components of variation in your data are relevant to construct validity.

Low construct validity is fundamentally construct "misrepresentation." It arises from some combination of missing aspects of your construct that are part of its essential meaning, and including aspects that are irrelevant or even contradictory to its essential meaning. In Chapter 2, we discussed the poor construct validity in John's research presented from Chapter 1. By operationalizing the construct of "dissociative institutions" by counting liquor stores in the phone book, for example, John missed much of the actual meaning of dissociative institutions at the same time that his data contained variation not relevant to the meaning of dissociative institutions. Similarly, biogeographers who operationalize the construct of "plant communities" by measuring only latitude and climate conditions will miss out on some aspects of the meaning of that construct at the same time they pick up variation in their data that is not relevant to which plant community is present in a particular place.

Of course, scientists attempt to increase construct validity by operationalizing variables (designing operations to measure constructs) in a thoughtful and carefully reasoned way that is informed by previous research and scholarship. A systematic approach to establishing and increasing construct validity is called **convergent-discriminant validation**. It is based on the fact that poor construct validity means that you are failing to capture some of your construct in your measurement, you are capturing some other constructs in your measurement, or both. To perform a convergent-discriminant validation, you attempt to measure multiple constructs with multiple measurement operations, that is, multiple manifest variables. Logically, variables that should conceptually tap into the same construct should largely agree, that is, should correlate highly across cases—this is the convergent part. By the same token, variables that should conceptually tap into different constructs should largely disagree, that is, should not correlate highly across cases—this is the discriminant part. In other words, you explore construct validity by measuring in more than one way and by measuring in ways that you do not expect to be relevant to your constructs of interest. Even if you do not apply a full-blown convergent-discriminant validation in your research, you should still avoid resting your research conclusions on a single way of measuring your most important constructs. No measurement technique is perfect, so measuring in multiple ways is the best way to avoid the dangers of drawing conclusions from any single technique—the so-called **mono-method bias.**

Statistical Conclusion Validity

Statistical conclusion validity concerns the truth of conclusions you draw from statistical analyses of data, a function of the appropriate use and interpretation of statistical tests. We discussed this in Chapter 9, but as we pointed out there, you should learn more about it in courses dedicated to statistical analysis. Issues of statistical conclusion validity include applying the correct descriptive indices to data at different levels of measurement; for example, the mean of a nominal variable is meaningless. Another issue concerns the truth of assumptions made in inferential statistics that allow the construction of the correct sampling distribution for determining probabilities about population inferences from samples. These include assumptions about the population distribution, the independence of scores, and correct and complete statistical model specifications.

There are statistical tests for the distributional assumptions. Some texts recommend applying these tests to your data before analyzing them. For instance, we can test the normality of a population distribution by examining the normality of a sample taken from that population. As we pointed out in Chapter 9, these distributional assumptions apply only to parametric statistical tests. Many researchers prefer to restrict themselves to nonparametric tests when these assumptions are shown to be dubious. Of course, these tests of assumptions about populations are based on data from samples and are therefore inferences themselves. Furthermore, several of the standard assumptions for parametric inferential tests turn out not to be very important; that is, a violation of population normality or homogeneity of variance across groups has very little distorting effect on the probabilities found as part of inferential statistical tests. In other words, the inferential tests are robust to violations of many of the standard assumptions.[4] Parametric tests can nearly always be used safely, even with samples that are fairly nonnormal or that reflect heterogeneous variances. Given this, we recommend using parametric analyses over nonparametric analyses in most situations—they are more flexible and typically more powerful.

There are statistical tests for the independence assumption too. In contrast to distributional assumptions, independence is something you should take quite

[4]Evidence for the robustness of statistical tests to violations of certain distributional assumptions comes from statistical Monte Carlo "experiments" like those discussed in Chapter 7 in which simulated sampling distributions are created by drawing a large number of random samples from hypothetical populations with particular properties. For example, the "central limit theorem" states that a population must be quite nonnormal and samples must be small before the sampling distribution of the sample mean will deviate much from normalcy itself and produce distorted probability values.

seriously. Nonindependence means that you actually have much less independent information than you might think. This can lead to radically mistaken estimates of inferential probabilities concerning what sample results indicate about population values. It is widely recognized that nonindependence in geographic research is often based on spatial relations, as we discussed in Chapter 9, but it is also frequently based on temporal relations or relations of identity, as we discussed in Chapter 7; two cities measured once provide more independent information than one city measured twice.

There are two other common threats to the validity of conclusions about statistical significance when using the inferential technique of hypothesis testing (Chapter 9). The first is **inadequate power**. It is possible whenever you make a decision to retain both the null and alternative hypotheses—that is, when you do *not* obtain statistically significant results. If you fail to obtain significance, it is always possible that you have made a mistake (a Type II error in Figure 9.5). You are likely to make this mistake when beta (β) is too high, or equivalently, when power is too low. So whenever you fail to obtain significance, there is always the question of whether your test had adequate statistical power to find an effect that is really there in the population. The statistical technique of power analysis introduced in Chapter 8 is the formal way to estimate how much power you have in a given hypothesis test, and how you can increase it to a particular level.

A second common threat to the validity of statistical conclusions about significance is, in a sense, the converse of inadequate power: the possibility of **alpha (α) inflation** whenever you conduct multiple significance tests on the same set of data. Alpha inflation is a potential threat to validity when you make a decision on some of your tests to reject the null and accept the alternative hypotheses—that is, when you *do* obtain statistically significant results. If you obtain significance, it is always possible that you have made a mistake (a Type I error in Figure 9.5). You are likely to make this mistake after multiple tests because the α-level per test may be .05, for instance, but the effective **studywise α** will be higher, much higher if a large number of tests are conducted. That is, the chance of making at least one Type I error in a set of tests is higher than the chance of making one on just a single test. To appreciate this intuitively, imagine that you are drawing a single card from a standard deck, and you are concerned about getting one of the two black Jacks (you harbor some superstitions). On one draw, the chance of getting a black Jack is only 2/52—less than .04. However, if you are forced to draw five cards all at once, your chance of getting hexed by at least one of the black Jacks is much higher—around .20. So whenever you conduct several significance tests on the same data, there is always the question of whether one or more of the significant ones reflect an inflated studywise α. This problem is much more serious when one conducts a large number of comparisons, even all possible ones, on a single data set without theoretical expectations to guide which tests to conduct and how to interpret their meaning. Such an approach is charmingly known as "going fishing." Unfortunately, such an atheoretical approach to empirical work is all too common in geography and likely has produced a great deal of apparent evidence for phenomena that are not true, or at least that are not very strong. The antidote to α inflation is using one of several possible techniques to reduce the α per test and therefore the overall studywise α (see references in Bibliography). Better yet, try to perform statistical tests that are theoretically guided—leave the fishing for seafood suppliers and recreational anglers. At the very least, phenomena discovered on a fishing trip must be replicated in new data sets.

Researcher-Case Artifacts

We finish with a brief discussion of certain special classes of validity threats that arise because research is an activity that can influence ongoing reality, and researchers are humans that can be influenced by their understanding of a situation as being research. In human geography, researchers' cases are often humans that can also be influenced by a situation as research. These **researcher-case artifacts** can produce biased data or data interpretations that reflect various expectancies or beliefs people, whether researchers or human research subjects, have about research situations, about outcomes, or about classes of people being measured or doing the measurement. They also derive from various aspects of the activity of the research situation. That is, researchers can act in ways that alter their phenomenon of interest, or they interpret that phenomenon in biased ways. These artifacts potentially threaten any of the four types of validities.

Researcher-case artifacts can be grouped into two classes, interactional and noninteractional. **Interactional artifacts** result from the interaction of the research situation (including the researcher himself or herself) with the case, providing an opportunity for the research situation to directly influence the cases and their values on the constructs being measured as data. That is, interactional artifacts actually change the value of the case on the construct being measured. In human geography, one usually considers the danger of interactional artifacts when the cases are sentient creatures, such as individual human subjects who are directly contacted by a researcher. Nonhuman animals studied by biogeographers could also show these influences, but interactional effects in general are undoubtedly rare in physical geography. An example could be a geographer who is trying to show that soil interflow depends on ground slope.[5] He or she might excavate a soil pit and insert measuring devices at various levels to intercept the water flow. The faces of the pit will act as "macropores," however, that alter the pressure relations experienced by the flow and possibly cause water that would have flowed downslope not to flow. This will lead to an altered indication of flux in the water.

The classic type of interactional artifact in human research occurs when research subjects change their behavior or their responses to explicit reports because of the way they interpret the meaning of the research situation, including its purpose, its risk, its relevance to their reputation or class grade, its political bias, and so on. In other words, the concept of reactance that we defined in Chapter 4 is an interactional artifact. This expression of interactional artifacts reflects a definition of research as

[5]This example comes from Reid (2003), referenced in Chapter 4.

seen from the perspective of a research subject[6]: Research takes place in "a context of explicit agreement to participate in a special form of social interaction known as 'taking part in a study.'" In this situation, artifacts can be brought about by the way researchers ask questions, the identity of the agency or organization sponsoring the research, or just the name or stated purpose of the study. Even the appearance of the researcher could affect subjects' responses if data collection occurs during an episode of direct contact between the researcher and the subject. Examples include a researcher's apparent sex, age, ethnicity, or physical attractiveness.

Noninteractional artifacts result from the researcher himself or herself, not from the interaction of the research situation with the case. They change the recorded values of data or the way they are interpreted but do not actually change the value of the case on the construct being measured. Instead, they change the values or interpretation of data as a result of bias or limitations in the way researchers perceptually and cognitively process information. To some extent, researchers see what they expect or hope to see (we discussed the "nonobjective" nature of human cognition in Chapter 6). Sometimes researchers might even intentionally bias the way they record or analyze data because of human foibles such as greed or hubris, which, we noted in Chapter 1, scientists suffer from as much as anyone else. Such intentional bias obviously violates ethics in research, which we return to in Chapter 14. A good example of an unethical noninteractional artifact would be making a decision that an observation is an outlier and deserves to be removed from the rest of your data only *after* you have seen that its inclusion weakens support for your hypothesis. Noninteractional artifacts potentially occur in much the same way and to the same degree in any area of geographic research. In general, they would be more common whenever data creation and analysis have less strict and predetermined rules for their conduct—that is, whenever the researcher's personal judgment plays a greater role determining data treatment. In some contexts, such as the coding of unstructured interviews, that would be more common in human geography. This is one of the important reasons for preferring systematicity in data treatment.

How can we reduce the occurrence and impact of researcher-case artifacts? With respect to interactional artifacts, it is important to recognize that they do not necessarily occur and do not occur equally to everyone. Just because a person is aware he or she is being questioned for a study, for example, does not guarantee that the person will deceive the researcher about the truth, as he or she understands it. Likewise, most of us change our behavior somewhat when we know are being watched, but skilled and experienced public speakers or performers often demonstrate the ability to avoid this change.

Having made that point, interactional artifacts such as reactance are a real possibility. Avoiding them is a major motivation for using so-called nonreactive measures in human geographic research (as we discussed in Chapter 4). Research subjects cannot change their behavior, thoughts, or feelings in response to the researcher or

[6]From Rosenthal & Rosnow (1991).

research situation if they are not aware they are in a research study, or if the data are not in fact produced as part of a research study. If nonreactive measures are not possible, such as when your research question deals with constructs like explicit attitudes, then you need to spend a great deal of effort during pilot testing and research design to produce unbiased instruments. In some situations, it would be worthwhile to show that the same data result when items are designed with alternative wordings. Sometimes it is appropriate to keep assistants "blind" to the specific hypothesis of the research or to the specific condition to which a subject has been assigned. This helps with noninteractional artifacts too, as when open-ended data coders are kept blind during coding. You should also consider the characteristics of assistants or interviewers, such as their sex or ethnicity, that could influence subjects' responses. Sometimes you should counterbalance these characteristics by making sure to use both male and female assistants, for example. In other situations, perhaps you should use only assistants with particular characteristics.

In the case of noninteractional artifacts, it is clear that new researchers need to be trained, through both explicit instruction and exposure to role models, in the ethical scientific values of neutrality, objectivity, honesty, and so on. We are not so naive as to think that such training, even when very high in quality and quantity, guarantees that researchers or their assistants will always treat data properly and interpret it objectively. Neither are we so jaded to believe that personal integrity in research is mostly a lost cause. We try to conduct our own research according to these values and believe that we almost always succeed at it. Ultimately, however, the social nature of science that we discussed in Chapter 1 comes to the rescue here. The fact that scientific results and conclusions are necessarily and appropriately subjected to critical scrutiny by other researchers at other places goes a long way toward muting the threat of researcher artifacts. Findings should be skeptically doubted (with justification) and should be independently replicated by other researchers with neutral or even opposing views.

Review Questions

Reliability

- What is reliability, and what causes high and low reliability of measurement?
- What are three specific approaches to assessing reliability, and why does each fall short of ideal for assessing reliability?
- What are some ways to increase measurement reliability?

Validity

- What is validity, and what are the names and meanings of the four major types of validity?
- What is regression to the mean, why does it happen, and how does it influence internal validity?
- What is ecological validity, and how does it relate to internal and external validity?

- How is poor construct validity different than low reliability?
- What are three assumptions generally involved in valid statistical inference (introduced in Chapter 9)? How important is it for researchers to attend to threats to statistical conclusion validity stemming from violations of these assumptions?

Researcher-Case Artifacts

- What are researcher-case artifacts, and how might they influence each of the four types of validity?
- What is the distinction between interactional and noninteractional artifacts, and what are some examples of each in geographic research?

Key Terms

alpha (α) inflation: threat to statistical conclusion validity when multiple hypothesis tests are conducted on a single data set, and some of them are significant; results from the overall increased studywise chance of a Type I error when multiple tests are conducted

construct validity: the truth of how measured (manifest) variables capture the meaning of the theoretical constructs (latent variables) they are supposed to represent

convergent-discriminant validation: systematic approach to establishing and increasing construct validity by measuring constructs with multiple variables, some of which are intended to redundantly measure the construct and others of which are meant to measure other constructs

ecological validity: the truth of research generalizations as a function of how well the context within which a phenomenon exists under natural conditions is present in a particular study; an aspect of external validity

external validity: the truth of research generalizations from the samples of cases, measures, settings, times, and so on, actually used in a study to wider populations of cases, measures, settings, times, and so on

inadequate power: threat to statistical conclusion validity when nonsignificant hypothesis tests may be due to power that is too low

interactional artifact: researcher-case artifact that changes a case's value on the construct being measured because of the interaction of the research situation (including the researcher) with the case

internal consistency reliability: approach to assessing the reliability of a complex measured construct that is operationalized as the combination of two or more separate but partially redundant variables (as in surveys); reliable measurement requires the separate variables to correlate with each other

internal validity: the truth of research conclusions about causal relationships

inter-rater reliability: approach to assessing the reliability of a measured or coded variable in which the data produced redundantly by separate observers or coders are compared for similarity

item-total correlation: way to quantify internal consistency reliability by averaging the correlation of each separate variable with all of the others in the set meant to measure a single construct

mono-method bias: shortcoming of research that relies on drawing major conclusions from a single way of measuring constructs

noise: another term for random error of measurement

noninteractional artifact: researcher-case artifact that changes data values or their interpretation, not the actual values of the cases on the constructs of interest, independently of any interaction of the research situation (including the researcher) with the case

random component: the component of a measured score value that is inconsistent across measurement events, reflecting nonsystematic influences on measurement operations; this random component causes lowered reliability and can be called noise or random error

reliability: the "repeatability" of measured values of variables; perfect reliability occurs when remeasuring a construct that is exactly the same on both occasions results in exactly the same measured value

remeasurement reliability: approach to assessing the reliability of a measured variable in which cases are measured twice or more at different times, and the resulting measurements are compared for similarity; sometimes called "test-retest reliability"

researcher-case artifact: threat to any of the four types of validity that arises because human researchers and/or research cases have various expectancies or beliefs about the research situation, outcome, or about classes of people being measured or doing the measurement

statistical conclusion validity: the truth of conclusions we draw from statistical analyses of data, a function of the appropriate use and interpretation of statistical tests

statistical regression to the mean: the phenomenon that extreme measured scores tend to be less extreme when remeasured, due to chance variation acting in a more likely manner after it acted in an unlikely manner; it is an especially intriguing threat to internal validity

studywise α: the α for making at least one Type I error among multiple tests conducted on the same data set; α inflation is due to the fact that the studywise α is generally larger than the α per individual test

systematic component: the component of a measured score value that is consistent across measurement events, reflecting whatever construct we are picking up with our measurement operations; when this systematic component is inaccurate relative to the true value of the construct, it can be called "systematic error"

validity: the "truth-value" of research results and interpretation

Bibliography

Cook, T. D., & Campbell, D. T. (1979). *Quasi-experimentation: Design and analysis issues for field settings.* Boston: Houghton Mifflin.

Hand, D. J. (2004). *Measurement theory and practice: The world through quantification.* London: Hodder Arnold.

Rosenthal, R., & Rosnow, R. L. (1991). *Essentials of behavioral research: Methods and data analysis* (2nd ed.). New York: McGraw-Hill.

CHAPTER 12

Geographic Information Techniques in Research

Learning Objectives:

- What is geographic information, what forms does it come in, and what are some ways it is used in geographic research?
- What are some important considerations when analyzing, mapping, and interpreting geographic information?
- What is spatially referenced remotely sensed imagery, and what are some ways it is used in geographic research?
- What are the major analytic capabilities of GISs?

The increased availability and sophistication of computers in the last several decades has enabled the development of geographic information systems (GISs). GISs have the capability of storing, manipulating, analyzing, and displaying **geographic information**. Geographic information is any information that is spatially referenced to the surface of the earth. Historically, geographic information was limited to verbal descriptions and increasingly sophisticated maps. Today, geographic information is stored digitally in simple and complex data structures on computer-readable media. The abstraction of spatial reality to these digital data structures both limits and extends our understanding of geographic processes and phenomena. Remotely sensed imagery from satellites, aerial photographs, and other sources provides a significant subset of geographic information used by geographers, both human and physical. This chapter overviews (1) geographic information and spatially referenced remotely sensed imagery, (2) the nature and capabilities of GISs, and (3) some of the spatial and nonspatial types of analysis that are significantly enabled by GISs.

Geographic Information

Fundamentally common to all geographic information is its spatial reference to the surface of the earth, as conceived and measured. Human conception of the earth has evolved from a large if not infinite flat plane to a sphere to an **oblate spheroid** to the **geoid**. As we mentioned in Chapter 1 and discussed further in Chapter 4, geodesy is the science of measuring the size and shape of the earth, and the distribution of physical and human features on its surface. Geographic information includes geodesic information but is more general in that it includes any information with a spatial reference to the sphere, oblate spheroid, or geoid. However, characterizing the nature of geographic information gets somewhat complex when we consider what is being represented, how it is measured (including uncertainty in those measurements), the spatial and temporal resolution at which it varies and is measured, and the data models and structures used to represent the information. These characteristics are interrelated in ways both subtle and not so subtle. They are very important because they influence how you can or should display, manipulate, and identify patterns in data. The nature of geographic information is an essential component of research methods and must be carefully considered before and during geographic inquiry. Furthermore, considerations about how geographic information has been handled constitute an important basis for critiquing the research of others.

For example, consider the four following kinds of geographic information: (1) The location and elevation of all mountain peaks in Colorado above 14,000 feet, (2) a 1 km^2 resolution grid (or raster) of the population density of the nation of India, (3) those areas of the conterminous United States where threat of exposure to the West Nile virus is high, and (4) the flight paths of all United Airlines flights on June 27, 1998. Our four examples might typically be represented, respectively, as points with elevation attributes; a grid of cells, each containing a population count; a polygon **coverage** (layer), each polygon scored for a threat level; and a set of lines (we discuss data models and structures below). All of these representations inherently contain some degree of spatial, temporal, and thematic error that must be considered, but the types and amounts of error associated with each vary dramatically. Global Positioning Systems (GPSs), discussed below, allow us fairly accurately and precisely to locate the 14,000-foot peaks in Colorado, whereas the population density of India changes daily in space and grows with time. Risk associated with the West Nile virus is not completely understood; therefore, maps of that risk are "fuzzy" at best or completely wrong. The United Airlines flights actually did take place along very specific routes; however, the accuracy of the instrumentation we use to map them is limited.

Although the comprehension of geographic information may seem daunting, you may find it somewhat comforting that the "children's" placemats that many restaurants give to kids of all ages often contain puzzles and activities that use the basic models of geographic information. A "connect the dots" puzzle on those place mats uses numbered points that can be used to create points, lines, and polygons. Even the "word search" puzzle is a raster representation of letters. Needless to say, commonly used geographic information is substantially more sophisticated but nonetheless comprehensible with a little effort. The following are important characteristics to consider when creating, analyzing, or mapping geographic information:

1. ***What is the nature of the phenomenon or entity being represented?*** As individuals and members of cultural subgroups, we conceptualize geographic phenomena (features and events) in terms of **conceptual models**. These are simplified mental or intellectual representations of phenomena in the world. The most fundamental conceptual distinction when representing geographic phenomena is that between **fields** and **objects**. This is the distinction we touched on in Chapters 2 and 8 between continuous and discrete reality. A clear example of a field is temperature, a continuously varying phenomenon that exists everywhere in space. Other examples include humidity, elevation above sea level, and barometric pressure. In contrast, a clear example of geographic objects would be fire hydrants. Fire hydrants are discrete entities, with fairly clear and precise boundaries, that exist in some places and not in others. Examples of geographic features often best suited to object representation include islands, water bodies, buildings, and streets.

Phenomena in human geography are very frequently modeled as objects, whereas those in physical geography are often modeled as fields. However, some geographic phenomena do not lend themselves obviously to object or field conceptualization. Often, the issues of object versus field are related to the spatial and temporal resolution of measurement—matters of scale. Consider the geographic phenomenon of population density. Population density is typically a number attributed to a polygonal area or a pixel: the number of people in that area divided by its areal extent. In actuality, however, population density is a temporally varying number of discrete humans in a given subset of continuous space. Imagine yourself chatting with a friend at a party—does the space between you and your friend have the population density of the room or the city you are in, or is it simply zero? For some applications, it might be reasonable to conceive of population density as a field; an example is a regional study of human impact on the environment. In other applications, an object-based representation would be preferred; an example is urban warfare. Similar issues are encountered in physical geography when studying vegetation—do you look at the forest, the trees, or a field of vegetation density? In any case, you should give careful consideration to the nature of the geographic phenomena you are attempting to represent in maps or data, and how those representations will be used and compared with other geographic information relevant to your particular inquiry.

2. ***Spatial scale or resolution of measurements.*** Questions about how information is measured are related to the previous question regarding the nature of the phenomena being studied. That includes the relationship of analysis scale to phenomenon scale (see Chapter 2). As we discussed in Chapter 9, geographic information is measured and analyzed at various spatial scales and is frequently aggregated from a larger set of measurements into a smaller set of measurements applied to the inquiry at hand. Terms such as **resolution** or **granularity** are often used as synonyms for analysis scale, especially when the information is in the form of digital

representations by means of a regular grid of small cells in a satellite image (**rasters**) or on a computer screen (**pixels**). Analysis scale would then refer to the area of earth surface represented by a single cell.

In Chapter 9, we reviewed many of the difficulties and ambiguities caused by the uncertain relation between analysis scale and phenomenon scale. However, scale difficulties are not limited just to issues of the proper size or resolution of measurement units, because geographers are often interested in linkages that occur across varying scales, such as "local to global" or "global to local"; these linkages are expressions of the hierarchy of scales, introduced in Chapter 2. Linkages of this nature can be independent of, or confound, questions associated with the spatial and temporal scale of measurement. For example, identifying a large-scale El Niño pattern (that is, over a large area) in the eastern Pacific Ocean during the spring may tell us a great deal about small-scale (over a small area) rainfall in many parts of Africa during the following summer. This is an example of identifying a potentially useful global-to-local linkage. In order to study this, you would probably want historical information about both the El Niño in the Eastern Pacific and the rainfall in Africa. The El Niño data could range from nominal or ordinal descriptions of the El Niño for several decades (for example, 1979 El Niño "yes," 1980 El Niño "no") to complex time-series satellite images of sea surface temperature. The African rainfall data could range from aggregate measures based on farmers' reports ("1981 was a good rainfall year around here") to a whole network of rain-gauge data from hundreds of stations in locations around the continent.

3. ***Temporal scale or resolution of measurements.*** All geographic phenomena vary as a function of time; no earth phenomena even existed a few billion years ago. Consequently, one must consider the rate at which geographic phenomena change. Clearly the location and elevation of Mt. Everest, the flow of the Mississippi river, the population of Yankee stadium, and the proportion of likely voters favoring presidential candidate Bob vary on different timescales. The temporal scale of measurement of geographic information is consequently another important aspect of research methods. Annual rainfall information is likely inadequate for understanding specific meteorological events like storms or hurricanes. Daily traffic counts for roads and highways in a street network may be useless for understanding traffic patterns that vary on a minute-by-minute or hourly basis. GISs still have room to improve with respect to handling temporal information. The development of dynamic models that describe urban growth patterns, for instance, presently constitutes an important current research focus in geography. One fundamental question associated with this kind of work concerns what temporal resolution is required for the geographic information represented in the models. The development of dynamic models describing two- and three-dimensional geographic processes will be an increasingly vibrant area of research in the future.

Temporal resolution is particularly interesting in the context of satellite remote sensing. The temporal density of remotely sensed imagery is large, formidable, and growing. Satellites are collecting a great deal of imagery as you read this sentence. However, most applications in geography do not require extremely fine-grained temporal resolution. Meteorologists may require visible, infrared, and radar information at sub-hourly temporal resolution; urban planners might require imagery at monthly or annual resolution; and transportation planners may not need any time-series information at all for some applications. Again, the temporal resolution of imagery used should meet the requirements of your geographic inquiry. Sometimes researchers have to delve into archives of aerial photographs to get information from the past that predate the collection of satellite imagery.

4. ***Thematic scale and classification of features.*** In Chapter 2, we pointed out that scale was not just relevant to space and time but also to theme. **Themes** are the non-spatial and nontemporal characteristics of human and natural phenomena that geographers measure as variables and store as attributes in information systems. Like space and time, themes can be aggregated or disaggregated to various levels. A biogeographer can study plant subspecies, species, families, or vegetation communities. An economic geographer can study the manufacturing of computer chips, computers, all electronic goods, all high-tech goods, or manufactured goods in general. As with spatial and temporal scale, you need to think about the proper thematic scale at which to study your phenomena of interest.

Research issues involving the thematic content of geographic information go well beyond the question of the proper thematic scale at which to study a problem, however. We claimed in Chapter 1 that nearly all scientists accept a philosophy of realism; they believe the universe actually exists and is not a mere construction of human minds. We also pointed out, however, that the patterned matter and energy that make up reality is organized into meaningful pieces—objects and events—by sentient beings. In fact, this organization, essentially a process of interpretation, is a fundamental mental activity carried out constantly in an informal way by everyone, including scientists. But scientists carry out this interpretation more systematically, with greater concern for the coherency, consistency, and reality correspondence of the ontologies that result. **Ontologies** are systems of concepts or classes of what exists in the world; they reflect the structure of objective reality as well as personal and social acts of cognitive organization.[1]

Geographers must also concern themselves with the nature of geographic reality, recognizing that the tools of their intellectual and empirical efforts are verbal, graphical, and numerical *models* of reality that necessarily depend on categories of

[1]In Chapter 1, we defined the study of epistemology as how scientists (and others) can know what they know. As traditionally defined in philosophy, and in Chapter 1, ontology deals with the question of the objective nature of that which actually exists. Ontology and epistemology together make up the traditional domain of metaphysics. We are using the concept of ontologies in a broader sense that has become quite popular in the information sciences, including geographic information science, since the 1980s. This broad sense more or less combines ontology with epistemology in recognizing that human concepts of reality, whether in human minds, cultures, or information systems, depend as much on human conceptualization as on the nature of reality. We'll leave the question of whether we can ever know the nature of reality independently of human conceptualization for you to ponder during your next midnight philosophy session.

earth and human structures and processes. Cities, mountains, rivers, and industries cannot be understood simply as elements of reality sitting out in the world for geographers to objectively assess. They are also products of what geographers choose to focus on in the world and how geographers organize their perceptions into units of reality. This is true when geographers study their phenomena directly via field and lab observations, and it is true when they study their phenomena indirectly via maps and remotely sensed imagery. Most people can readily see the ambiguity of trying to classify all forms of agricultural activity as "intensive" versus "extensive," or of trying to classify vegetated land cover as "grassland" versus "broadleaf forest" versus "needle leaf forests" versus "mixed forest"—surely there is grass to be found in all four land cover classes. But it is even ambiguous to define "lake" clearly enough so that the exact number of them in Minnesota can be counted accurately. That's right, no one knows *exactly* how many lakes there are in the "Land of 10,000 Lakes," except by making some partially arbitrary decisions about what constitutes a single lake.

The nature of geographic ontologies has great implications for the potential of different people, cultures, disciplines, and information systems to "communicate" effectively with one another about geographic phenomena. In the information sciences, this is known as the problem of **interoperability**. In addition to geography and computer science, linguistics, cultural anthropology, cognitive category theory, and artificial intelligence are all important to the study of the interoperability of geographic information systems, a cutting-edge area of research on geographic information.

Data Models and Data Structures

Our conceptual models of geographic phenomena are represented in simplified form in the computer in terms of a **data model**—a simplified computer representation of the spatial, temporal, and thematic attributes of geographic information. We represent most geographic information in terms of either the **vector** or **raster data models.**[2] The vector model consists of geometric objects formed by vector connections among node points in the data space, and includes features modeled as **points**, **lines**, or **polygons**, which are zero-, one-, or two-dimensional features (three-dimensional **volumes** are possible as well). Spatial information in a vector model is expressed in terms of the coordinates assigned to the vertex points and any geometric inferences that can be made from those coordinates. In contrast, the raster model consists of a regular grid of small cells, usually square, that **tessellate** the planar surface like a big checkerboard with values in each cell. A raster does not represent coherent objects as such. Spatial information in a raster model is expressed in terms of the coordinates assigned to the raster cells and any geometric inferences that can be made from those coordinates. The vector-raster distinction is often described as the data-model equivalent of the conceptual distinction

[2]There are other, less common geographic data models, such as "triangulated irregular networks" (TINs) and various hybrid models.

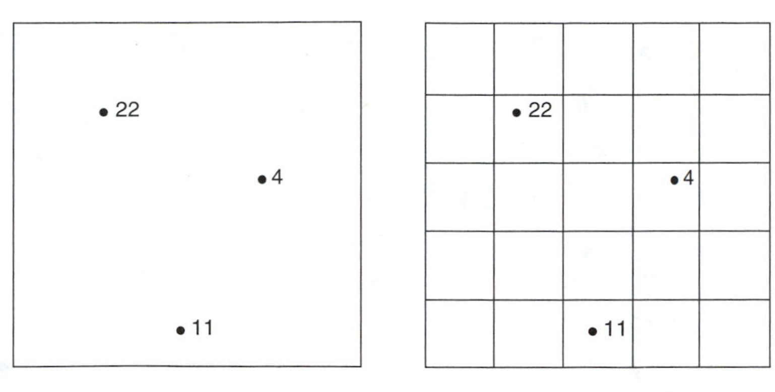

Vector Data Model
three points in space
with numerical attributes

Raster Data Model
with coordinate
system superimposed

Pt. ID	X	Y	Value
1	1.5	3.5	22
2	3.5	2.5	4
3	2.5	0.5	11

Vector Data Structure

0	0	0	0	0
0	22	0	0	0
0	0	0	4	0
0	0	0	0	0
0	0	11	0	0

Raster Data Structure

Figure 12.1 Data models and data structures.

between objects and fields, although this equivalence is inexact at least insofar as rasters digitally represent fields as collection of small grid objects, not as truly continuous entities.

In turn, our data models of conceptual phenomena are digitally expressed in the computer in terms of a **data structure.**[3] Consider a simple data set of thermometers measuring temperature at three locations in space. The data model is simply three points in space with temperature measurements attributed to them (Figure 12.1). A **coordinate system** could be imposed on this data model to facilitate describing the relative locations of the points. The data structure associated with this data model will be significantly different depending on which data model is chosen to represent the geographic information. In the case of a vector representation, the data structure would simply be three lines of numbers in addition to an ID code (*X-coordinate, Y-coordinate,* and *Temperature Value*), perhaps separated by commas, easily stored as an ASCII text file. In the case of a raster representation, the data could be stored as five rows of five numbers or *n* rows of *n*

[3]Our use of the term "data structure" conflates a distinction that is sometimes made between low-level data structures that express the fundamental machine code used to represent information digitally and "data formats" that connect low-level structures to high-level data models.

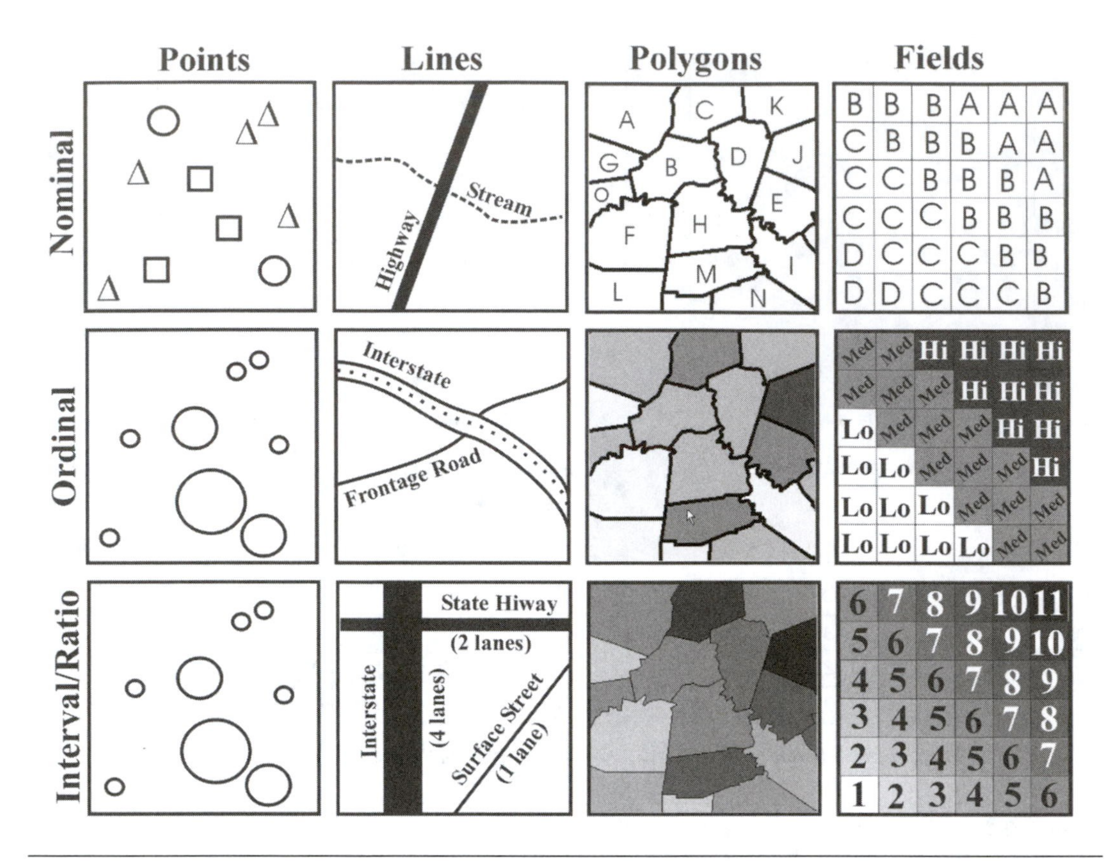

Figure 12.2 Schematic representation of entity-attribute spatial data types (adapted from O'Sullivan & Unwin, 2003).

numbers, depending on the spatial resolution of pixels or grid cells considered appropriate. Note that in the raster data structure example, zeroes represent "No Feature Present."

It is important to consider data models and structures because they influence how data can be processed and displayed and how certain analyses can be performed, as well as how errors in those analyses might propagate. Figure 12.2 depicts 12 different spatial data types that result from cross-tabulating measurement level (nominal, ordinal, interval/ratio) against the three major vector data models (points, lines, and polygons) and the raster data model (fields).[4] These 12 different data types are best displayed graphically in different ways, propagate error differently, and lend themselves to different kinds of spatial analysis. We often have choices as to how we want to represent geographic information; the data model we choose is an important characteristic of geographic information that we must carefully consider when doing research.

[4]Based on O'Sullivan & Unwin (2003).

Remotely Sensed Geographic Information

We have noted in Chapters 3 and 4 that geographers frequently collect data by observing physical traces of human and natural activity on the earth's surface. This is commonly done from above by image recording devices carried by airplanes or satellites. Although information derived from satellite images, aerial photographs, and other imagery is simply a subset of geographic information in general, such **remotely sensed geographic information** does warrant some special consideration. Remotely sensed images can record solar radiation reflected off the (near) earth surface, radiation emitted from natural and human features, and radiation bounced off near-earth surfaces that were emitted originally from the sensors themselves (for example, Lidar and radar); the first two are known as **passive sensing**, and the third is known as **active sensing**. Remotely sensed images of the earth surface were historically derived from optical photographs taken from airborne vehicles such as planes or balloons. Today the majority of remotely sensed imagery comes from satellites in outer space.

Aerial photographs were originally black-and-white images taken in the visible range of the electromagnetic spectrum. This naturally evolved into color photographs, which provide more spectral information—separate information from the red, green, and blue portions of the spectrum. Later, color-infrared film was developed that provided information from the infrared part of the electromagnetic spectrum, a profound development for assessing the health and water content of vegetation. The development of more sophisticated sensors has expanded the **spectral resolution** of remotely sensed imagery, an aspect of thematic scale, to a dizzying extent. There are now satellites and airborne sensors producing remotely sensed images of the earth in the radar, microwave, visible, near-infrared, thermal infrared, and ultraviolet parts of the electromagnetic spectrum. Detailed descriptions of the potentials of these different types of imagery are beyond the scope of this book, but the interested reader can look at the books we list in the bibliography, or better yet, take a course in remote sensing.

Multispectral remotely sensed imagery in some sense has the "layer-like" characteristics of a set of vector data layers (for example, road layer, soils layer, county layer) in which each spectral band is a layer. These data sets sometimes have hundreds of spectral bands and are referred to as **data cubes**. The size of these data cubes can become overwhelming. Imagine a 1,000 by 1,000 pixel image, about the size of a typical computer monitor, that has 256 separate layers representing narrow slices of the electromagnetic spectrum; that is, each pixel of the image has 256 **data numbers** or values associated with it. Analysis of a time series of this kind of data set actually starts to make computers work hard. The data volume of remotely sensed imagery can become staggering, really exceeding the comprehension capability of the human mind. What's more, many of the data are partially redundant; for example, several spectral bands sensed from the same area of ground surface will often be strongly intercorrelated and thus represent less information than they appear to. Consequently, the analytic techniques of remote sensing typically include methods to reduce the spectral, spatial, and/or temporal resolution of the imagery in order to reduce this redundancy and make the information interpretable by humans. The amount and type of resolution reduction applied to an image is preferably determined by the needs of the user.

Remotely sensed imagery comes in a large range of spatial resolutions. The spatial resolution of remotely sensed imagery must be appropriate to the research questions it is being used to investigate, but choosing the spatial resolution of imagery appropriate for your particular research can be challenging. Aerial photographs can be obtained in which the spatial resolution is measured in inches, whereas radar images can have spatial resolution measured in kilometers. *IKONOS* imagery provided by the Space Imaging Corporation is provided at 1-meter resolution. Despite the dramatic improvements in computational power over the last decade it would be very difficult to meaningfully display a composite *IKONOS* image covering, for instance, the state of Alabama. Global image products with spatial resolution less than 1 kilometer are pushing the computational capabilities of computer systems to some degree. In general, it is fairly easy to reduce the spatial resolution of imagery via aggregation to larger pixels; however, this is not always a good approach. In addition, it should be noted that finer-resolution imagery is not always the best means of getting "better" information. For example, consider an attempt to map and identify "exurban" areas of the United States. Exurban areas are residential developments outside of the traditional suburban boundaries of 20th- and 21st-century urbanized areas. They are rural, wilderness, and small town areas that contain residents engaged in traditional urban economic activities by commuting or telecommuting to urban workplaces. To identify these areas, a 1-kilometer-resolution nighttime image proves to be more useful than a 30-meter-resolution image derived from *Landsat* imagery (Figure 12.3). Lights from exurban development are captured in the nighttime image, although the buildings associated with those lights are lost in the forest of the 30-meter *Landsat* image. Although the fine-resolution *IKONOS* imagery captures some of the exurban development, it would be prohibitively expensive both financially and computationally to do this analysis across the United States with 1-meter imagery. Figure 12.3 presents several spatial resolutions of remotely sensed imagery that show both the benefits and drawbacks of finer spatial resolution.

Likewise, it is important to know the temporal resolution of imagery data. Production of the aforementioned nighttime imagery by the Defense Meteorological Satellite Program's Operational Linescan System (DMSP OLS) provides a good example of data reduction that utilizes temporal resolution to improve the quality of the data product. Global products of the DMSP OLS imagery are available at 1-km resolution for city lights, gas flares, lantern fishing, lightning, and forest fires. None of these products could be produced from a single night's observations because of clouds and the fact that different kinds of light are confounded with one another. The city lights product, one of the most popular of these data products, would not be possible without manipulation of the temporal characteristics of the DMSP OLS imagery. The city lights data product is produced from hundreds of orbits of imagery that occur throughout the year. The DMSP OLS system has a thermal band that allows for the screening of cloud impacted imagery;

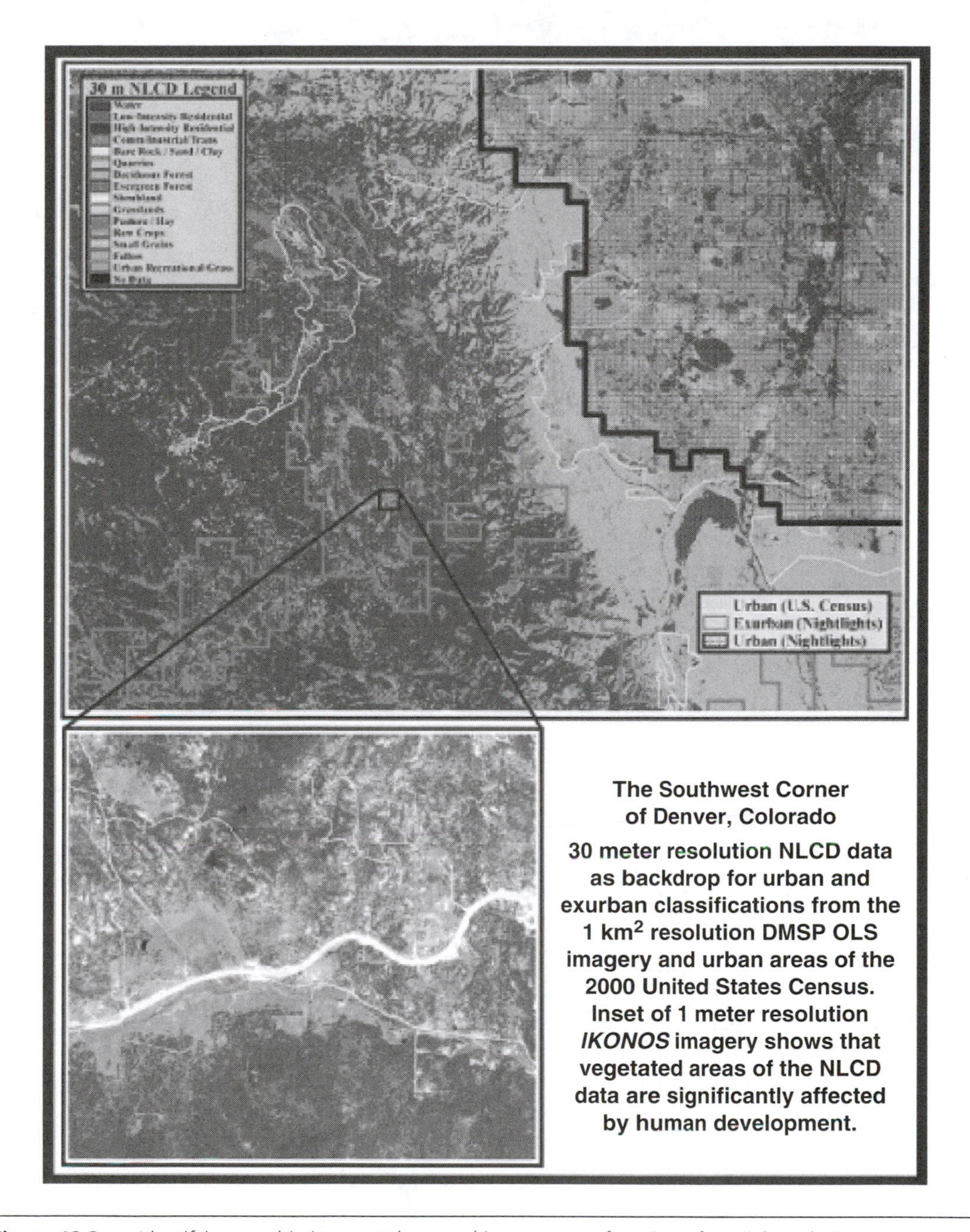
30 m NLCD Legend
High Intensity Residential
Bare Rock / Sand / Clay
Quarries
Deciduous Forest
Evergreen Forest
Grasslands
Pasture / Hay
Row Crops
Small Grains
Fallow
Urban Recreational Grass
Urban (U.S. Census)
Exurban (Nightlights)
Urban (Nightlights)
The Southwest Corner
of Denver, Colorado
30 meter resolution NLCD data
as backdrop for urban and
exurban classifications from the
1 km^2 resolution DMSP OLS
imagery and urban areas of the
2000 United States Census.
Inset of 1 meter resolution
IKONOS imagery shows that
vegetated areas of the NLCD
data are significantly affected
by human development.

Figure 12.3 Identifying exurbia in remotely sensed imagery, as a function of spatial resolution.

geo-location of the imagery allows for discriminating where the light sources come from—land or ocean, Middle East or Texas, and so on. City lights are generally nonephemeral and occur every night in a given pixel; fires and lightning are ephemeral and can be distinguished on the basis of their temporal behavior and spatial context. Thus, the city lights nighttime image data product is produced at an annual temporal resolution and enhanced in accuracy by collapsing the daily measurements made by the satellite from which it is derived.

Let's consider an example of how imagery resolution reduction works in the context of spectral resolution and thematic classification. *Landsat* imagery is provided at 30-meter spatial resolution (pixels contain information about 30 m by 30 m squares of earth surface) with seven bands of spectral information: ultraviolet, blue, green, red, near-infrared, mid-infrared, and thermal infrared. Typically, when a *Landsat* image is displayed on a computer, the user chooses to display three of the seven bands to make unique color combinations for each pixel. However, remote-sensing GIS software (for example, ENVI or Erdas Imagine) can apply mathematical techniques that use all the spectral information for each pixel to classify them into a number of unique categories. A typical application of these methods would be to classify a *Landsat* image into land use/land cover types, such as "deciduous forest," "grassland," "water," "urban," and so on. The classified image would then become a "single layer" image in which the pixel values take on nominal land use/land cover values (for example, 1 = water, 2 = urban, and so on). Classification of remotely sensed imagery is in essence a data reduction technique in which the spectral resolution of the image is reduced to increase its usefulness and interpretability.

Geographic Information Systems

Geographic information systems (**GISs**) consist of the hardware and software used to store, query, display, manipulate, and analyze geographic information. It is no exaggeration to describe the contribution of GISs to geography, and several other disciplines and professions, as revolutionary, but it is critical to recognize that the value of a GIS cannot be realized without human capital (including data collectors, analysts, and programmers) and social capital (including institutions and infrastructure). GIS technology began in the 1960s and has developed dramatically since then. Its future potential appears to be nothing short of incredible. In fact, GISs may become the standard for databases of all kinds in the future, when the capability of traditional databases is combined with the benefits of utilizing any spatial content that is part of the information in the databases. This could be considerable, of course, when we remember that so much information is from or about a *place.* Perhaps the abbreviation "GIS" will evolve from meaning geographic information system to meaning *general* information system.[5] In any case, GISs have become fundamental tools for conducting geographic research in many topical domains. In the rest of this chapter, we review the basic functions of GISs and consider their powers and limitations. We begin with a look at using GISs to store geographic information.

[5]Recently, "GIS" has frequently come to be defined as **geographic information science**. Geographic information science is the discipline, or collection of disciplines, that emerged during the 1990s as the scientific study of geographic information, including its representation and use in computers, human minds, and societies. To avoid confusion, geographic information science is often abbreviated as "GISci."

Information Storage

Most GISs can store information that is spatially referenced to an earth coordinate system (for example, latitude-longitude or some map projection), spatially referenced to itself (for example, features in the data are located relative to other features), or aspatial (not spatially referenced at all). Ideally, geographic information is best stored as spatially referenced to the earth surface. This allows one to spatially "cross-reference" it with other **geo-referenced information**. This information can be based on digitized maps, census data, remotely sensed imagery, data collected from observation in the field, surveys, or any other source. Storing information in a geo-referenced manner allows for many kinds of data manipulation, display, and spatial analysis that would otherwise be impossible. Consider the three temperature measurements described in the section above on data models and data structures. If these temperature measurements were stored with geo-referencing, they could be compared to other geo-referenced information such as elevation, population density, or rainfall. If they were stored without geo-referencing, they could be compared only to other information collected using the same "lost in space" coordinate system, and that information may not even have come from the same part of the world. If they were stored without any spatial reference at all (aspatially), then it would be difficult if not impossible to compare them with other geographic information, and simple calculations such as the distance between measurements could not be made.

Geographic information is stored as points, lines, polygons, or fields, and it is linked to associated tables containing one or more numbers, text units, or images. The spatial reference of geographic information manifests itself in many ways. It is now generally accepted that the earth is a somewhat "bumpy" oblate spheroid orbiting the sun. The development of this understanding of the size and shape of the earth co-evolved with the mechanisms we used to map and measure it. Today, one of the most sophisticated and accurate systems for measuring locations on the earth is the **Global Positioning System (GPS)**. The GPS consists of a suite of satellites that communicate via timed radio signals to receivers and base stations on the surface of the earth; the Bibliography contains references that explain this in detail. With a GPS receiver, one can fairly precisely map oneself or any other object onto our beloved bumpy and nearly spherical geoid. Consequently, a GPS receiver is an excellent way to obtain geo-referenced coordinates (typically latitude and longitude) for any entities or phenomena on the surface of the earth, and they are increasingly being used for this purpose.

Satellite images and aerial photographs provide information that is also best analyzed when geo-referenced, but such imagery is sometimes obtained without any geo-referencing at all. To geo-reference this "lost in space" imagery, ground control points with known geo-reference are used. Sometimes the imagery comes with complex orbital information about the location of the sensor that produced the image. This can also be used to geo-reference the imagery. However, a great deal of spatial information in geography, particularly human geography, does not use GPS or geo-referenced satellite imagery at all. Census data, economic data, and survey data can all be geo-referenced, although they often are not. Survey data could be geo-referenced according to the home or work address of the survey respondent. This would likely be done using a street network data set and the **address matching** functionality of a GIS. Economic data such as median household income could be geo-referenced by attributing the data to polygons that delineate the geographic boundaries of the regions from which the data were derived.

Data Query and Display

We pointed out in Chapter 10 that GISs provide powerful opportunities to query and display geographic information that far exceed not only traditional paper displays such as maps, but electronic calculators and other nongeographic computer databases. Displaying geographic information against the background image of its portion of the earth surface gives us many advantages over traditional databases when it comes to understanding data. It can help us identify errors or inconsistencies in the data, unanticipated spatial and temporal patterns, and previously unrecognized relationships that exist among data from diverse sources. These capabilities, when combined with the human ability to perceive spatial patterns, extend GIS functionality beyond that of traditional database systems.

A traditional database stored in a system like Oracle or Microsoft Access is typically a bunch of tables (you can think of them like Excel spreadsheets) that are usually linked together by common "key" columns. The functionality of these databases constitutes a powerful set of tools enabling the exploitation of the information they contain. A simple example of this functionality is the ability to query linked tables of information. Imagine a table that consists of names, social security numbers, and tax returns. Imagine another table that consists of the social security numbers, Visa card numbers, and addresses of people. These data could be queried via a traditional database system to identify the Visa card number of those people whose annual income exceeds $200,000. The same functionality is available with a GIS, but the results of the spatial query could then be mapped simultaneously with the tabular result of the query (in this example the addresses could be address matched and mapped). This operation would be useful to a researcher who wanted to contact people living in particular areas to solicit their participation in a research study.

Our example can be extended to show some of the profound benefits that can result from displaying geographic information with a GIS. Suppose you downloaded census data for the Denver metropolitan area. You load them into a GIS and display the population density of the census block groups of the metro area. Figure 12.4 shows the block groups of the Denver area displayed by both population density and housing value. Even without a legend, most people would be able identify the image on the left as being population density. If you displayed population density and the image looked like the image on the right, you would seriously explore the possibility that your data were corrupted in some way (or that you had not commanded the GIS to display them properly). Your ability to "see" that the left-side image "makes sense" and the right-side image "does not make sense" as a representation of population density could come from your knowledge of one or more of the following: (1) You could have firsthand knowledge of the Denver area; (2) you could know that in general the highest population densities in U.S. cities

Census data (block groups) for the Denver, CO, area mapped according to population density and housing value

Figure 12.4 Seeing spatial patterns in census block groups of Denver, CO.

tend to be in the city center, and that population density generally drops off with distance from the city center; or (3) you could know that census tracts try to encompass the same number of people, and that smaller ones are therefore more densely populated. Imagine trying to catch such an error if the output were simply a table of block group numbers and their associated population densities. The ability to display geographic information easily in map form is a profound advantage of GISs over traditional database systems. This ability of GISs to quickly and easily display spatial data begs the researcher to use this functionality for the purpose of quick and easy error checking, and in the process, to do some preliminary exploratory analyses that can help identify errors or inconsistencies in the data, unanticipated spatial and temporal patterns, and unrecognized relationships that exist among diverse data sources. These capabilities, used in tandem with the human ability to perceive spatial patterns, extend GIS functionality far beyond that of traditional database systems.

The profound capability of GISs for data query and display can be further demonstrated with this "over the top" example. Suppose you have a nighttime satellite image of Santa Barbara County, a polygon coverage of the census tracts, a street coverage of the county, and a point coverage of 700 Santa Barbara residents who took a survey on their attitudes about gun control and abortion. If you were courageous, you could use the black-and-white nighttime image as the background, display the population density of the census tracts as a semitransparent overlay in varying shades of red, and display the streets with symbols showing government jurisdiction, width of streets displaying street type (gravel, paved, interstate), and various shades of green showing traffic density. On top of it all, you could display the points with symbols showing survey respondent sex ("M" for male, "F" for female), the size of the symbol showing attitude about gun control, and the color showing attitudes about abortion. This map shows variation on at least eight variables but could be created from the data in a matter of minutes (interpreting this complex map is another story). In addition, a good GIS could virtually instantaneously calculate the correlation between population density and the measured attitudes about gun control, determine whether or not there is a statistical difference between the men and women in their attitudes about abortion, and determine whether or not this sample of survey respondents was likely to be spatially randomly selected from the population of Santa Barbara county. A GIS and a few data sets would allow you to look at a virtually infinite number of relationships, patterns, and phenomena. The challenge for the geographic researcher is finding the useful, relevant, effective, and interesting manipulations and analyses to perform.

Data Manipulation and Analysis

The distinction between data manipulation and analysis can be a little difficult to make in a GIS context. Often a simple manipulation of data (sometimes simply displaying the raw data) can enable analysis that takes place in the eye of the beholder. So we will not attempt to answer the question of where manipulation ends and analysis begins. There are several kinds of manipulations and analyses that are made dramatically easier when performed with a GIS:

1. ***Projection and coordinate transformation.*** As mentioned above, geographic information can come in various geo-referenced and non-geo-referenced formats. Most GISs are capable of converting geographic information from one coordinate system to another. Typically, the transformation of a data set from one coordinate system to another is just a matter of processing the X and Y coordinates through mathematical functions. This process is simplest with point data and gets more complex with line, polygon, and raster data. With line and polygon data, the transformation still operates on the points that make up the lines and polygons, then reconnects them using topological information in tabular form. In these situations, the density of points representing the lines and polygons can become a problem. For example, a minimum-bounding rectangle containing the conterminous United States in a geographic (lat-long) projection could consist of only four points. A reprojection of such a rectangle into an Albers Equal Area projection would result in a trapezoid with straight sides that should in fact have a curved top and bottom edge. This can be corrected by increasing the number or density of points making up the lines and polygons. When coordinate transforming raster data, you run into the issue of **resampling**. The space of a raster image is reprojected to a new coordinate system. This "space" is tessellated, typically with square pixels. The problem then is to determine what values those pixels should take. In resampling, the coordinates of the new pixels are sent through an inverse of the function used to perform the original transformation. This produces a location in the original image that is unlikely to land dead center in one of the original pixels. Consequently, one must choose from among several techniques available for determining the pixel value in the new image; three common resampling techniques are nearest neighbor, bilinear interpolation, and cubic convolution. If the raster data set is of a nominal

measurement level, then the nearest neighbor resampling method is the only appropriate choice.

2. ***Mathematical and logical manipulation of tabular data.*** Typically point, line, and polygon data consist of a geometric element (spatial reference and identification) and attribute information (tables linked to the geometric information). For example, line data representing roads might be associated with tabular information containing speed limits and traffic information, and polygon data of census tracts are associated with tabular information about population, areal extent, and median household income. This tabular information can be manipulated usefully, but also in erroneous and misleading ways. For example, the tabular polygon data with population and area attributes can be manipulated to create a new data attribute of population density by simply dividing the population column by the area column. These manipulations of tabular data can result from user judgments but also from mathematical and/or logical operations that produce new nominal, ordinal, or interval information.

3. ***Spatial aggregation and filtering.*** As we have suggested, it is sometimes desirable for various reasons to aggregate geographic information up to larger spatial units. Larger spatial units may represent a more appropriate resolution for a given geographic phenomenon, of course, but may also be preferable for display purposes, maintaining anonymity of survey respondents, more appropriate correspondence with other data, or ease of comprehension. Careful thought should be given to spatial aggregation, because GISs typically allow users to perform both appropriate and inappropriate aggregations. For example, suppose you have block-group census data with population counts, population density, percent of population that is Hispanic, and median household income as attributes. If you want to aggregate these block-group polygons to spatial units such as tracts or counties, you must give some thought as to what should happen to the block-group tabular attributes. You could sum the population count attribute to get an appropriate population count for tracts or counties. However, the other attributes are difficult if not impossible to retain at the aggregate level. You could recalculate population density if you either summed or recalculated the area of the new aggregate polygons. You could also recalculate the percent of population that is Hispanic with some effort. However, the median household income of the aggregate spatial units cannot be determined from the disaggregate information. In many GIS packages, this aggregation function (sometimes referred to as a "Dissolve") simply deletes the tabular attributes upon aggregation; other packages query the user as to what attributes to retain, and what logical or mathematical function to use to derive the new aggregate value (in the case of population in this example, it would be sum).

Raster data pixels can also be aggregated, and there are several ways in which to do this. Commonly the average value of the pixels being aggregated is used; however, there are many different ways to accomplish this task, including assigning the maximum or minimum value of any of the pixels being aggregated. **Spatial filtering** is an image processing technique that is typically applied to image or raster datasets. A filter (sometimes referred to as a **kernel function**) is an *n* by *n* window that is applied to each pixel of an image. The simplest example is a 3 by 3 pixel window that changes the data number of the pixel in question to the mean of the values in the 3 by 3 window. This is applied to every pixel in the image. A mean filter has the effect of "smoothing" an image. There are many different image filters, which use different mathematical functions, designed to do **edge detection**, remove a "salt and pepper" look, or serve other purposes.

4. ***Buffering and distance calculation.*** Measuring distance is one of the most fundamental capabilities of a GIS. It is a critical component of common functions like generating variograms, determining shortest paths on networks, and finding the nearest fire hydrant to the location of a given fire. A basic function of most GISs is the ability to create a **buffer** based on a fixed distance or a distance that could be derived from a variable attribute of points, lines, or polygons. An example is the way a logging company might buffer a road network in order to identify forest patches visible from the road. To minimize negative public perceptions of their clear-cutting activity, the company would avoid harvesting these patches. The buffer function of the GIS creates polygons at a specified distance from the roads. Of course, there has to be some thought as to how "thick" this buffer should be, or whether it should be a buffer of variable thickness that is larger for more heavily traveled roads. A GIS can calculate distances from points to points, points to lines, points to polygons, lines to lines, lines to polygons, and polygons to polygons. Any and all of these measurements can then be added to the tabular attributes of specified geographic features in the data set.

Another example of the utility of GIS distance calculations involves real estate prices and distance from the city center. Suppose you were interested in finding out if there was distance decay of real estate prices as they increased in distance from the central business district (CBD) of some city. You could overlay a point coverage of recently sold homes on a street network that contained the CBD. The GIS could calculate the shortest distance along the street network from each house to the CBD and add it to the attribute table of the house data set. A multivariate statistical analysis of the housing sales prices could then incorporate the new GIS-derived variable: distance to the CBD.

5. ***Overlay.*** Another fundamental function of GISs is the ability to compare two or more spatially referenced variables. There are numerous ways to conduct these kinds of comparisons, all of which are examples of the general concept of **overlay**. Given two raster data layers, most GISs can calculate the correlation between variables in the two layers and also add, subtract, or perform other mathematical functions on cells in the same geographic location that are expressed in the separate layers. A vector overlay may be performed between points, lines, or polygons.

The mathematical operations conducted as part of a raster overlay are part of the general concept of **map algebra**. The logic of vector overlay, in contrast, is derived from **Boolean logic**, the logic of combining sets. There are two major types of vector overlays, which represent the OR and AND statements of Boolean logic, respectively. The operation representing the OR statement is called a **union**. As shown in Figure 12.5, a union combines all of the geographic space of both layers, even if they are not co-incident in space, thus implementing the OR statement. The operation representing the AND statement is called an **intersect**. An intersect combines only the geographic space *shared* by the input and the intersect data layers, thus implementing the AND statement. When performed on polygons or lines, the layers are merged and new nodes are created at all crossings of polygon or line boundaries. The newly organized lines or polygons have the attribute characteristics of the two original layers. A specialized form of overlay with vector data is the **point-in-polygon** operation, which identifies spatial-set membership of a point layer within the polygons of a polygon layer. For example, one could perform a point-in-polygon operation on a U.S. "States" polygon layer with a point layer representing all the claimed "Elvis sightings" since his death. This could be accomplished in such a way that each point in the Elvis sightings data set has a new "State" attribute in its table of attributes, or it could be accomplished so that the attribute table of the States coverage has the new attribute of "Number of Elvis sightings" (with high numbers expected in Nevada and Tennessee!).

The details of how the principles of overlay are implemented vary between software systems. No matter the system, it is always worthwhile to read the documentation for each function to be certain that the theoretical logic expressed by your selected overlay operation matches the logic of your problem.

6. ***Spatial interpolation.*** As we discussed in Chapter 8, phenomena that are best represented as fields are measured only at a finite number of locations, typically points. This is especially common with physical geographic variables, such as rainfall, temperature, humidity, and ozone concentration. With such data, spatial interpolation is performed on the point coverage to generate the entire field of values—values intermediate to the sampled locations. A well-known example of this is the temperature map in the weather section of a newspaper. This is a field representation of temperature derived from a point network of temperature measurements (and a little bit of predictive modeling). There are many specific mathematical techniques for spatial interpolation, such as "inverse distance weighting," "kriging" (which comes in many flavors), "splines," and more. A sufficiently complete description of all of these techniques is beyond the scope of this book; references at the end of the chapter provide detail about them.

7. ***Spatial model building and spatial regression.*** Spatial model building and spatial regression analysis are some of the most sophisticated functions of GISs. Suppose you have a point network of rainfall measurements with information on rainfall, elevation, and temperature. You can build a simple regression model to predict rainfall from elevation and temperature. You can also apply the model to field data sets of elevation and temperature to produce a field representation of rainfall. Spatial model building often attempts to produce a representation of something that is not directly measurable. For example, suppose you wanted to produce a map of fire risk for the country of Portugal using variables like vegetation, slope, aspect, and distance to roads. You could obtain or derive a vegetation layer from the classification of a set of *Landsat* images, the slope and aspect layers from a digital

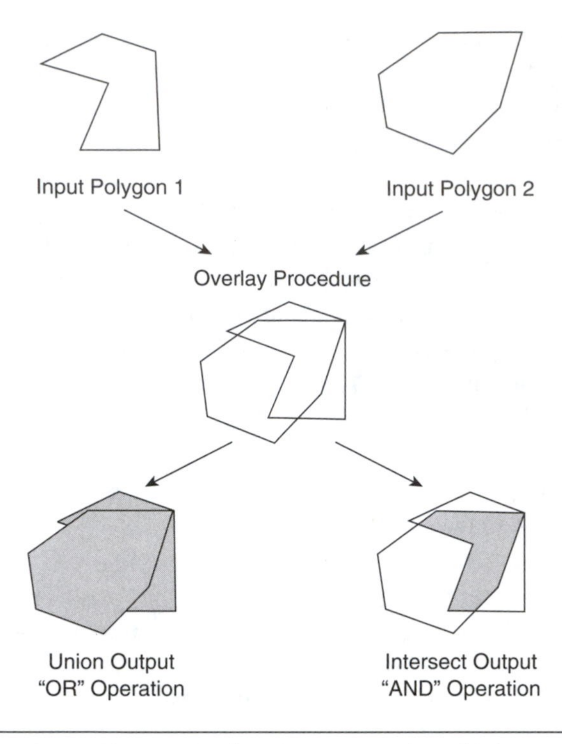

Figure 12.5 Union and intersection functions in a GIS. (Graphic by M. V. Gray. Reprinted with permission.)

elevation model (DEM—see Chapter 4) of Portugal, and the distance to roads from a vector coverage of the roads in Portugal. You could then calculate fire risk from any logical or mathematical combination of those input variables on a pixel-by-pixel basis. You could then input this model into a GIS to produce a raster representation of the abstract concept of "fire risk" for Portugal.

Review Questions

Geographic Information

- What is geographic information, and what are some sources of such information?
- What are some important characteristics of your phenomenon and the data you use to study it that you should consider when working with geographic information?
- How does the distinction between fields and objects relate to the distinction between physical and human geography?
- Why are issues of ontology important when working with geographic information, and what are some examples of ontological issues in geographic research?

Data Models and Data Structures

- What are data models and data structures?
- What are the vector and raster data models, and how does each fit different kinds of phenomena and data more or less appropriately?
- Why is it important when conducting research with geographic information to consider data models and data structures?

Remotely Sensed Geographic Information

- What are some different types of remotely sensed geographic information?
- How do spatial, temporal, and thematic scale express themselves in the context of remotely sensed geographic information? Provide specific examples of each.

Geographic Information Systems (GISs)

- What is a geographic information system, and what are its technological and human components?
- What are some methods and technologies for spatially referencing geographic information?
- What are some benefits of being able to query and display spatially referenced data with a GIS? Demonstrate your answer with a specific example.
- What are the data manipulation and analysis techniques of (1) projection and coordinate transformation, (2) mathematical and logical manipulation of tabular data, (3) spatial aggregation and filtering, (4) buffering and distance calculation, (5) overlay, (6) spatial interpolation, and (7) spatial model building and spatial regression? Provide specific examples of each.

Key Terms

active sensing: remote sensing of radiation that is originally emitted by the satellite and then reflected back from the earth surface; a good example is Radar

address matching: GIS operation in which a spatial database is used to georeference standard street addresses, ZIP codes, and so on

Boolean logic: the logic of combining sets, used as the basis for vector data overlay operations

buffer: GIS operation in which a band or zone of specified width is placed around selected vector features

classification: the creation of semantic systems, or ontologies, for organizing reality into meaningful types or units

conceptual model: mental or intellectual representation of geographic phenomena, including their spatial and nonspatial attributes, as understood by individuals or groups of people; the central conceptual distinction in geography is between "objects" and "fields"

coordinate system: spatial reference system of numerical units and rules that encodes location on the earth; latitude-longitude is the most common, but there are many

coordinate transformation: GIS data manipulation in which locational information in one coordinate system is mathematically translated into another

coverage: data layer in a GIS

data cube: large geographic information data set constructed from multiple spectral layers of remotely sensed information about the same portion of the earth, recorded at the same time

data model: simplified computer representation of a conceptual model of some phenomenon; the central distinction in geographic information is between "vectors" and "rasters"

data structure: the digital expression of a data model as represented in a computer

distinct number: the number of data points in a data cube stored for each pixel of earth surface; each data point is an intensity value in one narrow slice of the electromagnetic spectrum

edge detection: spatial filtering transformation designed to identify and highlight boundaries or discontinuities in a raster data layer

fields: fundamental conceptualization of geographic phenomena as continuous surfaces (usually two-dimensional); much more common in physical than in human geography, it is usually contrasted with "objects"

geographic information: information that is spatially referenced to the surface of the earth; nowadays, the term usually implies information in digital form

geographic information science (GISci): the discipline, or collection of disciplines, that emerged during the 1990s as the scientific study of geographic information, including its representation and use in computers, human minds, and societies

geographic information system (GIS): the computer hardware and software used to store, query, display, manipulate, and analyze geographic information

geoid: the average surface around the planet Earth that is used as its baseline shape; it is a little more accurate as a model of the planet's shape than is a strict oblate spheroid

geo-referenced information: information that is spatially referenced to an earth coordinate system

Global Positioning System (GPS): electronic technology for determining the two- or three-dimensional location of any point on or near the earth surface, based on a set of satellites and computer software that triangulates location on the basis of time information in the satellite signals

granularity: synonym for resolution

interoperability: the problem of effectively communicating geographic information among different people, cultures, disciplines, and information systems; typically used in the context of digital information systems

intersect: GIS operation applied to overlaid vector data in which common values are identified that are present in all layers

kernel function: the *n* by *n* window of cells in a raster data layer to which spatial filtering transformations are applied in order to convert the cells to a single common value

lines: entities modeled as one-dimensional features in a vector data model

map algebra: the set of mathematical GIS operations, such as addition or subtraction, that can be applied to overlaid raster data

objects: fundamental conceptualization of geographic phenomena as discrete entities (of any dimensionality); common in both physical and human geography, it is usually contrasted with "fields"

oblate spheroid: geometric term for a sphere that is "flattened"; it is very nearly the shape of the planet Earth, which has a pole-to-pole diameter 42 km (26 miles) less than its equatorial diameter

ontologies: systems of concepts or classes of what exists in the world; somewhat different forms are found in individual minds, cultural belief systems, and computer databases

overlay: GIS operation in which two or more data layers based on the same area of earth surface are superimposed in order to be compared

passive sensing: remote sensing based on recording solar radiation reflected from the earth surface or other types of radiation emitted from the earth; a good example of emitted radiation is that from fires or electric light bulbs

pixels: approximations of fields on digital image displays as checkerboard coverages of small regular "cells," usually square; the computer-screen analogue to remotely sensed image data rasters

point-in-polygon: GIS operation applied to overlaid vector data in which a layer of points is intersected with a polygon layer

points: entities modeled as zero-dimensional features in a vector data model

polygons: entities modeled as two-dimensional features in a vector data model

raster data model: data model that represents geographic information as a continuous two-dimensional field consisting of a regular tessellation of small cells, typically square; the typical data model for satellite remotely sensed imagery

remotely sensed geographic information: information about the (near) earth surface in the form of analog or digital recordings of patterns of electromagnetic energy in one or more portions of the spectrum; it includes aerial photography and satellite imagery

resampling: GIS data manipulation in which new values are determined for raster data layers that are coordinate transformed

resolution: term for spatial, temporal, or thematic scale, particularly used in the context of analysis scale with digital geographic information; synonym of granularity

spatial aggregation: GIS data manipulation in which information at one spatial scale is transformed into some larger scale by combining areas

spatial filtering: type of spatial aggregation in which raster information is transformed by modifying cell values within a particular defined area according to a mathematical or computational formula

spectral resolution: the degree to which remotely sensed information can pick up precise portions of the electromagnetic spectrum rather than broad portions; it is an expression of thematic scale because it greatly influences the specificity with which particular geographic features can be identified

tessellate: to "tile" a two-dimensional (planar) surface with a coverage of small regular cells; equilateral triangles, squares, and regular hexagons are the three regular polygons that can tessellate a planar surface exhaustively without overlap

themes: the characteristics of human and natural phenomena other than space and time that geographers measure as variables and store as attributes in a GIS

union: GIS operation applied to overlaid vector data in which values are identified that are present in any of the layers

vector data model: data model that represents geographic information as meaningful geometric objects of varying dimensionality

volumes: entities modeled as three-dimensional features in a vector data model

Bibliography

Burroughs, P. A., & McDonnell, R. A. (1998). *Principles of geographic information systems.* Oxford: Oxford University Press.

Chang, K.-T. (2004). *Introduction to geographic information systems* (2nd ed.). Boston: McGraw-Hill Higher Education.

Hurn, J. (1993). *Differential GPS explained: An exposé of the surprisingly simple principles behind today's most advanced positioning technology.* Sunnyvale, CA: Trimble Navigation.

Jensen, J. R. (2000). *Remote sensing of the environment: An earth resource perspective.* Upper Saddle River, NJ: Prentice Hall.

Longley, P. A., Goodchild, M. F., Maguire, D. J., & Rhind, D. W. (2005). *Geographical information systems & science* (2nd ed.). Chichester, U.K.: Wiley.

McNoleg, O. (1996). The integration of GIS, remote sensing, expert systems, and adaptive co-kriging for environmental habitat modeling of the Highland Haggis using object-oriented, fuzzy-logic, and neural-network techniques. *Computers & Geosciences, 22,* 585–588.

O'Sullivan, D., & Unwin, D. (2003). *Geographic information analysis.* Hoboken, NJ: Wiley.

Peuquet, D. J. (2002). *Representations of space and time.* New York: Guilford.

Quattrochi, D. A., & Goodchild, M. F. (Eds.) (1997). *Scale in remote sensing and GIS.* Boca Raton, FL: Lewis Publishers.

Worboys, M., & Duckham, M. (2004). *GIS: A computing perspective.* Boca Raton, FL: CRC Press.

CHAPTER 13

Scientific Communication in Geography

Learning Objectives:

- What are the purposes and major forms of scientific communication, including types of scientific literature?
- What are some recommendations for giving effective oral presentations?
- What is the peer review system for scientific literature, and how does it work?
- What is the structure of a one-study empirical research article?
- What are major library resources for scientific research, and how can they be found and used?

The results of research have little value if they are not communicated to people. These people are other researchers but also students, potential employers, journalists, administrators, business leaders, elected officials, and members of the general public. The act of conducting research itself necessarily involves sending and receiving communications—by reading the ideas of other researchers and the results of past research, by discussing ideas with collaborators and assistants, and by requesting financial support from funding agencies. Thus, scientific communication is critically important to the entire process of research, not just to its results, and it involves an assortment of communications that vary from the relatively formal to the relatively informal. Much scientific communication occurs in verbal form, both oral and written, but also in numerical and graphical forms.

In this chapter, we examine the various forms of verbal scientific communication; we discussed numerical and graphical communication in Chapters 9 and 10. We first look at the archive of written communications known as **scientific literature**. Perhaps the most important are scientific **journals**. These are serials published as frequently as once a week or as infrequently as once a year, but most often once every two or three months. Journals provide the primary forum for the communication of scientific ideas and empirical results. They are a bit like scientific magazines, except the editors of scientific journals are aided in their evaluation of manuscripts submitted for publication by a panel of peer experts; we discuss this peer review system below. And unlike the authors of magazine articles, authors of journal articles are not paid—they may even have to pay some charges for things like including color figures in their article. But scientists definitely get rewarded for publishing journal articles. There is an intrinsic value to having one's work put into print; most people find it quite satisfying for both egotistic and altruistic reasons. In addition, academic researchers, if not other scientists, must publish in order to get hired, promoted, and funded.

Journals vary in their topical domain, their intended readership, their methodological style, and their quality and status; status is largely a function of the number of subscribers and readers a journal has, and the rate at which it accepts submitted manuscripts for publication. Table 13.1 provides a list of some of the major English-language journals that publish research in geography and related fields. There are many other journals relevant to geographic research, however, and most geographers read outside the walls of the discipline of geography, strictly speaking. For instance, population geographers read journals in which mostly sociologists or economists publish, and biogeographers read journals in which mostly botanists or ecologists publish.

There are several kinds of journal articles. Some journals publish several of these kinds, whereas others primarily publish one or two. The most common kind of article reports one or more empirical studies; we discuss its basic structure in detail below. Another kind is the literature review, which summarizes all relevant articles presenting research on a particular issue; for instance, there are literature reviews on the interpretation of topographic symbols on maps and on the migration patterns of refugees. Ideally, authors of literature reviews do not just summarize, like a grade-school book report, but organize, evaluate, and synthesize the empirical results and ideas in the literature.[1] A third kind of article is the theoretical paper that presents new concepts, theories, or models but not new empirical results. The methodological/statistical article presents new methods for collecting or analyzing empirical data, or critiques existing ones. Finally, there are various short forms of journal articles, including research notes, commentaries, letters to the editor, and

---

[1]**Meta-analysis** is a quantitative review of research that makes conclusions about some phenomenon by statistically combining the empirical results of tens or hundreds of independently conducted studies on the phenomenon. State-of-the-art conclusions about the health effects of smoking, for example, are based on meta-analytic reviews of many independent studies of smoking. Meta-analysis is underused in geography.

Table 13.1 A Sample of Major English-Language Academic Journals in Geography

General Geography:

Annals of the Association of American Geographers
Applied Geography
Area
Australian Geographer
Canadian Geographer
Geoforum
Geographical Analysis
Geographical Journal
Geographical Review
Geography
Journal of Geography
The Professional Geographer
Transactions of the Institute of British Geographers

Physical Geography and Related Fields:

ASCE
Climate Research
Earth Science
Earth Surface Processes and Landforms
Geografiska Annaler. Series A, Physical Geography
Global Ecology and Biogeography
International Journal of Earth Sciences
International Journal of Environmental Studies
Journal of Biogeography
Journal of Climate
Journal of Geophysical Research
Journal of Hydrology
Journal of Soil Science
Physical Geography
Polar Geography and Geology
Progress in Physical Geography
Soil Science

Human Geography and Related Fields:

Antipode
Environment and Behavior
Environment and Planning A-D
Environmental Education and Information
Geografiska Annaler. Series B, Human Geography
Journal of Cultural Geography
Journal of Economic Geography
Journal of Historical Geography
Journal of Transport Geography
Location Science
Networks and Spatial Economics

(Continued)

Table 13.1 (Continued)

Political Geography
Population and Environment
Population, Space, and Place
Progress in Human Geography
Regional Studies
Urban Geography

Geographic Techniques:

Cartographic Perspectives
Cartographica
Cartography and Geographic Information Science
Computers & Geosciences
Geodesy, Mapping and Photogrammetry
Geographical Systems
GIScience & Remote Sensing
IEEE Transactions on Geoscience and Remote Sensing
International Journal of Geographical Information Science
International Journal of Remote Sensing
Photogrammetric Engineering & Remote Sensing
Remote Sensing of Environment
Spatial Cognition and Computation
The American Cartographer
Transactions in GIS

book and software reviews. Of course, these kinds are not pure; nearly every empirical report, to take an important example, contains a short review of literature and a discussion of theory.

Books are another form of written communication in science. Books contain some combination of concepts, theories, empirical results, literature reviews, and methodological discussions. They are less timely than journal articles (that is, they take longer to write and publish) and are less stringently peer reviewed, if at all. They may be written by one author (or one group of authors), or by several authors and an editor, in which case different authors write each chapter. The editor of a multiauthored **edited book** revises or organizes the chapters. Some editors write short overviews or commentaries for the book, based on the content of the various chapters; these overviews can help unify the chapters into a single book. Other editors do little more than remind authors of their deadlines and decide on the order of the chapters. This is typically what the editors of **conference proceedings** do; proceedings are collections of papers based on talks or discussions at scientific conferences.

Another important form of written communication is the **grant application**. A grant application is a request for research funding for a proposed program of research, made to sources such as public agencies and private foundations. The exact structure of a grant application depends on the particular source where the money is requested. In general, grant applications consist of the following: a summary of the proposed research, a description of its benefits (both to a scientific area and to society), a review of relevant literature (with references), a detailed plan for what and how studies and analyses will be conducted, a plan for dissemination of the research in publications and conference presentations, a timeline for research activities, academic and research biographies of investigators, a description of research infrastructure and equipment available, and a detailed and justified budget.[2] The major differences between grant applications and journal articles are that grant applications must be especially concise and are written to persuade funding agencies to award money. Thus, even when an application proposes a program of basic research, it must attempt to describe possible applied benefits of the research to society. Instructions for preparing a grant proposal to the U.S. National Science Foundation, which are fairly typical for any grant proposal, may be found at https://www.fastlane.nsf.gov/fastlane.jsp.

There are a variety of other forms of written communication in scientific research. The Internet supports a variety of written communications, stored and sent digitally: electronic journals, discussion lists, Web sites, and so on. High school, undergraduate, and graduate students write research papers for classes, honors programs, master's theses, and doctoral dissertations. Student papers are typically styled like journal articles, with the exception that their major purpose is to demonstrate academic proficiency to instructors or advisors. As a consequence, they tend to be more didactic in their presentation of material (just as texts like this one are). That includes summarizing research literature more comprehensively and describing in more detail the research materials (such as the satellite imagery or entire survey instrument) and research results (such as detailed information about all of the sites where data were collected). However, doctoral dissertations in particular often become actual journal publications eventually; when this happens, the material in the dissertation is restructured and condensed for publication after the doctoral degree is completed, sometimes years after.

Oral communication is important in science too. Most notable here are talks given at academic and scientific conferences. These talks are commonly called

[2]Grant applications typically request funding to cover the costs of some or all of the following: partial or complete salary for lead investigators, graduate students, staff members, and postdoctoral research assistants; data collection or acquisition costs, including direct fees paid for data, research materials and equipment (including computer software and hardware), payments made to research participants, and travel expenses associated with data collection; travel expenses incurred by attending meetings and conferences; miscellaneous supplies; and "overhead." The last category refers to the (rather large) percentage of the grant that university and research laboratories require to support the infrastructure costs of doing research at their institutions—room rental, utilities, custodians, and so on. Although it is difficult to know exactly how institutions like universities actually spend overhead, because administrators generally do not reveal that to research investigators, they apparently use some overhead money to help support activities at the institution other than funded research.

Table 13.2 Recommendations for Giving Effective Oral Presentations

1. Content
   - Have a single theme or main point, even for a long talk.
   - Think about your audience. What do they know? What do they need explained or defined?
   - Even for a sophisticated audience, it is always best to overview your talk briefly and define fundamental concepts clearly and unambiguously when first used. Explain acronyms and other abbreviations upon first usage.

2. Presentation Style
   - Your goal is to inform *and* entertain. You want to hold your audience's attention. Interesting content, well organized, is a big part of this, but presentation style is critical too.
   - Don't read your paper (although that is acceptable in some humanities disciplines).
   - Establish eye contact with the audience; speak to individual people (if you want to pretend some of them are naked, as is sometimes advised, be our guests, but we never found this to calm us down).
   - Speak loudly enough, vary your tone, and be enthusiastic and energetic. This may require that you fool yourself into believing you are fascinated by what you are discussing.
   - It's OK, even good, to move around some (see previous recommendation) but avoid nervous movements.
   - Avoid defensive responses to questions or comments. Be comfortable saying "I don't know" or "that's a good point."

3. Appearance of Presentation
   - Use words *and* graphics (photos, graphs, maps, videos, animations). Presentation software like PowerPoint makes this easy to do. If sound recordings are appropriate for your topic, by all means use them.
   - Avoid overly complex or busy slides that include too much information, whether verbal or graphical.
   - Use large font sizes. This is one reason to avoid overly complex slides. Some of your audience may have poor eyesight, but even those with 20/20 eyesight deserve to see your slides without straining.
   - Carefully proofread for typos and inconsistencies. The way you use words, symbols, and punctuation are sometimes less important than making sure to use them the same way each time.

4. General Recommendations
   - Stay within your allotted time. It is very hard to be adequately brief and concise, but very important. Leave enough time for questions—that time belongs to the audience, not to the speaker. Besides, speakers often learn from questions, and audience members usually enjoy listening to questions and their answers.
   - Know your room and audiovisual equipment beforehand. Be prepared for technical failures. Will the broken digital projector or Internet connection destroy your talk?

- Dress with an appropriate level of formality. This might be considered dressing for success. Or it might be dressing to conform, which not everyone considers a good thing. Either way, the need to appear casual or even radical, as admirably self-expressive as that might be, strikes us as a rather superficial reason for turning people off, having them ignore your brilliance, or having them refuse to hire you. Put another way, one dresses formally to communicate respect for the audience and the task at hand—it's a matter of etiquette.
- Practice beforehand for content, flow, style, and time (ask a friend or two to play the audience).

papers even though they are presented orally.[3] A **conference** is like an academic convention, where new and established researchers meet to present and listen to talks about research. Conferences are social events too; attendees meet and "network" with other members of the discipline, representatives of funding agencies, publishers and other sales representatives from relevant companies, and potential employers and employees. The largest geography conference in the United States is the annual meeting of the Association of American Geographers (AAG). About 5,000 members, affiliates, and guests of the AAG attend each year; as of 2005, the AAG had about 9,000 members. Something like 2,000 research presentations are given there.

Besides conference talks, other relatively formal examples of oral presentations include "job talks" given by candidates for faculty positions during job interviews and "colloquia" given by researchers visiting labs and departments. Of course, instructors lecture in class. In Table 13.2, we offer recommendations for giving effective oral presentations of a relatively formal nature. There are also a host of relatively informal oral and written communications among researchers, including e-mails, letters, phone calls, "hallway chats," and so on; although informal, they may be of great importance. Finally, one should not forget scientific communications with nonscientists (or scientists from other disciplines) that go on in various popular media, including TV, radio, newspapers, and magazines.

Peer Review System for Academic Publishing

In Chapter 1, we pointed out that science is a social activity and that social processes help reduce the potentially distorting effects that the human motivations of

[3]Although the orally presented paper is the primary format for formally communicating research ideas and results at conferences, there are other formats, such as panel discussions and **posters**. Posters are brief written and graphical presentations of research that are literally posted on bulletin boards (about 3 m^2 in area) and set up for a session of a couple hours in a room with several other posters. Conference attendees visit the poster sessions of their choice, walking around to view the posters and discuss the research with the poster authors. Although most senior researchers in geography avoid presenting in poster sessions, they do not avoid attending them. The poster session is a unique way to meet researchers at all career stages and discuss research more deeply and intimately than at a paper session.

Table 13.3 Issues That Reviewers and Editors Typically Consider When Reviewing Manuscripts for Publication

- Is the manuscript appropriate for the journal under consideration, considering the topical domain, target audience, and methodological style of that journal?
- Does the manuscript make an original contribution? Is the magnitude and relevance of the contribution sufficient to deserve publication?
- Is the appropriate literature cited and accurately described? Are appropriate links made with theory?
- Are the proper methods used? Are concerns about the validity of the methods addressed?
- Are the proper statistical or mathematical tests applied, and are they applied correctly? Are results interpreted correctly?
- Is the title appropriate and informative?
- Is the writing clear, unambiguous, and effective? Are punctuation and grammar correct?
- Can the manuscript be shortened without loss of important content, or does it need to be expanded in some places?
- Are the tables and figures necessary? Should there be additional ones? Are the tables and figures well designed? How could they be improved?

individual scientists can have on scientific beliefs. As we noted there, one of the most important examples of such a social process is the **peer review system** in academic publishing. The peer review system consists of authors, editors, and reviewers, and the process they use to determine whether a manuscript should be published in a particular outlet, and if so, what should be changed about the manuscript to improve it for publication. The criteria used to decide if a manuscript should be accepted for publication, and used as the basis for recommended improvements, include a great variety of issues concerning the way research studies were conducted, the way they were interpreted, the way the research was connected to other research that exists in the scientific literature, and the way the manuscript is written and otherwise communicated in tables and graphics (see Table 13.3). These criteria are more or less the same for every journal, although most journals have their own list of reviewer criteria that vary somewhat; the specific criteria are distributed to reviewers when their reviews are requested. These review criteria largely apply to the review of grant applications too. In addition, grants are reviewed for their potential benefits to a scientific area and to society, and for the likelihood that particular applicants can pull off approximately what they propose, given the particular timeframe and resources of the grant.

Peer review typically proceeds as follows. An author decides where he or she wants to publish the manuscript. This decision is ideally based on the topical domain and target audience of the journal. That is, one would hopefully not try to publish research on the diffusion of linguistic dialects in the *Journal of Retailing*, let alone the *Journal of Biogeography;* perhaps the *Journal of Cultural Geography*, a general geography journal, or a linguistics journal would be appropriate. In addition to topical appropriateness, authors usually try to maximize the prestige of the journals in which they publish, both because articles in more prestigious journals are read by more people and will thus have greater impact and because publishing in more prestigious journals inflates one's career reputation, if not one's salary. It can even inflate one's ego (there's that human side of science again). Once the author has decided where to submit the manuscript, he or she follows the "Instructions to Authors" printed in occasional issues of the journal and on the journal's Web site. These instructions discuss stylistic rules to follow in preparing the manuscript, how many copies to send for consideration, in what form to send them, and to what person and address. The instructions usually also spell out an important general rule of publication: Authors may submit manuscripts *to only one journal at a time.*

When the editor receives the manuscript, he or she usually makes an initial decision whether it is, on the whole, appropriate for publication in that journal. The editor sometimes decides the submitted manuscript is *not* appropriate for that journal and informs the author of that; the author decides whether to submit it to a more appropriate journal. But more often than not, the editor proceeds with the review by sending the manuscript to reviewers.

Each manuscript is reviewed by between one and five **reviewers**; three is a common number. Reviewers are researchers and scholars at universities and research labs who have expertise in the specific topical area of the manuscript, often having published in that area, but who do not have any clear conflicts of interest (we discuss ethical issues in research in Chapter 14). Reviewers decide if they want to review the manuscript based on whether they have the time and expertise to do so. They are not paid for reviewing; rather, it is considered an act of service to the discipline that is appropriate especially if a reviewer later wants to have his or her own research manuscripts reviewed for publication. The review is usually carried out **blind** or **double blind**. In blind review, either the author is not informed of the identity of the reviewer or the reviewer is not informed of the identity of the author—the author's name is cut out or obscured in the manuscript; in double-blind review, neither author nor reviewer is thus informed.[4]

[4]There is an ongoing debate in the scientific community about the merits of blind and double-blind review, and some journals have stopped using it or at least give reviewers the option to reveal their identity or not. Blind reviewing is intended to reduce or eliminate any residual bias in a review that might occur as a result of a reviewer knowing the identity of an author or an author knowing the identity of a reviewer. For instance, the work of prestigious authors likely gets reviewed with a more positive attitude at the outset, and reviewers may hesitate to be honestly critical if they think the author will learn their identities. On the other hand, some critics of blind review believe it is futile because an experienced reviewer can often recognize the work of particular researchers, and in any case, it is difficult to fully expunge signs of who authored a manuscript just, for instance, by removing the name on the title page. Also, critics argue that the anonymity of blind review gives license to reviewers to be unnecessarily harsh, even hostile or sarcastic (some people we know don't require anonymity to get this license).

Whether blind or not, the reviewer reads the manuscript carefully, often more than once. He or she writes a review of the manuscript based on criteria like those in Table 13.3, and often fills out a short survey that requires the reviewer to rate the manuscript on the criteria. Most reviewers consider their task to be twofold: to advise on whether the manuscript should eventually be published in that journal, and to advise on how to improve the manuscript if it is to be published. To these ends, the reviewer makes a single summary rating of the manuscript as his or her recommendation to the editor. This summary rating scale varies somewhat across journals but usually includes something like the following choices:

(a) accept for publication without revision

(b) accept for publication with minor revisions

(c) revise extensively and resubmit for review

(d) reject the manuscript but encourage submission to a different journal

(e) reject the manuscript as not publishable in any journal.

Upon receiving reviews, the editor makes the decision and informs the author. The editor usually reads the manuscript too, basing the editorial decision on his or her own evaluation combined with the evaluations of the reviewers. Some editors follow the advice of the reviewers very closely, especially if the paper is outside the topical expertise of the editor (and no editor is expert in every topic). Other editors sometimes veer from the recommendations of reviewers, and especially when the reviewers disagree about the quality of the manuscript (not uncommon), the editor has to step in and make a more independent decision about the manuscript. The editor sends the decision to the author, along with copies of the reviews.

Receiving an editor's decision can be a joyous event for the author but is unfortunately often frustrating. Rejection rates at journals can be quite high, especially at very prestigious journals. Such journals may accept for publication no more than 5–20% of the manuscripts sent to them (by "accept," we mean *eventually* accept). Some revisions are requested for virtually all manuscripts, even very good ones by top researchers. A summary rating like (a) above is exceedingly rare, perhaps never to be gotten in the career of a competent geographic researcher. So a rating like (b) is excellent and even (c) should often be considered fairly positive. In the end, editors frequently decide that manuscripts may be accepted for publication if the author carries out relatively extensive revisions, sometimes including collecting more data or conducting new analyses. Thus, manuscripts often get published only after they have been reviewed and revised twice or more. Furthermore, if a manuscript is rejected at one journal, authors often attempt to improve it and send it to a different journal. This is not a sign of incompetence; as in many areas of life, such persistence is critical to eventual success. But resubmissions to multiple journals should include interim revisions that attempt to improve the manuscript on the basis of feedback from editors and reviewers—this feedback usually contains mostly good advice for improving the manuscript.

The Basic Structure of a Journal Manuscript: The One-Study Empirical Article

There is no single style dictating the structure of a journal manuscript in geography. As we stated above, every journal provides style instructions to potential authors. Some journals instruct authors to follow the style presented in one of the widely known style manuals, such as Strunk & White's *Elements of Style*, the *Chicago Manual of Style*, or the *Publication Manual of the American Psychological Association* (see the Bibliography at the end of the chapter). Style manuals like these have been revised and updated over many years; they contain a wealth of useful and interesting material about how to write a research manuscript, ranging from manuscript structure, citations and references, word choice, scientific abbreviations, punctuation, table and figure design, and more. They are essential reference books for the research author and surprisingly good reads.

We do not have space here to cover exhaustively the structure of all kinds of journal manuscripts. We focus on providing a blueprint for the most common kind, the one-study empirical article (Table 13.4). We recommend that new authors try to follow this blueprint rather closely. As you become more experienced, however, you can introduce variations that help tell the particular story better and allow for the creative expression that potentially makes any writing more engaging, even scientific writing. But until you master the standard format, your attempts at creativity are not likely to work well, because you will not clearly understand how to introduce variations meaningfully and effectively. It's like any artistic act: Novelty can work only when it is expressed judiciously and intentionally, not excessively or accidentally.

The first page is the **Title page.** It lists the authors' names and institutional affiliations. The list of authors usually includes everyone who contributed "substantially" to the conception or design of the research, the interpretation of its results, or the writing of the manuscript. Research assistants who merely helped with data collection or routine analysis are probably not authors, nor are people who helped with routine tasks such as typing, computer administration, and grant management (not to imply that "routine" means unimportant or valueless). Also, the order of authors' names matters, because the scientific community, including hiring and promotion committees, typically gives the first author major credit for the manuscript. A general rule is that the first author is the person who initiated the research and made the largest contribution to its conceptualization, theoretical analysis, and write-up in the manuscript. However, opinions of individual researchers (and disciplines) vary somewhat in who should be included as authors and the significance of first authorship. It is sometimes standard practice always to include a student's advisor as an author on anything published by the student, and sometimes the first author, regardless of the advisor's actual contribution. This is an issue of research ethics to which we return in Chapter 14. We will say here, however, that we do not endorse this practice and emphasize that students have the same ethical right to first authorship as anyone else (not to imply that students and other new

Table 13.4 Basic Structure of a One-Study Empirical Article

a. Title Page
   (1) author names (order matters and must be discussed ahead of time)
   (2) departmental or institutional affiliation

b. Abstract
   (1) short summary; as few as 100 words, rarely more than 300
   (2) what you did, why you did it, what you found, what it means

c. Introduction
   (1) topic (first sentence should not refer to researchers or to literature)
   (2) citations to relevant literature; put your problem in context
   (3) specific research question or hypothesis at the end

d. Method
   (1) cases (who or what, how sampled)
   (2) materials (including equipment, survey questions, computer displays, and so on)
   (3) procedures (narrative of what you did to get data, make model)

e. Results
   (1) descriptive (especially) and inferential statistics
   (2) concise interpretation—avoid broad conclusions here
   (3) most central question first, if possible
   (4) data treatments, transformations, data problems and how you dealt with them

f. Discussion
   (1) results restated and interpreted specifically
   (2) possible problems or threats to validity
   (3) results interpreted more broadly; future work that should be done
   (4) final conclusions (usually no need to say that "more research is needed"—tell us what research is needed)

g. References or Bibliography
   (1) list, usually in alphabetical order by first author's last name, second author's last name, and so on (sometimes in citation order)
   (2) list in chronological order if there are multiple references by same authors

h. Tables and/or Figures

i. Appendices (occasionally)

researchers always have an accurate view of what contribution deserves first authorship). In any case, authorship credit is one of those issues that can potentially cause conflict within the "relationship" that is scientific collaboration. If you and your collaborators want to maintain a harmonious relationship, you should discuss this well ahead of manuscript submission.

After the title page comes the **Abstract**, often on a separate page. The abstract is a short summary of what you did and why, what you found, and what it means. Writing an informative and complete abstract is quite challenging, as they are typically limited to as few as 100 words and rarely more than 300.

Next come the four major sections of the body of the manuscript: Introduction, Method, Results, and Discussion. The **Introduction** section is more substantial than its name suggests. It does introduce the topic of the manuscript in the first paragraph or so, preferably with a first sentence about the topic area or specific phenomenon of interest.[5] After this short topical introduction, narrow your focus by turning to the specific research issue or question addressed in the manuscript. Provide a review and discussion of relevant literature, making sure to cite it, in order to put your problem in context. This is a big part of scholarship, as we discuss later. The Introduction should end narrowly with a specific research question or hypothesis. This hypothesis is often a prediction about what you will find in the study, but a prediction per se is not necessary and should not be forced unnaturally. Another important point in this regard is that you do not need to write your research manuscript according to the actual temporal order of what you did and what you believed. Many new research writers assume, for instance, that their Introduction must be based on what they knew and thought *before* they did the study. This is not true; write the manuscript in a logical and engaging manner, given what you know having completed the study. Writing the Introduction so that it does not give away everything in the Discussion, however, can contribute a little suspense and does increase the engagement of your story. It's OK to offer a hypothesis in the Introduction that you know will be discredited by your Results.

The second major section is the **Method** section, in which you describe in detail what you did to obtain your data. Do not describe what your data show; those are results, and they go in the next section. As several chapters have shown, geographers who study different subject areas use fairly different sorts of data, obtained in different ways. But all Method sections must describe the nature and number of cases, how they were sampled and contacted, the variables that were measured and how they were operationalized, any materials used to measure the cases (including recording or sensing devices, survey questions, and so on), and the procedure used to collect data or construct a model. Describe the cases and materials in a static manner; describe the procedures as a temporally ordered narrative with a beginning and an end.

After the Method comes the **Results** section. Present your data and any analyses you have conducted. Make sure to focus on describing the pattern of results, not just their statistical significance (see Chapter 9). Inferential questions about statistical significance speak to what sample patterns probably indicate about population patterns, which is not the first thing a reader needs to know to interpret your results. The Results section should focus on a correct and complete presentation of your data, but in order to help the reader understand the meaning of your data, you should concisely interpret your data in words as you present them. The key word here is *concisely*—save broad interpretations and conclusions for the Discussion section. Engage the reader by trying to present those results that speak to your most central question first, if possible, followed by the second most central results and so on. However, it is often necessary first to discuss manipulations you may have

[5]The first sentence of the Introduction should *not*, however, refer to researchers or literature, as in "Geographers who study river channel development disagree about . . ." It is the topic area or phenomenon of interest that we care about, not what geographers do or think.

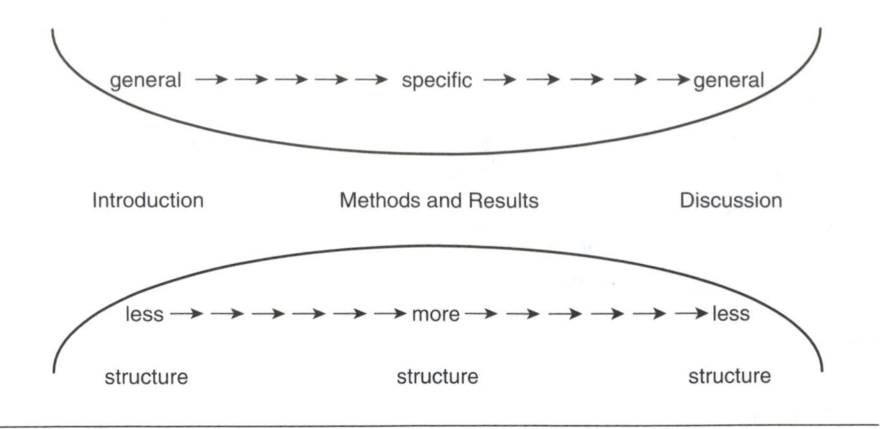

Figure 13.1 The "hourglass" structure of a one-study empirical research paper.

applied to your data (in the everyday sense of the word, not assignment control), such as omitting bad cases or variables, transforming or combining variables, or treating outliers. It is to be hoped that these manipulations are justifiable for a priori reasons, not because you find out afterward that the results fit your hypotheses better if they are manipulated. Always be forthright in describing data manipulations, a component of research ethics discussed in Chapter 14.

The final of the four major sections of the body of your manuscript is the **Discussion** section. Start narrowly by restating your major results, interpreting them specifically. Then broaden out by considering possible problems, limitations, or threats to the validity of your study (Chapter 11). Then interpret your results even more broadly, possibly suggesting implications they have for a research community, discipline, or even society at large. You can speculate a little more at this point. Recommend future research that should be done. However, there is little need to say that "more research is needed"; that's pretty much a foregone conclusion. Instead, say specifically what research is needed and why. Finally, you want to state your conclusions succinctly, often designating them with an explicit heading of "Conclusions" or "Summary and Conclusions."

To repeat, the four major sections of the body of the manuscript are the Introduction, Method, Results, and Discussion. You may have noticed that there is an "hourglass shape" to these four sections (Figure 13.1). That is, the two end sections, the Introduction and Discussion, are broader than the two middle sections, the Method and Results. Furthermore, the Introduction starts broad and narrows as it nears the Method section; the Discussion starts narrow at the Results section and broadens as it nears the end. Also, the Method and Results are fairly tightly structured, whereas the Introduction and Discussion are less structured—you have a little more creative leeway as to exactly how you structure these end sections.

After your concluding sentence comes the **References** section, usually starting on a new page. This is a list of your literature sources, occasionally including other types of sources such as data sets or media reports. Depending on the journal's particular style, references are listed in alphabetical order by the first author's last name and initials; sometimes they are listed in the order in which they were cited in the manuscript, particularly if footnote citation is used. After considering the first author's last name, use the second author's last name, and so on, to determine reference order; a single-author reference precedes a reference by that same author and a second author. If there are multiple references by exactly the same author, list the earliest references first. Unlike a bibliography, which lists general source material, all of your references must be cited in the body of the manuscript at specific locations where they are relevant.[6]

Citations are done in one of two ways; every journal will specify this in their instructions to authors. The more common way in geography journals is to list the last names of the authors, followed by the year of the publication (with or without a separating comma). Here is a (fictional) example: "Retail location is strongly market driven, as compared to manufacturing (Garcia, 1994)" or "Garcia (1994) showed that retail location is strongly market driven, as compared to manufacturing." Many journals prefer that three or more authors be shortened to "et al.," a Latin abbreviation meaning "and others." So Ruocco, Konty, and Ishikawa (2001) would be cited as Ruocco et al. (2001). Multiple citations to the same author and year would be distinguished by the addition of lowercase letters, which are also added in the References: Lovelace and Cova (1999a), and Lovelace and Cova (1999b). Multiple citations at one place in the text are ordered alphabetically or chronologically. The second way of citing references is to use numbers or letters as footnotes, such as "Retail location is strongly market driven, as compared to manufacturing.[1]" Either the footnotes are in sequential order (1, 2, 3, . . .), which means the References are listed in the order they were cited, or the References are in alphabetical order, which means the citation superscripts are not in sequential order.

Most manuscripts will include one or more tables and figures, including maps (Chapter 10). These are either printed on separate pages following the References or placed in the body of the text at the location where you want them to appear. If located after the References, a note is placed in the text to indicate where the publisher should place the table or figure ("Insert Table 2 about here"). Either way, all tables and figures must be cited in the body of the manuscript at least once: "Figure 3 shows a model for climate variability in the Tropics." In addition to tables and figures, a few manuscripts also include one or more appendices at the very end that present technical details some readers may wish to skip. An appendix might show a proof, the derivation of some formula, or a computer algorithm, for example.

Two Aspects of Style in Scientific Writing

There are two important aspects of writing style for scientific manuscripts that we want to discuss briefly; you should definitely consult style manuals and guidebooks like

[6]There are a variety of software packages (for example, EndNote, PAPYRUS, ProCite) that store and organize references and support the creation of Reference and Bibliography sections for manuscripts. We recommend you start your own reference database as soon as you start writing manuscripts—that is, while you are still a student. Oh, and make sure to keep a backup copy.

those listed in the Bibliography for numerous additional aspects of style. The first is that scientific research should be expressed in precise and straightforward language. Avoid flowery prose—*say what you mean clearly and simply.* Term repetition is more acceptable in this type of writing than in some others; using the same word or words to express the same concept helps the reader easily know that you are talking about the same concept. Avoid stilted, pedantic writing, including the dreaded "scientese" that employs passive and jargony sentences in order to sound scientific.[7] Although jargon is to be avoided, certain technical jargon should be used because its specific meaning is recognized within a particular research community, and its use can communicate more efficiently; make sure to define such terms when first used. It may be hard for some newcomers to believe (and we regret that some old-timers don't seem to realize) that writing in an engaging style is a good thing even in scientific writing. Interesting and appealing writing is not just for novels and short stories.

The second aspect of writing style we want to comment on is that scientific communication should avoid **sexist language**. Some see this advice as political correctness, whereas others argue that it is social and even semantic correctness. Rather than try to defend one of these positions, we will simply observe that current usage recommendations in nearly every field (as explained in style manuals) require avoiding what we describe here as sexist language. Specifically, writers should not consistently use masculine singular pronouns in a generic or default manner, that is, when the sex[8] of the person referred to could be male or female and is otherwise

[7]**Active** or **passive voice**—that is the question, at least with respect to writing style. We cannot fully treat this question here, but we note that you should not strictly adhere to the general advice "not to use passive voice" (the Microsoft Word grammar checker believes this). Passive voice, wherein the subject of the sentence expresses the goal of an action, does tend to reduce the flow and stimulation of writing; rewriting to reduce passive voice definitely tends to improve what is written. "We took samples from 56 vernal pools" is better than "samples were taken from 56 vernal pools." However, passive voice is appropriate when one wants to place focus on the goal of an action rather than on the agent bringing about the action. There are a few different situations when this might be true, including when the action and its target is important but the particular agent is not. "The vernal pool at the lowest elevation was found to have the highest concentration of organic solutes." We direct the reader to Williams (1990) for an excellent discussion of active versus passive voice, and in general, how to write for readability and engagement.

[8]People have both sex and gender. Sex is a dichotomous variable that corresponds to "female" and "male"; it is determined by genetics and biological development in the womb and after (actually, as we noted in Chapter 2, sex is not a perfect dichotomy). Gender is a continuous variable that corresponds to what is meant by "feminine" and "masculine" (gender is also applied to words; pronouns in some languages have gender rather than sex). Socialization and other experiential variables join with genetic and biological development to determine gender, a process that social scientists have studied extensively. We use the term "sex" rather than "gender" to refer to a person's membership in one of the two categories of female and male. This usage does not imply, however, that all variables correlated with a person's sex are caused by their biology. The income disparity between females and males working in the same profession is a sex difference, not a gender difference, but that doesn't mean it is necessarily caused by hormones or genitalia.

irrelevant. This poses a small difficulty for English speakers, as English has no gender-neutral singular pronouns for animate beings. Referring to a person as "it" generally does not fly. We recommend using the plural whenever possible, which is often. Make sure to maintain plurality agreement with verbs when you do this. Sometimes singular is appropriate, however, such as when you are making a point about the actions of a single person. Of course, if you are speaking of a particular person who is male or female, don't hesitate to identify the person's sex. It is correct to refer to the sex of people if that is known and relevant; don't write "Nearly every player on the Denver Broncos football team gives money to *his or her* favorite charity." If the person could be of either sex, but you want to write in the singular, we recommend "he or she" and "his or her" (the reverse orders sound less fluid in English). Some writers feel this is clunky, but we find it easily tolerable if done in small doses. Alternating the gender of the pronoun is another possibility, although we find it rather distracting. Besides pronouns, nonsexist language means referring to all people as "people," "humans," "humankind," or the like—not "man" or "mankind." Also avoid sexist assumptions about people's careers, traits, sexuality, and so on. For example, some engineers are women, and some nurses are men.

Using the Library for Scientific Research

In this final section of the chapter, we take a look at the major storehouse of scientific communication—the library. The library is not just the "brick-and-mortar" library building but increasingly refers to the incredible storehouse that is the Internet, including the World Wide Web. Libraries contain a variety of scholarly and research resources, including scientific literature, reference books, publicly available data sets (secondary data), maps and imagery, dissertations, newspapers, special collections, and more. Library resources are obviously invaluable for research in all areas. Most importantly, they provide the basis for **scholarship**. Scholarship is knowing what has and hasn't been said or done, what is and is not believed to be true, and so on. Scholarship is putting your research within the context of the history and intellectual development of your topical area. In addition to scholarship, library resources provide ideas that help you design and carry out your research, such as techniques you could use to collect or analyze data. Finally, library resources allow you to give recognition to people who have come up with new ideas or results, thus giving credit where it has been earned (see our discussion earlier about the rewards of publication). But more critically, this recognition allows you to avoid accidental plagiarism. It's up to you to know that someone has published a particular idea before.

The library is clearly the place to go to find relevant literature on a given topic. *Finding it* is the key. The library provides many tools to help find relevant literature. There are reference books and databases (almost exclusively electronic nowadays) built on journal and book titles, article titles, publication years, author names, abstracts, and citations.

There are a few databases specifically for geography and cartography, such as GEOBIB, GEOBASE, and Geographical Abstracts (in two parts: Human Geography and Physical Geography). You can find literature in various generic databases (such as Web of Science) or in more specialized databases that deal with one of the many topical areas with which geography overlaps (urban studies, environment, geosciences). If, as a newcomer to literature searches, this seems daunting to you, ask a reference librarian for help. It may not sound fashionable in this day of digital automation and budget cuts, but flesh-and-blood librarians have training and experience that can be very helpful—even to experienced researchers and scholars.

We can offer some specific strategies for finding relevant literature. Word searches are the first place most people go, including keywords, title words, and author names. **Keywords** are words or short phrases that have been attached to pieces of literature in an attempt to represent their content concisely. Word searches are carried out within a database like we just discussed. Word searches are obviously powerful, but we want to emphasize that they are generally not sufficiently thorough and authoritative in and of themselves to find everything relevant (remember scholarship?). A straightforward reason for this is that some relevant literature may not be indexed in particular databases or even in any database. But the more interesting and profound reason that word searches are not wholly sufficient, and probably never will be, stems from the way words relate to meaning. This is the same issue we discussed in Chapter 5, where we pointed out the major reason that coding verbal records is so difficult. The many-to-many and contextual mapping of words to meanings makes word searches inadequate in and of themselves, at least if one wants a truly scholarly and comprehensive search. Try searching with the word "map" sometimes and see how many disparate topics your search finds. Word searches can thus be very inefficient, finding many things that are not relevant to your problem. What they do find is not sorted very well for relevance, importance, validity, and so on. By the same token, you can readily miss a relevant piece of literature during a word search because the semantic content of the piece is not labeled with the word you use but by some near or exact synonym.

These caveats about word searches are especially potent in the case of searches carried out directly within the entire Web, via a search engine. This is becoming increasingly common; many students and, we fear, many scholars and research scientists don't search any other way. The Web is an incredible tool, but it is not sufficient for academic research by itself, and it may never be. All relevant and important literature is definitely not on the Web, although the proportion of literature that is not mentioned by someone somewhere on the Web is clearly becoming smaller. What *is* on the Web is comically (or perhaps tragically) inconsistent; we have often been bemused if not annoyed by the number of different answers to factual questions one can find on the Web. The Web is mostly unchecked for quality control, with many things posted there that are not put through gatekeepers like editorial fact checkers or the peer review system discussed earlier. We don't know if anyone has reliably estimated how much sheer baloney and hogwash may be found on the Web, but it must be shockingly high. It reminds us of the old joke that "42.6% of statistics are made up." The Web is a great place to apply some of that scientific skepticism we discussed in Chapter 1.

We therefore recommend supplementing keyword and title word searches, not only if you want to increase your chances of finding what's really been written, but also if you want to increase the efficiency of your search and its focus on the most central pieces of literature. One of our favorite ways to track down literature is to look over the references listed and discussed in some relevant book or article we already have, including a textbook. You can then look at the references in the new literature you examine, and so on; this strategy will obviously produce an explosion of possible literature before too long. To acquire real expertise in some topical area, scan all years of a key journal, although this strategy is probably too time consuming to justify unless you are a doctoral student. Or search for all literature by the name of key researchers. Finally, there is always the shortcut of asking established experts in a particular area. Well, it's a shortcut for you anyway; maybe it's good etiquette to do some work on your own before you request that someone who might be very busy give his or her time to save yours.

Review Questions

- What are the various written and oral forms of communication in geography?
- What are some specific types of journal articles?
- What is some advice for giving an effective oral presentation?

Peer Review System

- What are typical steps in the peer review system for publishing research literature, and what are some aspects of this system designed to promote objectivity in geographic research?
- What are blind and double-blind review, and what are pros and cons for using these forms of review?
- What are possible outcome decisions of the publication review process?

The Basic Structure of a Journal Manuscript

- What is an Abstract, and why is it important to write good Abstracts?
- What content should generally be found in the four main sections of the body of an empirical article: Introduction, Method, Results, Discussion?
- How does an "hourglass" structure describe the basic components of a one-study empirical article?

Two Aspects of Style in Scientific Writing

- What is the appropriate style of writing for scientific communication, and what are some ways to achieve this style?
- What are some examples of sexist writing, and what are some ways to avoid them?

Using the Library for Scientific Research

- What are some ways the library (including the Internet) is used by scientific researchers?
- What are strategies for finding relevant literature for a research project or paper?
- What are some limitations of keyword or title word searches?

Key Terms

Abstract: very short summary appearing after the Title page in a research manuscript

active voice: style of writing in which the subject of the sentence expresses the agent of an action rather than the goal or recipient of an action; although increasing its use generally improves writing, it is not appropriate in all writing situations

blind or **double-blind review:** peer review in which an attempt is made to keep either the author or reviewer, or both, uninformed of the identity of the other person

citation: statement of a literature source at a particular place in a manuscript, meant to give credit to specific ideas or data

conference: regular gathering (annual or less frequent) of new and established scholars and researchers in a given discipline or topical area where research is presented and discussed, social networking occurs, and other activities take place—essentially an academic convention

conference paper: talk given at a conference that may or may not end up in written form

conference proceeding: edited book comprising a collection of papers based on talks or discussions at a scientific conference

Discussion section: fourth and final major section of a research manuscript, it narrowly restates major results, interpreting them both specifically and broadly, considers problems and limitations, and suggests future research

edited book: book with chapters written by different authors; the collection of chapters is organized by one editor (or one group of editors)

grant application: written request for funding for a proposed program of research, made to a funding source such as a public agency or private foundation

Introduction section: first major section of a research manuscript, it introduces the topic of the manuscript, discusses previous data and ideas relevant to the topic (including citations), and ends narrowly with a specific research question or hypothesis

journal: primary form of scientific literature, particularly for initial communication of research results and conclusions; serially published usually once every two or three months

keywords: words or short phrases that have been attached to pieces of literature in an attempt to represent their content concisely; they support computer keyword searches, a common technique for trying to find literature

meta-analysis: quantitative review of research that makes conclusions about some phenomenon by statistically combining the empirical results of many independently conducted studies on the phenomenon

Method section: second major section of a research manuscript, it describes in detail what was done to obtain the data or create the model, including how cases were sampled and measured; subsections typically include detailed descriptions of cases, materials, and procedures

passive voice: style of writing in which the subject of the sentence expresses the goal of an action rather than the agent of the action; although reducing its use generally improves writing, its complete elimination is inappropriate

peer review system: important social process that increases the objectivity and quality of scientific literature by subjecting it to a gateway process in which other scientists read and review manuscripts submitted for publication

poster: brief written and graphical presentation of research that is posted on a bulletin board for viewing and discussion at conferences

References section: list of literature sources for a manuscript placed at the end, including only those sources that are cited in the body of the manuscript

Results section: third major section of a research manuscript, it presents data and their analyses, including both descriptive and inferential analyses; only concise statements about the meaning of data analyses are included here

reviewers: researchers and scholars who serve as the peers in the peer review system; their job is to carefully read a manuscript submitted for publication, and offer reasoned arguments for whether the manuscript should be published and how it should be modified to improve it

scholarship: knowing what has and hasn't been said or done in prior scientific and academic work; allows authors to put their research within its historical and intellectual context, credit people for their prior contributions, and avoid plagiarism

scientific literature: the archive of written research communication that includes journal articles, books, and so on

sexist language: the use of masculine pronouns or terms such as "mankind" to refer by default to people of either sex; also assumptions about people based on their sex or sexual orientation that are at most *probably* true, not definitely true

Title page: first page of a submitted manuscript, it lists the authors' names and institutional affiliations

Bibliography

American Psychological Association (2001). *Publication manual of the American Psychological Association* (5th ed.). Washington, DC: American Psychological Association.

Davis, M. (2005). *Scientific papers and presentations* (rev. ed.). Burlington, MA: Academic Press.

Fink, A. (2005). *Conducting research literature reviews: From the Internet to paper* (2nd ed.). Thousand Oaks, CA: Sage.

Northey, M., & Knight, D. B. (2004). *Making sense: A student's guide to research and writing: Geography & environmental sciences* (2nd ed.). New York: Oxford University Press.

Strunk, W., & White, E. B. (2000). *The elements of style* (4th ed.). Boston: Allyn and Bacon.

University of Chicago Press (2003). *The Chicago manual of style* (15th ed.). Chicago: University of Chicago Press.

Williams, J. M. (1990). *Style: Toward clarity and grace.* Chicago: University of Chicago Press.

CHAPTER 14

Ethics in Scientific Research

Learning Objectives:

- What is ethics, and what are the three guiding "ethical directives" in scientific research?
- What are some reasons for scientific researchers to act ethically?
- What are some guidelines for treating the natural and cultural world ethically when doing research?
- What basic rights do human research subjects have?
- What are institutional review boards, and how do they function?

We conclude this book with a discussion of ethics in scientific research. **Ethics** can be defined as the study of moral or proper action. Like all domains of human activity, scientific research involves a variety of ethical considerations. Ethical issues in research express themselves as an ongoing, dynamic attempt to achieve balance among three basic sets of rights or "ethical directives":

1. The right of scientists to pursue knowledge, using methods of scientific inquiry
2. The right of people and other sentient beings to be free from harm
3. The right of society to gain benefit from research while avoiding harm.

Each of these directives must be understood as contingent upon the other two; each must be exercised while maintaining respect for the rights specified in the other two. This leads us to realize that scientific ethics involves *responsibilities* as well as rights.[1] That is, the right of society to avoid harm means that scientists have a responsibility not to do net harm with their research. Similarly, the right of scientists to pursue knowledge is a responsibility on the part of society to allow open scientific inquiry and communication.

These directives essentially provide a principled basis for determining what you *may* and may not do, as well as what you *should* and should not do. Why should you, as a research scientist, follow ethical principles? From a utilitarian point of view, you need to realize that principles like the ethical directives provide the basis for a variety of specific legal and professional standards of conduct that you must follow if you want to avoid losing your license or your job, being sued or jailed, and so on. Aside from the utilitarian, however, most people who are not psychopaths find that acting ethically provides positive emotional benefits—it feels good. At the same time, we do not deny that acting ethically may be justified for spiritual or religious reasons. However, although many people may see these various justifications for ethical behavior as self-evident or natural, we recognize there is a certain amount of cultural variation in what is considered ethical. Certainly the philosophical literature on ethics is not filled with great consensus on exactly what is ethical and why. Nonetheless, we do believe that roughly common ethical considerations are universally part of human nature and human society under nonpathological conditions. We think doing the right thing will resonate in similar ways with most of you, even if, like us, you sometimes fail to meet the highest ethical standards you normally endorse.

Let's consider the right of scientists to pursue knowledge. This suggests that scientists are to be motivated by honesty, sincerity, and an open search for truth. They should be free to share data and other information with each other and with society. They have the right to make a living as researchers, as long as some entity is willing to pay for it; there is no ethical guarantee per se that someone has to pay you for any particular job or career. Given that a person works as a scientist, he or she should be allowed to freely investigate any topic desired; censorship of research topics or methods is not legitimate. We also believe this directive suggests that scientists should pursue what interests them, what excites them, what inspires them to get out of bed and get to work. Scientists should not simply "follow the money" by choosing to focus their research on whatever someone is willing to fund. By the same token, research institutions should not become inappropriately beholden to corporate funders who are willing to "donate" large amounts of money in return for new buildings, endowed positions, and so on.

Our second directive concerns the right of people and other sentient beings to be free from harm. As we mentioned in Chapter 1, sentient beings think and feel—they can recognize the consequences of acts, they can feel pain, and so on. It's not obvious or universally agreed who or what this includes. The two of us do not

[1]Ethical responsibilities, as opposed to rights, may be seen to stem in part from the fact that humans are creatures who have agency, mentioned in Chapter 3. Nothing can "do the right thing" unless it has the power to choose to do the wrong thing.

entirely agree with each other. It's safe to say that humans are sentient. Clearly some nonhuman animals are sentient to a degree. This definitely includes "higher" mammals, perhaps all mammals. It likely includes many or all vertebrates. How about crustaceans, insects, and viruses? It seems unlikely that unicellular entities, especially those without cellular nuclei, have adequately complex physiologies to produce sentience. It is not clear that plants, no matter their complexity or evolutionary heritage, are sentient. Surely the material world of rocks and water and air is not sentient. Or is it? As we consider further below, the argument is increasingly made that entities deserve ethical consideration even if they are not sentient in our mental or experiential sense.

Uncertainties about who or what deserves ethical protection aside, let's return to some specific principles that follow from the second ethical directive that people (and others) have the right to be free from harm. As we discuss in more detail below, this directive guides scientists in their treatment of research **subjects**, the human and nonhuman animals who are measured by geographers and other scientists working in particular topical areas. However, this directive has much broader implications. It also suggests that research assistants and other colleagues should be treated ethically. They should not be harassed, overworked, undercompensated, and so on. Attribution of credit for work and for ideas should go to the people who have earned it. People who deserve credit as authors should be so credited, and those deserving first authorship should get that. If you take ideas or data from someone, you need to cite them or recognize them in footnotes (we discussed authorship and citation in Chapter 13). Don't **plagiarize**. If you treat data fraudulently (**data cooking**), your unethical behavior compromises the rights of other individuals, and society at large, to honest communications. If, because of past personal experiences, you have a positive or negative assumption about someone or their work that would significantly hurt your ability to judge their work impartially, you have a **conflict of interest** and should not review that person's article or grant application. We refuse to review manuscripts by colleagues in our departments, for instance. You also have a conflict of interest if the success or failure of another person stands to benefit or impair your publishing success, reputation, or finances.

Finally, let's consider the ethical directive concerning the right of society to gain benefit from research while avoiding harm. That implies society has the right to be told the truth about what scientists believe their research actually means. Society has the right to choose whether to pay for research and which research to pay for. But society also has a responsibility in this respect—the responsibility to fund scientific work in proportion to what scientists as a whole believe is most appropriate for understanding and controlling a particular problem, rather than what is politically expedient. Scientists have a responsibility not to pursue research they could reasonably anticipate will cause more harm than good. And as far as this directive goes, scientists should choose research topics that will potentially benefit society, and they should disseminate their research to society in a beneficial way, although one might argue that society has a right to beneficial research only if it specifically pays for the research. Various principles we have already noted, such as avoiding data cooking and conflicts of interest, also follow from the directive to benefit society, not harm it.

Treating the Natural and Cultural World Ethically

In 1964, a graduate student in geography traveled to Wheeler Peak in the eastern Great Basin of Nevada, near Utah, to study glaciers of the last Ice Age. Near the upper tree line, he came upon a scattering of bristlecone pine trees. The great age of some of these trees in the western Great Basin had only been discovered about a decade before by the dendrochronologist Edmund Schulman. The student took tree cores (Chapter 4) and saw that some of the trees appeared to be at least 4,000 years old, making them among the oldest living things known at that time. Unfortunately, his coring tool broke, and with summer coming to a close, the student felt he could not wait until next spring to return for more cores. So he asked U.S. Forest Service Rangers for permission to cut down a tree that had been named "Prometheus." They granted permission and down came the ancient one. The student discovered that Prometheus was even older than he had expected—it had over 4,800 rings. The young geographer had just killed what was then believed to be the oldest living creature on earth. Eventually it was determined that Prometheus had actually died at the age of 4,950 years.[2]

We noted above the argument that entities deserve ethical consideration even if they are not sentient. Was the felling of Prometheus an unethical act, and if so, why was it unethical and who was primarily responsible for it? One could argue about whether the student or the Forest Service ranger bore more ethical responsibility for the felling; possibly the student's advisors or Forest Service policy bore some responsibility too. One can also recognize that the ethical harm of this act was committed against Prometheus itself, against the scientific community, and against society at large, including future generations who had not yet been born in 1964. But whoever we decide was to blame and whoever (or whatever) we decide was harmed, we find that this episode points to some of the ethical issues geographers must consider when collecting data in the field.

Our list of ethical directives states that people and other sentient beings have a right to avoid harm from research; increasingly, this ethical philosophy has been expanded to include all aspects of the natural world, whether sentient or not. This provides a rationale for the ethical treatment of the natural world and of local cultures, in addition to society at large. At the same time, scientists have an ethical right to pursue knowledge, and society has a right to benefit from research. The research and expertise of geographers has often helped make the world a better place—in a myriad of ways—and certainly will continue to do so in the future. If it is ethically directed, geographic knowledge can produce economic prosperity, medical breakthroughs, cleaner

[2]This incident attracted worldwide attention. The bristlecones on Wheeler Peak were eventually protected when their territory was included in the Great Basin National Park in 1986. The current titleholder of oldest bristlecone tree is "Methuselah," about 4,700 or 4,800 years old. This is not necessarily considered the oldest living thing anymore, as "clonal" trees, shrubs, and ferns have been found that are tens of thousands of years old. Bacteria that have been in suspended animation for millions of years have been reanimated.

environments, increased food production and energy efficiency, more livable cities, and greater peace and security. It can also entertain us, stimulate our curiosity and sense of beauty, increase justice in the world, and bring us closer to our spiritual ideals in any number of ways. Just one important example is the fact that the continued existence of ecologically valuable and vulnerable habitats probably depends in part on research by geographers and other scientists. Geographers catalog and document characteristics of places, they explain their creation and their continued development, they relate characteristics of places at particular locales to those at other locales, they educate the public and policy makers, and more.

Geographers often collect data in field settings that have sensitive natural or cultural characteristics deserving (and requiring) ethical treatment. These places and their characteristics deserve to be preserved, because people have the right to experience the benefits of these places, because cultures have the right to their traditional ways of existence, and because the natural world has the right to exist. Actually, a good argument can be made that any place deserves this respect, not just unusual, unspoiled, or "exotic" places. These ethical considerations suggest that researchers should strive to make a minimal impact on the places they study. Leave them as you find them—or better. But our existence on the planet makes these ethical goals challenging, even impossible in the strictest sense. There is no way to avoid all impact when visiting a place. The very existence of humans on the planet cannot help but change it, and that includes humans who are members of technologically undeveloped cultures as well as members of the mechanized world, the information world, and the world of consumerism. This applies not just to research, of course, but to all human activities and modes of existence.

However, our impact varies a great deal depending on the choices we make and the choices we urge others to make. Impact is a matter of degree. Killing two of the oldest trees around is worse than killing one of them. Geographers collecting data in the field can make a great many choices that minimize their impact on the natural and cultural environments in which they do their work. Making such choices is the ethical thing to do. Table 14.1 presents a list of guidelines for treating the natural and cultural environment ethically.[3] Of course, these guidelines apply to all people who are members of a research team; everyone must take responsibility for their actions, not just project leaders. The guidelines speak to all three of our ethical directives and include the nonhuman and inanimate parts of the environment as deserving of ethical consideration. During field research, geographers should strive to minimize the impact of their movement and access to sites; for example, to the extent possible, stay on established trails instead of walking over untrodden ground. Geographers should minimize the impact of their campsites; for example, locate and construct latrine sites carefully. Geographers should promote good community relations; for example, inform local peoples thoroughly about your project and involve them as much as possible. Finally, geographers must conduct responsible fieldwork; for example, conduct important and nonredundant research and disseminate it to others in an effective way.

[3]Table 14.2 is adapted from Smith (2002).

Table 14.1 Guidelines for Treating the Natural and Cultural Environment Ethically (adapted from Smith, 2002)

1. Reducing the impact of movement and access
   - traveling on foot or with pack animals
   - traveling by vehicles
   - traveling by boats
2. Reducing the impact of campsites
   - choosing a location
   - siting the camp and its facilities
   - using fires and firewood
   - using other fuels
   - reducing and disposing waste
   - performing actions at departure
3. Promoting good community relations
   - establishing and maintaining social relationships with the local community
   - establishing and maintaining awareness and respect for cultural differences
   - administering medical treatment to community members
   - taking photographs
   - carrying out economic exchanges with the local community
   - carrying out community projects
4. Responsible fieldwork
   - choosing locations
   - establishing liaisons with local community, scientists, and government
   - obtaining permissions and permits
   - carrying out valuable projects with competence and care
   - publishing, disseminating, crediting sources and contributors

Treating Human Research Subjects Ethically

Dismaying details of Nazi medical experiments revealed at the Nuremberg Trials of the late 1940s provided an early impetus to develop ethical principles for the treatment of human research subjects participating in medical and behavioral research. The principles that emerged were codified over subsequent decades into formal standards and rules that today guide and enforce the ethical conduct of research with human subjects. These rules have been formulated to prevent ethical transgressions against human subjects, whether minor or major. During the 1950s, for example, consternation was generated over a fascinating and important series of studies conducted by the psychologist Stanley Milgram. In these studies of obedience to authority, Milgram insisted that research subjects assigned to the role of "teacher" administer electric shocks to a "learner" who sat in another room (later variations of the study put the teacher and learner in the same room). The learner was actually a confederate of Milgram's, and no electric shocks were actually administered. However, the subjects didn't know this; a tape recording of the learner screaming and complaining about his heart made the situation sound shockingly realistic. In spite of the sounds and an ersatz "shock generator" that indicated the subjects were administering dangerous shocks over 300 volts did not stop many of them from eventually pushing the levers all the way to the highest voltage level.[4] Its contribution to our understanding of social influence aside, critics objected to the experience this study forced subjects to go through in their role as teacher and expressed concerns about its long-term effects, even given that the confederate came out after the study and met each subject with a reassuring smile.

In 1974, the federal government of the United States first published regulations to protect human subjects, basing them on recommendations that had been developed during the 1960s. There have been a variety of revisions over the years to these regulations, including the addition of regulations for research on nonhuman animal subjects, but the basic guiding ethical principles upon which they are based have remained fairly constant. These principles were explicitly stated in detail in 1979, in a publication put out by the U.S. Department of Health, Education, and Welfare, specifically the National Commission for the Protection of Human Subjects of Biomedical and Behavioral Research. This is the **Belmont Report**, a key document outlining the philosophical basis for the ethical treatment of human research subjects. Its full text can be found at the Web site of the Office of Human Subjects Research of the U.S. National Institutes of Health at http://ohsr.od.nih.gov/index .html.

The Belmont Report outlines three ethical principles that express aspects of the general ethical directives we discussed above as they apply to research with human subjects, especially the second one, which requires respect for the right of people and other beings to be free from harm. The three principles of the Report are

- **Respect for persons.** Individual people should be treated as autonomous agents capable of deliberating about personal goals and acting according to those deliberations. Special protection is to be accorded to individuals with diminished autonomy.
- **Beneficence.** Benefits to individuals should be maximized while potential harm is minimized.
- **Justice.** The benefits and burdens of research should be distributed fairly.

Several specific rules for the treatment of human research subjects follow from the principles of the Belmont Report. The first is **confidentiality** or **anonymity**, which express aspects of a person's right to privacy. Notice that the two don't mean exactly the same thing. Confidentiality means researchers know who the subjects are and what their measured data values are, but they will not give that information to anyone and will take steps to make sure no one gets a hold of it (like using a locked file cabinet or computer). Anonymity means researchers do not even record

[4]The irony is that Milgram's research began within the context of establishing that Nazi atrocities were committed because of some "inherent" flaw in the German character that made them exceptionally malleable to authoritarian control. His studies in the United States were so "successful" that he never bothered to take them overseas.

subjects' names or any other personally identifying information. From an ethical standpoint, anonymity is preferable to confidentiality and should be used whenever viable. Another specific rule is the requirement to provide **informed consent** to subjects before they agree to participate in a study. There are three components to informed consent: information, comprehension, and voluntariness. Potential subjects have to be informed up front about the procedures, purpose, and so on, of the study. They must also be informed about risks, both physical and psychological. Risks include injuries or diseases, negative moods (anger, sadness, fear, embarrassment), boredom or exhaustion, invasion of privacy and social harm, and litigation or criminal penalties. To balance this, subjects should also be informed about potential benefits. Comprehension and consent must be obtained via a communication between researcher and subject, a process that is often documented with a signed consent form. Finally, participation must be voluntary, not mandatory or coerced, and subjects must always be free to stop participation in any study at anytime without penalty.[5] To repeat: *Subjects are free to quit your study at any time for any reason with absolutely no penalty.* If they want, subjects can also request that their data be removed from the study and/or destroyed.

It may be difficult to meet all standards of the rules of informed consent with certain groups of people. In particular, certain groups may have difficulty comprehending or volunteering without explicitly or implicitly being coerced. In fact, there are specific groups who have special ethical rights, according to the implications of the Belmont Report. Minors—people under 18—cannot be studied unless their parents permit it; the minor must assent too (they cannot give legal *consent*). People with certain disabilities that might decrease their capacity to comprehend have extra ethical protection. Prisoners are thought to be at greater risk of coercion. Interestingly, because of the special benefit to society that is thought to follow from the accountability of elected officials, they are given less stringent ethical protections than other groups.

A particularly intriguing issue when it comes to the ethical treatment of human subjects is the issue of **deception in research**. Deception is sometimes used in research, most commonly deception about the purpose of a study or about all details of what will happen in it. Deceiving subjects in these ways is less common in geographic research than in some other behavioral sciences, but it definitely happens. The reason to deceive subjects is that you don't want their responses to change as a result of knowing they are in a study, or knowing exactly what the study is about. In many studies, if you are too forthcoming in informing subjects ahead of time, you run the risk of giving them information that would help them respond to your questions or direct their attention to things they would not otherwise think about. In other words, deception is used to reduce the biasing effects of reactance and other forms of interactional artifacts (Chapter 11). The principles of the Belmont Report in fact allow deception if it is necessary to the successful conduct of the research, and the research is otherwise seen to have value. That is, ethical

---

[5]Given this voluntariness, it is sometimes considered more correct to refer to human research subjects as "participants."

decisions about research are to be based on a balance between risks and benefits, as we stated above, and deception is considered a risk factor. Its use has to be justified by evidence of its need. If deception is used, however, a careful **debriefing** of subjects must occur as soon as possible after data collection is complete (typically at the end of each data collection session). Debriefing should (a) remove the deception, (b) express regret and explain why the deception was necessary, and (c) inform and educate the subject about the purpose and anticipated value of the research. In fact, all research studies with human subjects should at least provide the final component of debriefing, that of explaining the purpose of the research. The debriefing is also a good opportunity to learn more from research subjects about their experience of the study, any further thoughts they might have, and so on. This "empirical debriefing," a less systematic inquiry than the study proper, is a bit like a pilot study (Chapter 6) done after the fact.

Institutional Human Subjects Review

In response to the ethical principles of the Belmont Report and related documents, institutions such as universities and private research labs created **institutional review boards (IRBs)** to oversee the ethical treatment of humans and other animals that serve as subjects in research conducted at that institution. IRB committees at these institutions are charged with protecting the ethical rights of subjects. These committees typically consist of researchers from departments that commonly conduct research with human subjects (primarily medical and behavioral science). Students, staff, and individuals from the community are also represented. Research projects utilizing human subjects must receive approval from this committee before data are collected. To be more precise, federal regulations require that all human-subjects research "conducted, funded, or otherwise subject to regulation by any federal department or agency" must be reviewed and approved by an IRB before data are collected. Research institutions generally have **Federal Wide Assurance** with the federal government, so that *all* research involving human subjects done at that institution needs to be formally reviewed by an IRB, even unfunded research. The repercussions of skipping this can be quite severe, including loss of grant funding to the specific researcher or the institution as a whole. If skipping review is seen to be willful or continuing, it can be considered scientific misconduct.

The extent of the review process varies quite a bit, however, depending on the ethical issues a particular study invokes. It also varies somewhat across institutions. Research involving considerable risk to subjects, or involving minors or other especially protected groups, must be subjected to **full review** by the entire IRB committee. If full review is required, the researcher must submit a detailed application of at least several pages in length to the committee. In contrast, research involving anonymous or confidential surveys of adults about topics that are not personally controversial (for example, not about a person's criminal activities, sexual behavior, or suicidal intentions) might qualify as **exempt from review**. This status means the research can be certified as ethically acceptable even though it might qualify as human-subjects research (some institutions do not exempt any human-subjects research). Research qualifying as exempt is typically reviewed by one or two people, such as the chair of the IRB committee or an IRB administrator, instead of the entire committee. In between these two extremes is an **expedited review** that requires the full application form but can be reviewed by fewer people than the entire IRB committee. When researchers require subjects to do something like walk around a neighborhood, expedited review would probably be called for. Most human-subjects research in geography probably qualifies for exempt status; for example, most research using surveys or publicly available archival records would be exempt. Full review is quite rare, perhaps common only when research is conducted with children or considerable deception is employed.

However, most research conducted by geographers does not qualify as human-subjects research to begin with. Federal regulations technically define **human-subjects research** as "any systematic investigation, including research development, testing and evaluation, designed to develop or contribute to generalizable knowledge that uses (1) data collected through intervention or interaction with a living human subject, or (2) identifiable private information about a human subject." A variety of types of data collection employed by geographers, even human geographers, would not be considered human-subjects research in this technical sense, including the secondary analysis of aggregate data that are completely stripped of identifiers such as names, social security numbers, residential addresses, and so on. Thus, using census data does not require IRB review because the Census Bureau does not release data at the individual level. Any activities that are not intended to contribute to generalizable knowledge do not count as human-subjects research. Teachers may administer exams in class without permission, for example. Finally, anonymous public observation is also permissible, "anonymous" and "public" being the keys to deciding that techniques like counting cars with automatic traffic counters do not require review.

A Case Study for Geographic Research: The Ethics of Tracking People

Geographers who incorporate the study of human subjects in their research usually have only minor concerns regarding their ethical treatment to worry about. This is not always true, however. A new technique of geographic data collection emerging in the 21st century raises some fascinating and substantial ethical issues. That technique is the use of GPS transmitters to track people's travel activities, whether the transmitters are built into cell phones or installed as separate devices. Scientists have tracked the travel of nonhuman animals with radio transmitters and other technologies for decades, of course, but that has never been thought to raise ethical concerns. As we suggested above, however, people are considered ethically special; they are granted status as autonomous beings with rights of privacy over what other people may know about their activities, especially when they could suffer harm from that knowledge.

It has been recognized for some time that measurements of people's activities in space and place, a form of behavioral observation, can provide a rich source of data for geographic researchers working in a variety of topical areas. But these data have not been easy to collect. Even the most ardent researchers realize that stalking people without their permission is probably inappropriate, although we learned above that "anonymous public observation" is considered exempt human-subjects research for ethical review purposes. Archival records, such as those based on credit card, cell phone, or hotel use, could provide useful information in some contexts. But it's difficult or impossible to get permission to use these records, and they do not necessarily provide the most useful data for a given researcher's needs (presumably people travel quite a bit in between uses of their plastic). We have mentioned the use of activity diaries, transportation surveys, and other explicit reports in Chapter 6. Their administration requires informed consent by respondents, of course. They are quite expensive to administer and analyze, however, and suffer from the problems of explicit reports, such as memory fallibility and intentional deception by respondents. Attaching a GPS transmitter to a person, or having the person carry one around, could produce a source of activity data of unprecedented completeness, validity, and precision.

Thus, the possibility of acquiring a large number of continuous space-time records via an automatic technology has great appeal to geographic researchers—but it raises serious ethical concerns. How much could a researcher actually infer about a person's activities on the basis of partial or complete tracking records of their movements? After all, such a record would indicate, perhaps rather precisely, where a person is located at any time of the day or night, at least the street address of their location. Matching these addresses with the identities of places could be quite compromising, to say the least. You could reasonably infer if someone was visiting a paramour, which you might want to do if the person's spouse paid you enough. Consider how interested law enforcement agents would be to get a hold of your tracking data if they suspected one of your subjects was involved in a crime or even just a material witness to a crime. Research records could be subpoenaed for civil suits or child-custody cases. The threat to privacy this raises applies not just to committing criminal or clandestine activities, however. People have a basic right to control information about their activities, including *where* the activities occur.

Of course, the principles of the Belmont Report do not become moot just because satellites are involved. Informed consent would still be required of anyone who is tracked. Perhaps the tracked person would have the freedom to "shut off" the tracker whenever he or she wanted privacy. All that is fine, as long as only that individual is being tracked. What about tracking people who are driving or riding in automobiles, something of considerable interest to transportation geographers and others? Someone who has not given informed consent might well ride in the car or even borrow it to make their clandestine trips. What if such a person were a minor? We learned above that youngsters have extra ethical protections, according to IRB rules; for example, the youngster's parents must also provide permission.

The ethics surrounding the tracking of people with GPS technologies provides just one example of the privacy concerns raised by various geographic information technologies.[6] From the perspective of researchers, it would be desirable to be able

[6]These are discussed in the book by Monmonier (2002).

to promise people who have signed consent forms that their data will never be released to anyone else in a way that leaves the person identifiable. There is a federal "certificate of confidentiality" that aims to achieve just this; it has held up in court recently, but that could change with new judicial decisions. The future will surely bring more and more situations like this, where compromises will have to be fashioned among the ethical prerogatives of the various stakeholders in geographic research.

Review Questions

- What is ethics, and what are three "ethical directives" that guide ethical considerations?
- What are some reasons, both practical and philosophical, for acting ethically when doing scientific research?
- What are some specific ethical principles in scientific research that follow from each of the ethical directives?

Treating the Natural and Cultural World Ethically

- What are some ethical considerations suggested by the episode of the felling of Prometheus, the ancient bristlecone pine in eastern Nevada, by a graduate student in geography?
- What are some guidelines for treating the natural and cultural environment ethically during field research?

Treating Human Research Subjects Ethically

- What is the Belmont Report, and what are the three ethical principles outlined therein?
- What are institutional review boards (IRBs), how do they generally work, and which institutions are required to have them?
- What are the review categories of "full," "expedited," and "exempt," and what types of research projects in geography generally qualify for each?
- What are some ethical considerations suggested by geographical research in which people's travel activities are tracked and recorded?

Key Terms

anonymity: means of protecting privacy by not collecting or recording personally identifiable information about a subject

Belmont Report: key government document outlining the philosophical basis for the ethical treatment of human research subjects in the United States.

confidentiality: means of preventing inappropriate disclosure of personally identifiable information collected about a subject by not revealing it to anyone else

conflict of interest: when a reviewer has trouble evaluating a person's work fairly because of prior history with the person, or because the evaluation will have positive or negative implications for the reviewer

data cooking: the fraudulent analysis and/or communication of data

debriefing: informal interaction with subjects after a study session is completed, designed to inform them about the purpose or findings of the study, to remove any deception, and sometimes to collect more information relevant to the interpretation of the study's results

deception in research: any form of dishonesty a researcher intentionally uses in communicating to research subjects about any aspect of a study; an ethical issue that requires justification for its approval by IRBs

ethics: the study of moral or proper action

exempt from review: the least extensive of three IRB human-subjects review categories, required of human-subjects studies that invoke the least severe ethical issues

expedited review: the intermediately extensive of three IRB human-subjects review categories, required of human-subjects studies that invoke some ethical issues but not very severely

Federal Wide Assurance: agreement with U.S. government that all research involving human subjects done at a particular institution will be reviewed by an IRB, even if it is not directly funded by the federal government

full review: the most extensive of three IRB human-subjects review categories, required of human-subjects studies that invoke the most severe ethical issues

human-subjects research: technically defined by federal regulations as "any systematic investigation, including research development, testing and evaluation, designed to develop or contribute to generalizable knowledge that uses (1) data collected through intervention or interaction with a living human subject, or (2) identifiable private information about a human subject"

informed consent: IRB requirement that potential subjects be informed of the procedures and purpose of a study, that they indicate they understand this information, and that they participate voluntarily, without coercion

institutional review boards (IRBs): administrative units that oversee the ethical treatment of subjects as at research institutions

plagiarize: to take someone's ideas or work without proper attribution of credit

subjects: humans and other animals who are observed and measured by geographers and other scientists

Bibliography

Committee on the Conduct of Science, & National Academy of Sciences (1995). *On being a scientist: Responsible conduct in research* (2nd ed.). Washington, DC: National Academy Press.

Craig, E. (Ed.) (1998). *Routledge encyclopedia of philosophy.* London and New York: Routledge.

Monmonier, M. (2002). *Spying with maps: Surveillance technologies and the future of privacy.* Chicago: University of Chicago Press.

Proctor, J. D., & Smith, D. M. (Eds.) (1999). *Geography and ethics: Journeys in a moral terrain.* London and New York: Routledge.

Sigma Xi (1999). *The responsible researcher: Paths and pitfalls.* Triangle Park, NC: Sigma Xi, The Scientific Research Society.

Smith, M. (Ed.) (2002). *Environmental responsibility for expeditions: A guide to good practice* (2nd ed.). London: British Ecological Society and the Young Explorers' Trust.

Index

About the Authors

Daniel R. Montello is professor of Geography and affiliated professor of Psychology at the University of California, Santa Barbara (UCSB), where he has been since 1992. Before that, he was a visiting assistant professor at North Dakota State University in Fargo, and a postdoctoral fellow at the University of Minnesota in Minneapolis. Dan received a Ph.D. in Psychology (Environmental Psychology area) in 1988 from Arizona State University in Tempe and a B.A. in Psychology in 1981 from the Johns Hopkins University in Baltimore. He is a member of the Association of American Geographers, the Psychonomic Society, and Sigma Xi Scientific Honor Society. He has published widely in the areas of spatial and geographic perception, cognition, and behavior; cognitive issues in cartography and GIS; and environmental psychology and behavioral geography. Dan has taught courses in a great variety of topics, including research methods, introductory human geography, statistical data analysis, behavioral geography, environmental perception and cognition, cognitive issues in geographic information science, regional geography of the United States, introductory psychology, child and lifespan developmental psychology, cognitive development, perception, environmental psychology, and cognitive science. He lives in Santa Barbara, California.

Paul C. Sutton is associate professor of Geography at the University of Denver. He received his Ph.D. in Geography from the University of California, Santa Barbara, in 1999. His teaching and research interests are primarily in the domain of the human-environment problematic. Much of his work focuses on developing methods for improving our ability to estimate and map the human population using nighttime satellite imagery. Future work is focusing on using remotely sensed imagery and ground truthed sampling to map and estimate other socioeconomic variables such as impervious surface area, ecosystem services, income, energy consumption, and CO_2 emissions. These areas of research are an amalgam of work in remote sensing, geographic information science, and statistics. Paul has published in many journals, including *Nature*; *Remote Sensing of Environment*; *Photogrammetric Engineering and Remote Sensing*; *Population and Environment*; *Geocarto International*; and *Computers, Environment, and Urban Systems*. Prior to his life in academia, Paul worked as an engineer at the Santa Barbara Research Center and as a high school science teacher at the Anacapa School in Santa Barbara. He lives with his wife, Élan, and son, Paris, in Conifer, Colorado.